METHODOLOGICAL VARIANCE

BOSTON STUDIES IN THE PHILOSOPHY OF SCIENCE

VOLUME 131

G. L. PANDIT

Department of Philosophy, University of Delhi, India

METHODOLOGICAL VARIANCE

Essays in Epistemological Ontology and the Methodology of Science

KLUWER ACADEMIC PUBLISHERS

DORDRECHT / BOSTON / LONDON

Library of Congress Cataloging-in-Publication Data

Pandit, G. L., 1945-

Methodological variance : essays in epistemological ontology and the methodology of science
p. cm. -- (Boston studies in the philosophy of science ; v. 131)
Includes bibliographical references and indexes.
ISBN 0-7923-1263-5 (acid free paper)
1. Science--Philosophy. 2. Science--Methodology. 3. Ontology. 4. Knowledge, Theory of. I. Pandit, G. L., 1945- . II. Series.
Q175.M5418 1991
501--dc20 91-653

ISBN 0-7923-1263-5

Published by Kluwer Academic Publishers,
P.O. Box 17, 3300 AA Dordrecht, The Netherlands.

Kluwer Academic Publishers incorporates
the publishing programmes of
D. Reidel, Martinus Nijhoff, Dr W. Junk and MTP Press.

Sold and distributed in the U.S.A. and Canada
by Kluwer Academic Publishers,
101 Philip Drive, Norwell, MA 02061, U.S.A.

In all other countries, sold and distributed
by Kluwer Academic Publishers Group,
P.O. Box 322, 3300 AH Dordrecht, The Netherlands.

Printed on acid-free paper

Printed in the Netherlands

To Sir Karl Raimund Popper
and
all those physicist-philosophers who could
very clearly recognize that the glory of
science lay not in any kind of game of
playing safe in the situations of interaction
with nature but in *its* tortuous search-cum-discovery
procedures of exploring the subjects of objective
knowledge *without* one's rational belief in this
or that scientific problem ever coming in the way.

TABLE OF CONTENTS

PREFACE

For a philosopher with an abiding interest in the nature of objective knowledge systems in science, what could be more important than trying to think in terms of those very subjects of such knowledge to which men like Galileo, Newton, Max Planck, Einstein and others devoted their entire lifetimes? In certain respects, these systems and their structures may not be beyond the grasp of a *linguistic* conception of science, and scientific change, which men of science and philosophy have advocated in various forms in recent times. But certainly it is wrong-headed to think that one's conception of science can be based on an identification of its *theories* with *languages* in which they may be, alternatively, framed. There may be more than one place in my own book (1983: 87) where they may seem to get confused with each other, quite against my original intentions. The distinction between the objective knowledge systems in science and the dynamic frameworks of the languages of the special individual sciences, in which their growth can be embedded in significant ways, assumes here, therefore, much importance. It must be recognized that the problems concerning scientific change, which these systems undergo, are not just problems concerning language change. Unlike those problems which any enterprise preconceived on the model of language may raise, the most challenging problems here – those that can be of the greatest possible interest to humankind – are the recurrent problems of *methodological variance*, as we might call them. Most interesting, though neglected, among these relate to scientific changes in the search-cum-discovery procedures of a science. Such changes are a likely result of the scientists themselves working their way backwards from entanglement in those rules of method and strategy which become widely known once they are violated in the course of the developmental history of the sciences. For the *causal dynamics*, which is at work in methodological variance in a science and which accelerates the growth of its objective knowledge systems, we must turn to the methodologically significant problem-situations of *rule-entanglement* not just in its horizontal but vertical dimension. These can arise as and when the subject-specific *rules* of problem-determination (or problem-formulation) and theory-construction no longer work when extended to the qualitatively *novel*, unfamiliar or potentially more dense and opaque areas. Whenever it becomes possible, within science itself, to decide concerning a given science that it is, in fact, so entangled, it will be in need of methodological variance. Such a need will be fulfilled by resort to an appropriate enrichment in terms of novel rules of subject-specific relevance to it, in a manner which unfolds no methodological incommensurability-like after-effects (= a loss of commensurability) whatever. Thus, at any given time, the extent of methodological variance in a science resulting in the growth of its objective knowledge systems will depend upon how deeply entangled it is in the extant framework of its own rules of method and strategy at that time.

I think that it remains a central task of the foundational discipline of *epistemological ontology*, as we might call it, to explain how, in the course of the historical development of the individual sciences, objective knowledge systems grow, or can grow, in the form of theory-problem interactive systems without any kind of radical incommensurability coming in between them. In my earlier work (1983), I have argued for a general interaction-theoretic conceptualization of the nature of this growth, essentially in terms of the principles of methodological variance in the sciences, as being not only possible but feasible as a better alternative to various types of accounts which have been in fashion for quite sometime. Thus, it is possible to explore it as a function of – and, therefore, dynamically inconceivable without – methodological variance at various levels in the individual sciences. In a somewhat stronger sense, we might say that *it* follows a unique interactive pattern of development in which rule-governed creativity – in the form of the variable canonical systems of scientific description (= *CSD*s), their applications and their strategies in each science – intimately interplays with rule-enriching creativity (= *CSD*-variance). Its rationality lies, then, in this interactive pattern of growth itself instead of either in a Darwinian process of trial-and-error and natural selection extended to science (as Popper has suggested) or in the disciplinary-matrix-driven puzzle-solving activities of the highly closed society of normal scientists (as Kuhn has suggested).

But what is it that makes methodological variance in the sciences in this objective sense possible? It is, of course, inconceivable without those methodologically invariant structure-property correlations which operate at a deeper level, constraining it in various ways. I have here in mind significant, but neglected, correlations between the epistemic structural codeterminants of the objective knowledge systems in science and their holistic properties (= theoretical universals) of growth, rationality, causal dynamics, simplicity and so forth. The methodologically invariant structure-property correlations are generative of the general, subject-neutral, concept of a scientifically permissible canonical system of description (*CSD*) – the general concept of method and strategy – in science. Thus, they also make epistemic appraisal in science possible. They do so not just at the level of choice from among the competing *CSD*s in the special sciences undergoing revolutionary scientific change but at the level of the objective knowledge systems themselves which are triggered in the course of *CSD*-variance. I have here specifically in mind not only individual *theories* and *problems* but theory-problem interactive systems as growing objective knowledge systems in science. Thus, it is the structure-property correlations themselves which not only constrain but make methodological variance in the individual sciences possible without the ghost of incommensurability appearing at any relevant level of epistemic appraisal.

In this context, it is not possible just to ignore the strategies of scepticism, such as those of Thomas Kuhn and Paul Feyerabend, which seek to undermine the very objectivistic epistemological framework for analysing the structure and growth of scientific knowledge. But they have been either distorted or ignored in those philosophical discussions which find them admirable for their closeness to the history of scientific practice. Thus, hardly any attention has been paid to the socio-linguistic model of scientific rationality implicit in Kuhn's strategy of changing the subject of appraisal in science from the problem-solving structure of a scientific theory to the puzzle-solving closed community of normal scientists. The much discussed doctrine of incommensurability in Kuhn's and Feyerabend's senses is, as I have argued in my earlier book (1983), a simple consequence of the same strategy. This has resulted in considerable distortions in the role of epistemological ontology in generating not just interesting historiographies of science but an adequate account of the growth of scientific knowledge, both as a historical subject of epistemic appraisal and as a subject of future discovery. The double standards of the type illustrated in the Kuhnian and Feyerabendian approaches to science are a sure symptom of such distortions having taken place in so far as these have no basis, whatever, in epistemological ontology. I have here in mind particularly their common strategy of characterizing science in terms of complex pragmatical features of family resemblance between *it* and other types of activity (= forms of life) while seeking to undermine the methodological commensurability of the rival theories by invoking much simpler (= internal) considerations of semantic meaning variance of the individual scientific terms as introduced by such theories themselves

Some readers of my first book (1983) have complained about its difficult style. In respect of places where this has really hampered the very availability of the ideas I wanted to project, I wish to say, quite unapologetically, how unintendedly the book itself had assumed the structure and style which it did while all along I had had my eyes modestly set on an essay good enough for a scholarly journal of philosophy. The only compensation which it is possible for me to offer the reader of that book is, then, the present collection of essays as an offshoot thereof. They are essays in epistemological ontology, as I have here chosen to call the foundational discipline which studies, both synchronically and diachronically, the objective knowledge systems in the sciences, particularly in respect of their methodologically invariant structure-property correlations. Within the framework of a general interaction theory (Pandit, 1983), they address themselves to this very task by elucidating the view that *science* is an enterprise comprising the negative feedback controlled problem-solving (and problem-posing) systems – the *NFCPS* systems as I like to call them. They do so by assuming a basic principle of embedded methodology which says that a sound methodological model of epistemic appraisal of objective knowledge systems, and states of scientific change they can undergo, must be embedded in a sound model of their epistemic structural identity and growth. In other words, it must be embedded in well-identified structure-property correlations in these systems. While our models of the structure and growth of knowledge can serve, in this context, as a testing ground for the realism of an adequate epistemological theory, the principle of embedded methodology as a whole serves to ground the methodology of science in the basic principles of epistemological ontology itself. The result is a *two-fold* view of science upholding objectivism and realism but dispensing with both the evolutionary paradigm of Popper and the sociological paradigm of Kuhn:

(1) The structure of objective knowledge systems in the sciences is

always codetermined by their type-distinct structures as abstract as *theories* and *problems* in mutual interaction.

(2) These systems show interesting structure-property correlations. As a result, their *growth* follows an interactive, if unpredictable, pattern of its own.

The methodology of epistemic appraisal which this view, then, implies is called the methodology of theory-problem interactive systems as negative feedback controlled problem-solving (and problem-posing) systems, at once distinguishable from other types of systems of a cognitive as well as natural order. Of course, the term 'negative feedback' receives, in this context, an interpretation which deviates constructively from its currently standard cybernetic sense.

The concept of an *interactive system* is here, obviously, central to this view. The definition of such a system, as given in my book (1983: 67), has provoked criticism and is in need of an important correction in respect of its restriction there to deterministic systems only. These are characterized by determinate sets of relations of interaction between their successive states, with every state S having a unique successor state S'. But we must also include, as it has been pointed out to me by James H. Fetzer (personal communication), indeterministic systems "such that, for every S of such a system, there is a unique range (or class) of possible successor states S^1, S^2, . . . among which one or another occurs, with a certain probability distribution across the class of possible states". This is all the more necessary in the context of objective knowledge systems in the sciences that we have been just talking about. The states of scientific change, such as those of growth of knowledge, which they undergo are essentially unpredictable with respect to their detailed nature if not their probable direction.

From this point of view, it becomes possible, then, not only to reject as highly one-sided but to explain the type of approach to science and scientific change which looks to the historical scientific practice itself as the sole source and the chief arbiter of the methodology of the sciences. Thus, it can be traced to a general tendency among the historiographers and philosophers of science to ignore the following:

(1) An objectivistic enterprise of epistemological ontology, with the methodology of science embedded in it, is not only possible but necessary to any systematic attempt to understand that philosophical framework of the sciences which operates not as something

external to them but as a very vital component in their *internal life*.

(2) An adequate methodology, and philosophy, of science should itself be generative of non-trivial rational reconstructions and historiographies of science and scientific change, recovering as much structural and dynamical complexity as there may be to the growing objective knowledge systems in the sciences. Of particular importance are here those structure-property correlations which are methodologically invariant and, therefore, crucial to (i) the methodological commensurability of different theories, and problems, *within* the individual sciences and (ii) the incommensurability between science as a *whole* on the one hand and non-science on the other.

(3) An adequate methodology and philosophy of science should, moreover, not just recover but clarify those pervasive distinctions which are presupposed by historical scientific practice at every point and with which it is recurrently impregnated. This applies particularly to the distinction between *language* and *science* with the latter invariably embedded in but never identical with the former.

(4) As a matter of historical fact, a newly discovered (revolutionary) theory, which replaces or supersedes an older theory in its field, is always seen throwing fresh light on the latter, thereby making a better undestanding of both its past successes and failures possible. It is, therefore, quite reasonable to require that a sound epistemology, and a methodology embedded in it, should not only take into account but explain how – and by virtue of which type of principle – is it possible for a novel scientific theory to do all this to its own rival or predecessor in their field. Thus, for example, if we take Einstein's special theory of relativity, what calls for explanation is how it enabled the physicists at the turn of the century to gradually reconcile themselves with the null result of the 1887 Michelson–Morley experiment, and thereby to have a better and a decisive understanding of the Maxwell–Lorentz theory of the electromagnetic phenomena, of wave optics, of Galilean transformation and indeed of space-time concepts in classical mechanics; and all this without invoking any *ad hoc* ether-theoretic hypotheses or hypotheses of the type aimed at rescuing those already in difficulty.

(5) As a corollary of (1) – (4) above, neither the philosopher nor the historiographer of science should, in so far as their interest in its rationality and realism is concerned, look to those arbitrary conventions or decisions which the whole scientific communities may adopt over different historical periods. For issues of rationality and realism are matters of deeper principles based on the structure-property correlations in the objective knowledge systems in the sciences. At least on the present view, empirical science, as against non-science, is definable not on the analogy of the types of activity or forms of life as simple as *games* (= the puzzle-solving activities), whose character might be safely traced to the community-wide conventions or decisions, but on essentialist lines, where *essentialism* should be understood in a broadly conceived methodological, not metaphysical, sense.

Berlin, Zehlendorf
November, 1990

G. L. P.

ACKNOWLEDGEMENTS

The impetus to write this book in order to further develop the view of science set forth in my first book (1983) came from Robert S. Cohen and Marx W. Wartofsky in the Fall of 1981. Some early fragments of the present work (Chapters 1, 2, 4, 5, and 8) were tried out on audiences at Visva-Bharati, Santiniketan (March 1982), Punjab University Chandigarh (February 1983) and Delhi University South Campus (March 1983). A first full version became ready for refereeing in the Fall of 1983. It is to the critical reactions of the referees that I owe many of the improvements made later in the text. First of all, I am deeply indebted to James H. Fetzer. In respect of the structural reorganization of its contents, the book has benefited enormously from his careful reading of the text and his helpful criticism. My discussions with him (all in the form of personal correspondence in Winter 1985/1986) have resulted in quite a few essential improvements, especially in Chapters 1–4, 8 and 10 (particularly §§ 1.2, 2.6, 3.1, 8.2.). Criticism by an anonymous referee has enabled me to greatly improve, if not to perfect, the many formal and stylistic aspects of the text. The shortcomings that persist are admittedly my own.

The task of bringing the book to completion would have been for me a far too difficult one without the warm hospitality and support that I have been very fortunate to receive during the tenure of a Fellowship of the Alexander von Humboldt-Stiftung, Bonn (January 1986 – July 1987) from Lorenz Krüger on the one hand and Erhard Scheibe on the other. This enabled me to spend the period of my Humboldt Fellowship first at the Institut für Philosophie, Freie Universität Berlin, where research for § 8.2. (as also for §§ 6.2., 9.2. and Chapter 10) was partially carried out, and then at the Philosophisches Seminar der Ruprecht-Karls Universität in Heidelberg. The § 8.2. has benefited greatly from separate discussions with James H. Fetzer, Elliot Sober (by way of personal correspondence), Lorenz Krüger, Hans Poser and Don Howard; as also from discussions on a paper presented to a meeting of the Philosophischer Kreis, Institut für Philosophie, Freie Universität Berlin (May 1986). I am thankful to all those who participated and asked questions on that occasion, especially to Lorenz Krüger, Michael Heidelberger, David Pearce, Wolfgang Spohn and Jan von Plato. On the other hand, I am also highly thankful to Erhard Scheibe and Manfred Stöckler for lively discussions and conversations at our regular Colloquium at the Philosophisches Seminar der Ruprecht-Karls Universität in Heidelberg (April–July 1987) on subjects of common philosophical interest relating to topics I have dealt with in §§ 8.2.5. and 9.1. To my delight, they have paved the way to my last minute reformulations at several places in the text just alluded to. I see it clearly that had I not stumbled across the fascinating, and at times challenging, thoughts and texts of so many men to whom my debt is recorded in the bibliography, I could not perhaps think of human knowledge and human interaction in the way of the present book. I must also record my deep indebtedness to the following scholars and institutions: John Stachel (1986a, 1986b), Don Howard (1986) and David Pearce (1987a, 1987b) for making accessible to me their works while on the way to the press; the staff of the Fachbereich Physik Bibliothek, Freie Universität Berlin, where one could still see, and be inspired by, the picture of young Einstein hanging on a wall, and the Institut für Theoretische Physik Bibliothek, Ruprecht-Karls Universität in Heidelberg, for their hospitality; to the Hebrew University of Jerusalem, Israel, for permission to reproduce, at several places in the text, from the work of Albert Einstein (1905, 1934, 1948, 1949, 1960); and to the Goethe Institute Berlin where I learnt my German. Above all, I feel deeply beholden to the readers of my first book (1983) in different parts of the world, as also to my colleagues and friends in Germany. Their reactions, always well-meaning, have left an abiding impression on my way of thinking.

Heidelberg: Am Pferchelhang 4– July 1987 G.L.P.

NOTATION AND ABBREVIATIONS*

T_{EP}	the explanatory power of a scientific theory
T_{RP}	the resolving power of a scientific theory
STR	the special theory of relativity
GTR	the general theory of relativity
NGT	Newtonian gravitation theory
NCT	Newtonian corpuscular theory of light
NFCPS-system	negative feedback controlled problem-solving/ problem-posing system
CSD	canonical system of description
ES-variance	variance of the epistemic (= developmental) structural codeterminants of objective knowledge
T_{EP}-evaluation	evaluation of the explanatory power of a scientific theory
T-simplicity	theoretical simplicity of a scientific theory
P-simplicity	pragmatic simplicity of a scientific theory
SE	subjectivistic epistemology
OE	objectivistic epistemology
K-state	knowledge-state
TE	traditional epistemology
TE-theories	traditional epistemological theories
PI	*Philosophical Investigations*
TLP	*Tractatus Logico-Philosophicus*
CM	classical mechanics
$<$	is less than
$>$	is greater than

* Abbreviations and symbols (particularly in Chapters 8 and 9) such as, e.g., $T_{RP}(T_1)$, $T_{EP}(T_2)$, T_{RP} (GTR), T_{EP} (NGT), and T_{RP} (GTR) $>$ T_{RP} (NGT) should *respectively* be taken to mean: the resolving power of the theory T_1, the explanatory power of the theory T_2, the resolving power of GTR, the explanatory power of NGT, *and*: the resolving power of GTR is/as greater than the resolving power of NGT, depending on the particular context of its use.

$\geqslant$	is greater than or equal to
$=$	is equal to/equals
$\rightarrow$	triggers off/is in interaction with
$\Rightarrow$	strict implication
$\sim$	negation ('not')
$\cdot$	conjunction ('and')
$\vee$	disjunction ('or')
$\equiv$	is identical with/equivalent to
$\not\equiv$	is not identical with
$\leftrightarrow$	material equivalence ('if and only if')
$\mid$	given that
$\vdash$	explains (= deductively entails)
$\in$	is a member/species of
v	velocity of a point or a particle
c	the (constant of) velocity of light *in vacuo*

PART ONE

Was für eine Stellung nimmt das Weltbild des theoretischen Physikers unter all diesen möglichen Bildern der Welt ein? Es stellt die höchsten Anforderungen an die Straffheit und Exaktheit der Darstellung der Zusammenhänge, wie sie nur die Benutzung der mathematischen Sprache verleiht. Aber dafür muß sich der Physiker stofflich um so mehr bescheiden, indem er sich damit begnügen muß, die allereinfachsten Vorgänge abzubilden, die unserm Erleben zugänglich gemacht werden können, während alle komplexeren Vorgänge nicht mit jener subtilen Genauigkeit und Konsequenz, wie sie der theoretische Physiker fordert, durch den menschlichen Geist nachkonstruiert werden können. Höchste Reinheit, Klarheit und Sicherheit auf Kosten der Vollständigkeit. Was kann es aber für einen Reiz haben, einen so kleinen Ausschnitt der Natur genau zu erfassen, alles Feinere und Komplexere aber scheu und mutlos beiseite zu lassen? Verdient das Ergebnis einer so resignierten Bemühung den stolzen Namen „Weltbild“?

Albert Einstein (1934: 108).

CHAPTER 1

ON THE OBJECTS OF OUR SUBJECTIVE KNOWLEDGE

> . . . it is symptomatic of the confusion of philosophical language that the thing the painter refers to as his subject is what the philosophers call the object.
>
> C. H. Waddington (1977)

1.1. WHAT IS WRONG WITH TRADITIONAL EPISTEMOLOGY?

The problem of the nature of the objects of human knowledge was most clearly raised for the first time by Plato. For him, our very understanding of the nature of knowledge depended on an adequate solution of this problem. Four things, among others, are noteworthy in Plato's approach: (1) his sole concern with knowledge in a strictly subjective sense of this term, appearances to the contrary not-withstanding; (2) his firm commitment to infallibility as a necessary mark of (subjective) knowledge; (3) his view that perception is not a species of such knowledge; and (4) his view that the best candidates for objects of such knowledge are the (*eternal* and *abstract*) *individual* entities that he designated as 'Forms'/'Ideas'. In the course of the subsequent historical development of traditional epistemology (TE), it is only the doctrines (3) and (4) which have undergone radical modification, with the underlying basic *assumptions* staying intact and unquestioned. Thus while no philosopher from Aristotle onwards has ever subscribed to a Platonic doctrine of 'Forms' as objects of knowledge, or rejected perception as a *pseudo-species* of knowledge, there is discernible in TE a common commitment not only to the doctrines (1) and (2), but to the assumption that it is the *individual things/processes* of a suitable kind which alone can be the objects of our (subjective) knowledge. These are generally identified with, to borrow the Popperian categories, the world$_1$ of physical things on the one hand and the world$_2$ of mental processes/states on the other. This can be seen more clearly if we assume an ontological framework as wide-ranging as follows:

world$_1$: of physical things/processes
world$_2$: of minds/mental states of consciousness, belief, etc.

world$_3$: of epistemic structures $\equiv_{Df}$ linguistically neutral structures as abstract as propositions or theories (and problems).

There remains, nevertheless, a characteristic tension in TE between what the standard analysis of (subjective) knowledge *implies* as to the nature of its objects *and* the generally held belief in the world$_1$ and world$_2$ individuals as constituting these objects. I believe that the reasons why this is so have to do with the following features inherent in TE:

(1) TE has never been pursued within a clearly drawn framework methodologically sound enough to distinguish subjective from objective knowledge. (That the complexities inherent in human knowledge themselves warrant such a framework is argued by me elsewhere).[1]
(2) Plato's insight to the contrary not-withstanding, perception has always been viewed in TE rather dogmatically as a species of knowledge, the belief in *perceptual objects* of various kinds as objects of knowledge being a natural corollary thereof.
(3) At least one reason why the problem of perception haunts TE always bedevilling its discussions is traceable to a total absence in TE of a tool or a framework sound enough to distinguish *human knowledge* from *human interaction*, the former being in my view an orientating variable of the latter. It is not surprising that without this distinction it is not possible to realize how mistaken it is to admit the world$_1$ and world$_2$ individuals themselves as objects of subjective knowledge, these being the very things with which human beings interact in as many ways as may be demanded by our theoretical as well as non-theoretical problem-solving activity.[2]

1.2. THE NATURE OF SUBJECTIVE KNOWLEDGE: TRADITIONAL ANALYSIS

Before we may proceed with our discussion of the question of the nature of the *objects* of our subjective knowledge as it will be raised here, it is desirable, I think, to make a few clarificatory remarks concerning the central concept *knowledge* itself in the present context. When I first introduced the term 'K-claim' ('knowledge-claim') in my (1983) I did not make my intention explicitly clear. I had then intended to use it in a manner which was *neutral* between the subjective context of the knowledge-states of a given knowing subject *and* the intersubjective or objective context of systems of scientific knowledge (= theory-problem interactive systems). The term 'K-claim' appeared to me then quite suitable for purposes of referring to knowledge-states (K-states) in these two different contexts. Thus in the subjective context, I took 'I know that *p*' as (expressing) a K-claim in its own right as much as 'S knows that *q*'; each invariably involving the presumption of the knowing subject as a fixed frame of reference. Similarly, in the objective context, I took all our theories, whether true or false, as our objective K-claims. Another reason why I used the term 'K-claim' so liberally is this: I wanted to use a term which did not rob us of *fallibility* in both these two contexts. The underlying problem with which I was concerned might still be formulated thus: When and under which conditions could we say of a K-claim, made in the subjective context, that it involves genuine K-states, and of a K-claim, made in the objective context of science, that it involves genuine epistemic (= developmental) structures as structures of objective knowledge?

James Fetzer[3] has, in this context, drawn my attention to the importance of the distinction between a K-state and a K-claim. The point of the distinction between the former (for which it is true, say, that S knows that *p*, . . .) and the latter (say, S claims to know that *p*) is then this. For any number of pragmatic reasons (e.g., to deceive, to pretend, etc.) S might claim to know that *p*, without it being true of him/her that he/she knows that *p*. Thus it might be *true* that S claims to know that *p*, even though it is *false* that S knows that *p*.

As will be clear from my arguments in the present and the following chapters, I now think it proper to speak of K-states where I spoke of K-claims previously (1983). The concept of K-states will, therefore, retain the same neutrality and fallibility assumptions that characterized my earlier use of the concept of K-claims. Thus in the objective context of scientific knowledge, whose structural identity is, I believe, codetermined by theories and problems in mutual interaction, it would be quite legitimate to analyze the parameters of the structure and growth of knowledge in terms of K-states and dynamics of K-state-transition or K-state-variance, not only in the sense of movement from one asymmetrically ordered pair of theories (T_1, T_2) to another such pair but in the sense of movement from one complex theory-problem interactive system to another such system. The advantage of an approach like this would, then, lie in this: even most complex problem-situations in the individual sciences could be analyzed in terms of K-state-variance, since a new *problem* would, on this approach, signal a new K-state as much as a new theory is traditionally taken to do.

On the other hand, in the subjective context, it is proper to speak of a knowing subject S as being in particular successive or intervalled K-states, $S(K_1)$, $S(K_2)$, $S(K_3)$, . . .$S(K_n)$, at different times t_1, t_2, t_3, . . .t_n, such that if we were to analyze the structure and rationality of these K-states we would have to assume their minimal initial identity as *abstract* epistemic relations (= attitudes) of the following general form:

(a) S knows_1 that *p*
(b) S knows_2 that *q*
(c) S knows_3 that *r*
(d) . . . .
(e) S knows_n that *w*

where '*p*', '*q*', '*r*', '*w*' each ranges over a finite set of propositions.

I say 'abstract' epistemic relations, since it is not possible, even if we wanted, to take the entire range of complexities that must accompany the subjectively organized K-states and K-state-transitions into account. Even an analysis as excellent as James Fetzer (1981) provides of what he refers to as the knowledge context *Kzt* must employ, I am afraid, abstractions if rationality of the subjectively organized K-states in such contexts is to be at all analyzed in terms of a concept as central as that of a *minimally* and *maximally rational set of beliefs*.[4]

There can hardly be any doubt that TE is mainly concerned with knowledge in the sense of the subjectively organized K-states. That is, it is concerned with those subjectively organized K-states whose *minimal initial* structural identity involves epistemic relations of the general form (a)–(e) above. One of the central problems here is the problem of distinguishing the genuine from the non-genuine cases, a problem which is generally posed thus: How is knowledge different from mere belief? The underlying reasoning here, which is concerned with deeper questions of structural identity and rationality of knowledge, is as follows: For one to be able *to know* that such and such is the case (= to be in a specific and genuine K-state at a particular time and in a given knowledge-context) it is not sufficient, though it may be necessary, that one *believes* that such and such is the case (= is in a specific relevant Belief-state). Consider in this context the traditional standard analysis of subjective epistemic relations of the general kind above. On this analysis, a K-state, of the minimal initial identity 'S knows that *p*', in order to be genuine, must on proper scrutiny exhibit the following general structure:

S believes that *p*;
p is true; and
S has sufficient ground for
his belief that *p*.

Or in the case of another such K-state, 'S knows that ~*p*':

S believes that ~*p*;
p is false; and
S has sufficient ground for his belief that ~*p*.

What has gone unnoticed in the epistemological discussions generally is the philosophical significance of this analysis from the point of view of the very distinction between subjective and objective knowledge which it assumes quite implicitly. The distinction is bound to remain implicit and unclear unless a corresponding distinction is made between the two enterprises of objectivistic and subjectivistic epistemology.[5] Given the subjectivistic character of TE, one might say that the enterprise of subjectivistic epistemology is the first to develop historically. Objectivistic epistemology, which is concerned with objective knowledge, is on the other hand relatively of recent origin and is yet to receive universal recognition.

It can however be argued that there is an intimate relationship between the two enterprises. One of the ways in which the nature of their relationship might be clarified is to address oneself to the problem of the nature of the objects of subjective knowledge in the sense of TE: what is it precisely that one knows when one is in a subjectively organized K-state? One answer to this question is already implicitly suggested by the traditional analysis of subjective knowledge as follows: That it is invariably *propositions* (over which the variable '*p*' ranges) which are the objects of our subjective K-states of minimal initial identity 'S knows that *p*'.

This simple and straightforward traditional answer is however considerably disguised and complicated by what the essentialist Plato or, later on, the subjective idealist Berkeley, the empiricist Hume, the phenomenalist Ernst Mach and others had, one after another, to say on this subject. Also consider how, in this century, the logical empiricists Schlick, Carnap and Neurath remained divided for a long time as to whether the 'given', the 'rockbottom' of all knowledge, could even be captured in the language of the primary propositions. Russell's phenomenalism, distinguishing knowledge by acquaintance from knowledge by description, appears as a variation on the same theme that had started with Plato's doctrine of 'Forms' as the only objects of our (subjective) knowledge.

For Plato it is the abstract world of 'Forms' or 'essences' which provides the true objects of our knowledge; no other objects such as the ones we might perceive in the physical world could ever assume this privileged role. In what follows it will, I hope, become clear that the basic though unrecognized insight underlying the doctrine that the objects of perception cannot serve as the objects of our subjective knowledge is a deeply significant one. Plato's doctrine of 'Forms', however, cannot be accepted in its original form. No doubt, Platonic 'Forms' are abstract entities of a sort, a feature which they share with propositions. But it does not seem possible to suggest that Plato would have willingly equated what he called 'Forms' with 'propositions'. We have already considered briefly how on the traditional standard analysis of knowledge propositions tend to emerge as the abstract objects of subjective knowledge. Plato's doctrine of 'Forms' in its original form is thus already ruled out as a possible answer to our problem above, unless the doctrine is so amended as to imply that reference to 'Forms' is possible only by means of propositions and that, therefore, it is ultimately these propositions themselves which form the true objects of subjective knowledge.

Descartes' case is altogether different and, hence, must be singled out as of unique significance from the point of view of the problem concerning the objects of our subjective knowledge. What Cartesian essentialism or rationalism appears to accomplish is to shift the *objective* frame of reference of epistemology from Plato's world of 'Forms' to the physical and the mental worlds, respectively. Accordingly, one could now once again, after Aristotle, speak of these two worlds as what provide the objects of our knowledge. A closer study would however reveal that the traditional standard analysis of knowledge is traceable to Descartes' philosophy which employs the distinction between subjective and objective knowledge rather quite implicitly. There is indeed something of deeper philosophical significance about Descartes' 'Cogito, ergo sum' – something with consequences far-reaching in character but difficult to anticipate in his time. In order to bring this aspect out let us consider the question:

> What is it exactly that is *known* by Descartes when he asserts: 'Cogito, ergo sum'/'Dubito, ergo sum'?

In so far as the traditional theories of knowledge generally assume the objects of perception/intuition as the objects of our (subjective) knowledge, it is pertinent to reframe this question as follows:

> Is it some particular kind of *individual object* or its properties which Descartes claims to know or is it a set of *propositions* which he is thinking of?

It is interesting that at the very outset Descartes employs a *reductio ad absurdum* type of scepticism to establish the proposition allegedly expressed by the sentence 'I exist' as indubitable/knowable. *Any* attempt to deny one's knowledge of the proposition is shown to lead to inconsistencies and absurdities of an existential nature.[6] Thus what Descartes demonstrates when he asserts 'Cogito, ergo sum' is this: that there is a (synthetic) proposition which is *indubitable* and hence which each of us can *know* once for all; and that the propositions which can be deduced from it would be equally *knowable* to one in an act of subjective knowledge. Having already asked the question 'What can I know?', Descartes' 'Cogito, ergo sum' thus serves to build a whole corpus of *propositions as the clearly and distinctly knowable objects of our subjective knowledge*. Descartes' strategy may now be restated as follows:

> Wherever necessary, performative subjective K-states of the general form 'I know that p' must be subjected to the performatory version of the *reductio ad absurdum* test: 'I doubt that p' or 'I think that ~p'. If in this way the attempt to think the contrary of what one believes to be true leads to absurdities of the kind Descartes arrives at, it demonstrates the genuine character of the subjective K-state in question.[7]

It would thus seem to follow that Descartes' philosophy allows the distinction between the subjective and objective knowledge, with the former presupposing the latter. The fine structure of objective knowledge is in this context identifiable with a structure as abstract as that of propositions. That is, once we recognize this distinction, it is propositions themselves, whether true or false, which would have to be admitted as constituting, at least partially, the structural identity of objective knowledge.[8]

1.3. RATIONAL BELIEF, OBJECTIVE KNOWLEDGE AND HUMAN INTERACTION

Our understanding of TE is, however, considerably complicated if we consider its characteristic attempts to raise and deal with the problem of knowledge, particularly in the form of the problems of perception. All such attempts are associated with one form of empiricism or another. There is a popular philosophical doctrine which says that all our (subjective) knowledge is knowledge of objects – their properties and relations – belonging either to the world$_1$ of physical objects or the world$_2$ of inner mental processes/states. 'Perceptual knowledge' is, I think, often cited in this context as the paradigm of (subjective) knowledge, with physical objects/sense-data serving as the obvious *objects* of knowledge. As already seen, the view that it is the individual objects themselves – whether abstract or concrete – which are the objects of subjective knowledge is traceable to Plato who takes a highly abstract view of these objects and calls them 'Ideas' or 'Forms'. This view, however, runs into serious trouble if we want to (a) formalize/define the properties of a genuine (subjective) K-state and (b) distinguish knowledge, both subjective and objective, from human interaction – i.e. interaction between the humans and the objects rooted in the world$_1$ and world$_2$ respectively. In what follows, I shall argue that we must, therefore, reject this view in favor of some better alternative to it. My arguments will pave the way for an alternative view to the effect that not the deproblematized 'inmates' of a Platonic world of 'Forms', nor those of the world$_1$ and world$_2$ but the 'inmates' of the Popperian world$_3$ – the structures such as theories and problems (or propositions and problems) – are the objects of subjective K-states of the minimal initial identity: 'S knows that *p*', or 'S knows that $\sim p$'.

The very manner in which the problem of knowledge is formulated and discussed in TE reduces it to one of distinguishing the belief-states (B-states) such as 'S believes that *p*' from the K-states which have the general form 'S knows that *p*', where '*p*' is a propositional variable. I think that this is true of most of the important philosophers from Plato through Descartes to those of this century. Consider, in this context, how Descartes sets out to establish his subjective K-state 'I know *that* I exist' as a basis of all knowledge. Now, once the problem is formulated in this manner employing the concept of *belief*, etc. one cannot escape certain consequences of crucial importance to an adequate formalization of the properties of a (subjective) K-state. If knowledge is to be distinguished from belief one must ask:

> What is a belief? And, what is it that one does or does not believe in? Or what kinds of objects must be appropriate to the intentionality of B-states (as also to that of K-states)?

One might answer by saying that a belief is one's (epistemic) attitude towards a given *proposition* no matter whether true or false. What this would entail is the admission of propositions as what one may or may not believe and of beliefs themselves as propositional attitudes. If we follow this line of reasoning we would have to accept its further consequences: that it is not the individual things/processes themselves, whether in the $world_1$ or $world_2$, which are the direct *objects* of our beliefs.

Now if beliefs are propositional attitudes, and if the problem of *knowledge* is one of distinguishing *it* from mere belief, it seems a fruitful line of inquiry to study knowledge itself as a kind of propositional attitude or to analyze/formalize it in terms of such attitudes. And this is precisely what the traditional analysis of a (subjective) K-state seems to assume. Thus, it takes or requires a K-state to be of the minimal general form 'S knows that *p*' where

> 'S knows that *p*' $\equiv_{Df}$
> 'S believes that *p*';
> '*p* is true'; and
> 'S has sufficient grounds for his belief that *p*'.

However, what has received little or no attention in the epistemological discussions is the fact that this analysis of the structural identity of a (subjective) K-state is incompatible with the quite dominant view that it is the $world_1$ and $world_2$ things/processes themselves which serve as the objects of our (subjective) knowledge. For what the former presupposes is just an ontology of *problems*, *propositions* and *knowing subjects*; provided, of course, 'S has sufficient grounds' is interpreted as 'S can legitimately invoke a certain set of his *beliefs* that *q*, *r*, etc. in support of . . .'.

What is thus ruled out by the traditional formulation of the problem of *knowledge* is its analysis in terms of our relationship with the $world_1$ and $world_2$ things/processes themselves. That is, what is ruled out is a formalization of a kind such that our K-states may be of the form 'S knows *X*' where '*X*' ranges over the $world_1$ and $world_2$ things/processes themselves. That it makes no sense to impose a general form such as this on our (subjective) K-states seems to follow from the fact that it is not possible to impose on those of our beliefs, that involve the propositional attitudes, the following general form: 'S believes (in) *X*'. This only shows how philosophically significant the traditional formulation of the problem of knowledge is, and how deep the tensions that are, as it were, built into TE.

Now if it is correct to seek formalizations of the (subjective) K-states in terms of an ontology as rich as that of knowing subjects, their B-states with such properties as intentionality, and with *propositions* (and problems) for their objects, while ruling out any formalization in terms of just an ontology of knowing subjects on the one hand *and* that of the $world_1$ and $world_2$ things/processes on the other, it is then correct to go a step further so as to rule out *perception* as a serious candidate for a K-state. For to admit perception as knowledge would, I think, only amount to re-admitting the $world_1$ and $world_2$ things/processes as the objects of our (subjective) knowledge. And we have already seen how this is ruled out by the traditional formulation of the problem of knowledge. This particular consequence is of far-reaching philosophical significance to be considered presently. Before we may do so, however, it is necessary to have a closer look at the formalization of (subjective) knowledge as implied by its standard traditional analysis. According to this analysis, as we have seen already, all our (subjective) K states of the general form 'S knows that *p*' imply three things that form the following conjunction:

'S believes that *p*';
'p is true'; *and*
'S has sufficient grounds for his belief that p'.

And, correspondingly, the same should hold for K-states of the general form 'S knows that ~*p*':

'S believes that ~*p*';
'p is false'; and

'S has sufficient grounds for
his belief that $\sim p$'.

What is clear from this analysis is, I think, this: that it is propositions *about* the $world_1$ and $world_2$ things/processes and not these things/processes themselves which serve as the appropriate objects of our (subjective) knowledge and belief. However, what is not as clear as one would wish is this: while propositions as objects of our (subjective) knowledge may themselves be true or false, the fact that one's belief in the *truth* of a proposition is or appears quite rational, implies nothing as to whether that proposition is actually true or false. That this is so becomes amply clear if we consider a scientist's rational belief in the truth/falsity of a scientific theory as our example. What does a scientist's (subjective) K-state, viz., 'S knows that T_1, T_2, T_3, . . .' imply, where T_1, T_2, T_3 stand for a set of theories/hypotheses? I think that it is this:

> That he/she believes in the truth of T_1, T_2, T_3, etc.; and that he/she has grounds available in the form of his/her beliefs in the truth of some other relevant theories T_1', T_2', T_3' together with (new) experimental results which are strong enough to be invoked in support of his/her belief that T_1, T_2, T_3, . . .

But what it does not imply is that T_1, T_2, T_3 are *true*. If this brief analysis of a scientist's rational belief in the truth of T_1, T_2, T_3 is correct, the traditional analysis of our knowledge stands exposed as one which is quite faulty. It is, I think, faulty in that it assumes dogmatically and wrongly that knowledge implies not just *rational belief* in the truth of a proposition but the truth of the proposition itself. The real value of the traditional analysis lies then in its basic insight that knowledge implies rational belief in the truth of a proposition. And, I think, this in its turn implies that our knowledge is always knowledge of propositions. The aspect, however, in which it needs to be corrected is this: *that one's knowledge of a proposition being true in the sense of a rational belief in its truth does not imply the truth of the proposition itself*. If it did, true (and false) propositions would be so easy and so simple to produce. The same should be applicable to one's *knowledge* that $\sim p$. I don't see how one's rational belief in the falsity of a given proposition can ever imply the falsity of the proposition itself, except in those extremely simple situations in which one might be able to demonstrate the falsity of a synthetic proposition, using *modus tollens*. I suggest that a (subjective) K-state of the general form 'S knows that *p*' might be better analyzed afresh along the following lines:

1. 'S knows that *p*' $\equiv_{Df}$ 'S *understands* (what) *p* (says);
S believes that *p*;
S also believes that *q*, *r*, etc. – (background beliefs) – all of which can be invoked in support of his belief that *p*.[9]
2. 'S knows that ~*p*' $\equiv_{Df}$ 'S *understands* (what) *p* (says);
S believes that ~*p*;
S also believes that *q*, *r*, etc. – (background beliefs) – all of which can be invoked in support of his belief that ~*p*.'

Analyzed along these lines, this analysis enables us to characterize the structure of a (subjective) K-state as one of complex rational belief, with propositions emerging as the sole objects of belief and knowledge. This consequence is however quite disturbing, particularly from the point of view of the tension in TE between the view that it is the $world_1$ and $world_2$ things/processes themselves which serve as objects of (subjective) knowledge on the one hand *and* what the traditional analysis of knowledge implies on the other. Since the former does not stand closer scrutiny it is possible to resolve this tension in favor of the traditional analysis of subjective knowledge (in its by now amended form above) as follows. What has gone unrecognized in epistemology is, I think, this: that it is as important philosophically to distinguish human knowledge, both subjective and objective, from human interaction as it is to distinguish (subjective) knowledge from mere belief or rational from irrational/non-rational belief. There can hardly be any doubt that human beings are highly complex problem-solving beings. But the human problem-solving activity is the kind of activity which involves an enormous range of significant interactions between the humans and the $world_1$ and $world_2$ things/processes on the one hand *and* between one human being and another on the other. The most important question that arises in this context is this: What is it precisely that distinguishes human interaction in this sense from non-human (animal) interaction? On a closer look it would seem that all significant human interaction is *informed* by (subjective) knowledge. More precisely stated, (subjective) knowledge is, I think, what might be called an *orientating variable* of all such interaction. And it is, I think, this, among other things, which distinguishes the human from non-human interaction on the one hand

and knowledge itself from processes such as perception, sense-experience, etc. on the other. Thus, on this view, to perceive a world_1 or world_2 thing/process *X* is to *interact* in significant ways with *X* but not exactly to know that *X*. That is, while it is quite true to say that human beings perceive things/processes of various kinds in their environments it is also true, though unrecognized, that perception is a special case of significant human interaction. And certainly the difference between human perception and animal perception must be admitted as a serious problem worth exploring in terms of concepts such as the concept of a knowing subject (as against that of a perceiving organism), of subjective knowledge, etc.. It should suffice here to remark that what distinguishes the one from the other is the fact that the former, unlike the latter, is prompted by the human problem-solving activity in which (subjective) knowledge in the sense of rational belief plays a dominant role in its capacity as an orientating variable of all significant human interaction.

1.4. ONTOLOGICAL PRESUPPOSITIONS OF TRADITIONAL EPISTEMOLOGY

There can thus hardly be any doubt that the deeprooted tension[10] in TE needs to be resolved in a satisfactory manner, recent attempts to set up radical alternatives to it – which only radicalize the hidden tensions – notwithstanding. I think that the only reasonable way to resolve the tension is to recognize the vast complexities that are specific to the domain of K-states and to distinguish these K-states from human interaction along the lines indicated above. These complexities can be properly taken care of in a serious philosophical study only if we allow two distinctive enterprises of what I like to call subjectivistic and objectivistic epistemology (SE and OE, respectively) to operate non-dogmatically. Once we do so it will be possible for us to study human knowledge at various levels of abstraction. Already, the relationship between SE and OE is tentatively conceived, in Pandit (1983), as one of parallelism with possibilities of their interaction left wide open. I now think that a more precise conception of this relationship can be indicated thus: (i) SE ontologically presupposes OE[11]; and (ii) what cannot be defined within SE may be definable within OE. More precisely, what SE presupposes is this:

(1) There is objective knowledge capable of objective linguistic for-

mulation which does not involve any reference to the knowing subjects; and knowledge in this sense grows according to a rational pattern, with structures as abstract as problems and propositions giving it its structural identity.

(2) There are knowing subjects who are capable of being in states of subjective knowledge in the sense of rational belief; and capable of doing so by setting up various types of pragmatical relations with the structures of objective knowledge: relations such as those of belief, doubt, assertion, denial, confirmation, etc..

Subjective knowledge being a matter of one's epistemic attitude to propositions and questions – to theories and problems – reduces in ultimate analysis to pragmatical relations ranging from rational belief in a proposition/problem to (irrational/non-rational) ideological commitment to a theory-problem complex. What I am urging here is that besides subjective knowledge conceived as propositional attitudes, we must also recognize subjective knowledge as a range of attitudes to problems – as rational belief in problems.[12] Indeed at the level of objective knowledge, the propositions and corresponding questions, or the theories and corresponding problems, can always be seen forming interesting complexes of their own. I am thus arguing for the view that in so far as subjective knowledge (= the domain of K-states) ontologically presupposes objective knowledge in the form of problems and propositions, SE can be said to presuppose OE as an enterprise which deals with objective knowledge. As an interesting consequence of the present view, while TE/SE can be seen to lose the objects of knowledge traditionally conceived as rooted in *the* world itself, OE loses the knowing subject which is what SE centres around. It is quite clear that SE has the knowing subject as its fixed frame of reference in the sense that every K-state in its domain must have a structure which refers back to a knowing subject; for without such a reference, no subjective K-state will be complete by itself.

It follows that in TE (as a species of SE) there are no K-states which can exist independently of a knowing subject. There are no *absolute* K-states in TE. From the point of view of TE or SE a K-state is a state with intentionality which must be attributable to a knowing subject. Any K-state, of which this is not possible, is rendered impossible relative to TE. It turns out that the so-called self-knowledge states, as traditionally conceived, are precisely such states. If this is so, it follows that the possibility of TE or SE entails the impossibility of self-knowledge in the traditional sense of one's (subjective) knowledge of one's own self. And for self-knowledge states in this sense to be possible it is necessary, on the contrary, that TE/SE be dismantled.[13]

It may be seen now more clearly how the popular traditional conception of knowledge as *justified true belief* presupposes an ontology of its own. That is, it can be seen to rest on an ontology of abstract propositions on the one hand and that of knowing subjects with their B-states on the other, among other things. Given this ontological basis of SE, knowledge is traditionally viewed as a matter essentially of propositional attitudes expressible by means of complex pragmatical predicates: 'knows', 'believes', 'doubts', 'is certain', etc.. This shows TE in an interesting but hitherto ignored aspect as follows: although TE itself is not concerned with objective knowledge, it presupposes it. That is, it presupposes a whole pragmatically or ideologically neutral ontological order of epistemic structures, as abstract as *propositions* and *problems* in mutual interaction, with respect to which alone a knowing subject can adopt any one attitude out of the vast range of possible epistemic attitudes. Thus it is hardly surprising that TE should analyze/explicate the subjective K-states of the following general form:

'S *knows* that *p*'
'S *believes* that *p*'
'S *doubts* that *p*'
'S *is certain* that *p*

Thus, it is clear from the very complexity of its own problems how much TE borrows from OE. For subjective knowledge, as we have been accustomed to calling the very subject matter of TE, is another name for these subjectively organized K-states. They have to do, at their core, with our basic epistemic attitudes, whether to *propositions* generally or to *theories and problems* in mutual interaction, expressing fundamental relationships between the knowing subjects and the given sets of relevant epistemic structures. The epistemic structures I have here in mind, in their turn, codetermine the structural identity (= the fine structure, as we might call it) of objective knowledge systems as those systems which grow and develop in time both in their historical and methodological dimensions. (For discussion, turn to Chapters 5–10, Part Two, below).

CHAPTER 2

HUMAN KNOWLEDGE AND HUMAN INTERACTION

> All perception of beauty involves the feeling of being affected by the beautiful property as if by contact. The feeling of being contacted can be experienced in virtually any part of the body . . .
>
> Guy Sircello (1979)

> Each of us believes himself to live directly within the world that surrounds him, to sense its objects and events precisely, to live in real and current time. I assert that these are perceptual illusions, for each of us confronts the world from a brain linked to what is 'out there' by a few million fragile sensory nerve fibres. These are our only information channels, our lifelines to reality. These sensory nerve fibres are not high-fidelity recorders, for they accentuate certain stimulus features, neglect others. The central neuron is a story-teller with regard to the afferant nerve fibres; and he is never completely trustworthy, allowing distortions of quality and measure, within a strained but isomorphic spatial relation between 'outside' and 'inside'. Sensation is an abstraction, not a replication, of the real world.
>
> V. B. Mountcastle (1975)

> If the ultimate aim of the whole of science is indeed, as I believe, to clarify man's relationship to the universe, then biology must be accorded a central position, since of all disciplines it is the one that endeavours to go most directly to the heart of the problems that must be resolved before that of 'human nature' can be framed in other than metaphysical terms.
>
> Jacques Monod (1972)

2.1. OF THE PHILOSOPHER'S OBSESSION WITH PERCEPTION

'Perception' is a general term of wide connotation. From (what might appropriately be called) an interaction-theoretic view of perception it is plausible, I think, both to look at its range as a continuum of interaction between the perceiving subject/individual and its environment on the one hand *and* at the whole range of its endproducts, on the other hand, as a range of *apparent perceptual (anschaulichen) products* subject to variable interpretation. Thus, as a continuum of interaction, perception ranges from the most intimate (types of) contact with objects/persons in one's environment (including one's stumbling on such objects, say, in poor visibility, etc.) to the most distantly related visual or auditory experience of them. The perceptual products correspondingly range from one's visible excitement, tears in one's eyes, cries of joy, sketching, painting, uttering sentences (including one-word sentences), *reporting* what one sees or hears and bodily *movement* towards objects, and so on.

Now, each of the elements in the range of apparent perceptual-products will require interpretation, if only for the purpose of inter-subjective communication – no matter whether it is an uttered sentence, a tear in an eye, or a painting. Recognition of the fact that this is necessary will go a long way to show that perception (Wharnehmung) is not a species of knowledge – that perception by itself is not an epistemologically self-sufficient cognitive process or product. It is necessary since, as a rule, all our *interaction* requires a complex rule-governed conceptual setting without which it would be rendered impossible/insignificant or random, and its products unintelligible. And it can be shown how it is not only incorrect but absurd to think otherwise.

But, strangely enough, it is precisely an obsession with perception pre-conceived as an epistemologically self-sufficient cognitive process/product which has considerably bedevilled philosophy ever since antiquity. This deep-rooted obsession has had far-reaching consequences, making in-roads into the vital spheres of our day-to-day practical life. Thus, e.g., it can be shown how it has done greatest harm even to our (inherently imperfect) legal systems of punishment as embodied in different systems of criminal law. For, knowingly or unknowingly, there are built into these systems *infallibility assumptions and requirements* emanating from such concepts of evidence as are deep-rooted in our (ordinary and philosophical) pre-conceptions of perception.[1]

2.2. THE PROBLEM OF PERCEPTION

Consider a perceptual situation as a problem-situation involving a problem-solving human subject who not only 'sees' things and events taking place around him but does so self-consciously, and reports what he 'sees'. What do 'self-consciousness' and 'reporting' signify in such a context? If anything at all, these point to human problem-situations as highly complex situations of interaction of which perception is just a species. The very fact that a human perceiving subject is able not only to have perceptual experience but to understand and produce perceptual 'statements' – indeed a potentially infinite set of such 'statements' – suggests that he must be embedded in a much wider and highly complex framework which makes human interaction possible. Since, as I shall argue later on, the idea of the complexity of human interactions can be understood precisely only in terms of the idea of multi-variable interaction, human interaction must be impossible without such an embedding framework.

Depending on the degree of complexity of the perceptual situation, the perceptual reports of the human perceiving subject may range from relatively simpler 'statements' about a table in a room, a book on a table, a red flower in a flower-vase, a sweet tasting apple, to relatively more complex 'statements' about an over-crowded bus, a sudden traffic jam, a sudden failure of the vehicular traffic-control system or a clear sky after many days of cloudy weather and the like. Traditionally, such 'statements' of perception have been of undisputed interest and significance to the philosopher grappling with the problem of knowledge, the psychologist's study of perception being a relatively newer development. Thus, such 'statements' invariably figure in the traditional discussions on the problem of knowledge. In these discussions, perception has, I believe, always been considered a strong competitor of science as regards the latter's claim to its being a unique *search* for *knowledge*. That is to say, in a perceptual situation of the kind that involves a human perceiving subject, perceptions or the reports of perception as end-results have generally been considered epistemologically significant. Their epistemological significance is thought to consist in the 'fact' that they express knowledge-states of a simpler and least disputable kind, which are attributable to the human subject. Empiricist theories of knowledge go a step beyond this somewhat naive commonsense view to elevate them to the privileged position of *the very empirical basis* of all human knowledge. Taking perception to be a paradigm of knowledge, the whole problem of knowledge – of our knowledge of the external world as well as of other minds – has generally been thought by the empiricists to centre around the problem of perception. Thus, there has always been the problem of determining as to how much of the totality of our claims to knowledge can be made sense of in terms of perceptual/

observational 'knowledge'.[2] Placed as we are between these two views – i.e. the naive commonsense view *and* the empiricist view and its further variations – we have, I think, yet to do justice to that lone insight which historically we owe to Plato. This is the insight that knowledge is not perception *and* that perception is not a species of knowledge.

Now, how are the scientist's and the philosopher's theoretical interests in perception to be distinguished from one another? I think that one significant difference between their enterprises has to do with the fact that while the former aims at theory-building with a view to reconstructing in empirically significant and testable ways the system of *interactions* leading up to conscious perception as the end-result, the latter is simply so overwhelmed by the end-result of perceptual experience that it never occurs to him, in his state of epistemological worries, to overcome his intellectual excitement or to take a more detached view of perceptual experience. By studying perception as an end-result of the vastly complex processes of interaction between the human subject, himself a highly complex being, and the environment, the psychologist and the neuro-physiologist attack the problem of perception as part of their research programme to unravel important and deeper aspects of man's relationship with the world.[3] What is thus noteworthy is the significant fact that the problem of the nature of man's relationship with the world – of the structural and functional aspects of this relationship – itself provides the rationale for the psychological and neuro-physiological studies in the field of perception, among other fields. But what is the rationale behind the philosopher's persistant concern with perception? The most obvious and the least disputable answer would seem to be this: that it is his concern with the basic problem of the nature and foundations of human knowledge which provides the rationale for his habitual appeal to perceptual experience invariably with a view to removing his epistemological worries. But, I think, that is really no answer, unless it is beyond doubt and controversy from the very beginning that perception or the end-result thereof is a species of knowledge and that it can be studied to illuminate important aspects of human knowledge as a whole. I think that any appearance of a parallelism between the problem-situation of the scientist and that of the philosopher in the field of perception, in the present context, is bound to be illusory and deceptive. Thus it is quite reasonable to hold that, perception being a species or an active mode of man's complex relationship of interaction with the world, the scientist can quite fruitfully undertake to clarify the nature of this relationship by clarifying the nature and dynamics of perception. On the other hand, however, it cannot be granted, without conclusive argument, that perception is a species of human knowledge, or that it might be studied to illuminate the nature of the latter. Hence, a fundamental difficulty of the philosopher's rationale for his whole epistemological enterprise centering around perception must be faced and resolved.

Let us now consider situations involving man-made systems with built-in remote-sensing and central self-regulatory mechanisms or artificial systems like sophisticated computerized servo-mechanisms. As an example of the former type of systems one can think of a sophisticated strategic war plane on a bombing mission with a built-in remote-sensing device designed to signal warnings about the enemy's air offensive or counter-moves in the air. The signals are so fed back into the central self-regulatory system that it keeps changing the course of movement of the plane as part of its strategic performance. On the other hand, as an example of the latter type of systems that closely resemble the former type of systems, one may think of machines designed to perform certain specified functions, say in a factory. The performance of the machines 'is registered in a series of electric impulses which are fed into a computer'. The computer is so programmed as to produce specific signals whenever there is a discrepancy between the expected and the actual performance of the machines. These signals are then fed back into the central self-regulatory system for correcting the discrepancy. Now, important similarities between the signals of warning/discrepancy fed into the central self-regulatory mechanisms of such artificial systems *and* human perceptions/perceptual 'statements' are discernible as follows:

(a) The signals of warning/discrepancy are vital aspects of the complex processes of interaction between a remote-sensing device and the relevant parameters in its 'environment' or between a performing machine and its computerized monitoring system. Likewise, human perceptions/perceptual statements are vital aspects of a complex process of interaction between the human problem-solving subject and the surrounding world.

(b) Both, the signals of warning/discrepancy and human perceptions, have a strategic functional significance in the context of the larger systems in which the artificial systems and the human subjects may respectively be embedded. As is obvious from the fact that both the signals of warning/discrepancy and human perception are fed back into the central self-regulatory sub-systems of the artificial systems and the human subject respectively, their strategic functional significance is often full of negative feedback consequences for the subsequent processes of interaction at different levels. Thus, the signals of warning/discrepancy or of performance play an indispensable role in maintaining the complex problem-solving artificial systems in their strategic interaction with the relevant parameters in their 'environment'. Similarly, human perception/perceptual language plays an indispensable role in maintaining the human problem-solving subject in its strategic multi-variable interactions with the world at various levels.

(c) Without arguing for the point here, we may add that the human perception as well as the signals of performance, warning or discrepancy are characterized by varying degrees of *resolving power*. For example, the perception by a human subject intending to cross a busy road at a given point *resolves* the problem-situation for him into the chances for and against his crossing the road at a given moment of time. Similarly, we might say that the signals of performance, warning or discrepancy resolve the problem-situation for a performing artificial system into the chances for and against its expected performance. Resolving power in this sense can in ultimate analysis be understood as a function of the background conjectures, beliefs or theories in the case of the human subject or of the background plans and programmes of performance in the case of the artificial systems.

Now, given these interesting similarities, in what way are the two kinds of situation different? What is the philosopher's rationale of singling out human perception for the designation 'knowledge', when in its nature and function it seems essentially similar to the man-made problem-solving systems' signals of performance, warning or discrepancy? Why do we say in the case of the human perception that '*X* knows (perceives) that *p*', and in the case of the artificial monitoring system that 'the radar *shows* the air-space violation by the enemy plane at such and such altitude', where the latter serves as a warning signal which activates the air-defence system for an appropriate action? Any simple man-versus-machine argument cannot be invoked to justify the seeming asymmetry between the human perception and the performance signals of the artificial systems. Any such argument could be suitably framed and used instead to show that this seeming asymmetry is only an illusion: that the real distinction between the two lies elsewhere and is not brought out by calling one a case of knowledge while simply rejecting or ignoring the other.

It might be argued that human perception is distinguished by the fact of consciousness – i.e. by the unique phenomenon of *human awareness*. Human beings not only perceive what they do perceive, but they also show consciousness of their perceptions/perceptual activity. Thus, human beings are not just perceiving organisms; among other things, they are *self-conscious* perceiving subjects. But I think that to admit this is to recognize the vast complexities of human interaction as a multivariable interaction: If in perception man interacts with the world including the social world, in self-consciousness man interacts with his perceptions among other things. It might be argued that all this has been admitted in the discussion above in which the artificial problem-solving system was likened to the human perceiving subject. It might be said that the monitoring system of the former responsible for signals of performance/warning is similar or approximates to the self-consciousness of the human perceiving subject. And from here it might seem tempting to extend the analogy to the whole realm of the lower animals so as to render their problem-solving activity indistinguishable from that of the human subject. Any such argument is bound to involve over-simplifications including the one which models self-consciousness after the simple monitoring systems of either the artificial systems or of the lower animals. It may be correct to say that it is to self-consciousness that we must appeal as a first step of an affirmative answer to the question: what *unique* features are there in human perception or in the very *conditions* of human perception taken as a relationship of interaction between man and the world, which would warrant a philosophical investigation over and above a proper scientific one? The assumption that there must be some such unique features is made plausible by the fact that the philosopher's rationale for his enterprise concerning perception cannot be the same as that of the scientist for a subject-specific scientific investigation.

The very fact that we are trying to define human perception in terms of the concept of awareness/self-consciousness shows that the two concepts must not be confused with each other: generally, perception does not imply awareness. It implies interaction between the perceiving being and its relevant environmental parameters in the precise sense of "registering inputs as stimuli and behaving in relation to these promptings".[4] On the other hand, we may say that in awareness one interacts with oneself. But what does that mean? Awareness implies interaction at a higher level involving the knowing subject and presupposing the lower-level interactions between the human problem-solving subject and the world.[5] Thus, to be aware/conscious of what is taking place is to be a knowing subject in interaction with these lower-level interactions and the products thereof. It follows, as a special case, that to be a self-conscious perceiving subject is to be a vastly complex problem-solving being in interaction with oneself as a knowing subject. (For further discussion on self-consciousness turn to Chapter 4, § 4.3.)

2.3. ON INTERACTION WITH WHAT OUR THEORIES INVARIABLY SINGLE OUT AS OBSERVABLES

What makes our procedures of (non-random) *observation*, one might ask, so significant to us? Does their significance in any sense derive from the fact that we are ourselves the kind of complex problem-solving and problem-posing beings that are *capable* of participating in the interactive organization of the whole universe? It is possible, I think, to conceptualize these procedures, say in science, as a species of our significant interactions with nature, particularly those aspects of it that are capable of identification beforehand as its 'observable' aspects. These interactions are, in their turn, invariably traceable in their orientation and organization to our problem-ridden or problem-posing background systems of knowledge. All human interaction is, one might say, essentially a problem-solving and problem-posing activity. Observation – scientific and non-random, no matter whether aided by the sophisticated technology or not – cannot be an exception to this. It may thus be quite safe and reasonable to assume that, for man in his capacity as the problem-solving knowing subject, to engage in the observation of any kind of entities, their properties and relations in the field of the 'observables' is to *interact* with those entities in significant ways full of possibilities of negative feedback. But all problem-solving and problem-posing activity that entails our significant interaction with the observable aspects of nature would be impossible if we did not *learn* first to delineate, or single out, as the observable aspects of the world, those very aspects of it with which we could interact in selective and significant ways. Thus, every step that we may take in order to orient our problem-solving activity in a particular field will presuppose both an identification and a delineation of the observables of an appropriate kind for us to interact with. And as a result of our actual interactions with the observables already so delineated, our problem-posing activity, like its problem-solving counterpart, is bound to acquire a dimension of complexity demanding ever more comprehensive delineations, thereby leading to a constant widening of the world of the observables. More specifically, all our theoretically significant problem-solving (and problem-posing) activity presupposes, one might argue, the singling out of certain types of entities, processes and relations as those relevant *observable* aspects of the world with which we can interact according to our varied human interests. Once clearly identified, they can serve as an important source of constraint on the ongoing enterprise of theory-construction on the one hand and theory-change on the other. To recognize this is, then, to accept the primacy of *knowledge* as an orientating variable of significant human interaction. It is also to accept, I think, the primacy of the knowing subject in all our significant interactions with an ontically complex world[6] that we will never be tired of knowing more and more about. But what is it which enables us to identify and delineate as the 'observables' of the relevant type those aspects of the world with which we might choose to interact in significant and instructive ways? No doubt, this is the type of question which we should ask not just in the human-interests-constituted contexts of ordinary day-to-day life but in the context of science realistically conceived as an enterprise comprising relatively autonomous theory-problem interactive systems (being considered in Chapter 8 below).

Historical development of science shows how often novel theoretical problems have arisen as subjects of scientific discovery, demanding novel theories which can yield solutions that are not within the power of the relevant background theory. Adequacy of the kinds of solution sought will depend on the power of the underlying theory to explain the explananda – its T_{EP} – in the domain already delineated, or singled out, as one of the appropriate type of observables. Not only this, novel problem-determination itself shows a crucial dependence on interactive systems of background knowledge. It is well-known about these systems that it is the basic assumptions they employ which are responsible for singling out certain aspects of the world as its observable aspects. If we argue that these assumptions must themselves be either theoretical or empirical in nature, it would simply shift the analysis from one level to the other. But the important question which we should consider in some depth is this: How does a science achieve all its delineations with which it singles out relevant aspects of the world as its 'observable' aspects?

I think that it is possible to answer this question in terms of the principle of a theory's resolving power (= T_{RP}). Roughly, this is its power to resolve, or differentiate, the relevant entities, properties and relations in its field into (hitherto unsuspected) 'observable' and 'non-observable' types respectively. A theory is able to achieve this differentiation by virtue of the theoretical identifications it itself postulates. It is to these identifications that it owes its power to shift the scene of problem-determination from the field of the background theory to its own field. More precisely, then, a theory's resolving power may be defined as its power to give rise to new problems on the one hand and radical transformation of the familiar problems on the other. Only a theory which possesses T_{RP} in this sense, one might argue, can differentiate the relevant entities, processes and relations in its field into 'observables' and 'non-observables' respectively. Thus the delineation of the 'observables' as against the 'non-observables' in the relevant domains of scientific investigation must be a function of a theory's T_{RP} in the sense just defined. Bergmann's following remark serves, I think, to illustrate the whole point:

> What appears to me significant is that to single out the motion of the distant masses as something observable is an integral part of the structure of a physical theory that we might identify as classical celestial mechanics. Other theories involve different assumptions as to what is 'observable', and usually these assumptions are non-trivial.[7]

If it is correct to hold the identification, and delineation, of the appropriate type of observables in the domain of one's interest as a function of T_{RP}, it follows from this that all observation in science is T_{RP}-determined. At any time, and in any domain of our scientific interest, relevant type of observation (as a species of significant human interaction) can take place within the limits already set by the resolving power of our extant background theory. In a given domain of our scientific interest, therefore, there are conceivably no observations which are not T_{RP}-determined. There follows from this a special consequence for the present day quantum mechanics, itself a center of debate for more than half a century now concerning the EPR, measurement-interaction, locality, separability and so forth. It is necessary that its theoretical background and extant structure are spelt out and elaborated more exactly if the permissible range of observations, and experimental situations, appropriate to its field are to be specified and delineated without any conceptual paradoxes whatever. Time and again we have been here reminded of Heisenberg's uncertainty principle that we can choose to measure accurately either one or the other aspect of an experimental situation – e.g. either the *momentum* or the *position* of an electron – but not both *simultaneously*. In other words, " . . . there is no experimental arrangement conceivable by which we can gain knowledge simultaneously of all the characteristics of an atomic system which separately can be ascertained without difficulty".[8] On most orthodox interpretation of the formalism of quantum mechanics, known as Copenhagen interpretation, this is generally given an ontic interpretation so that what appears at first as an uncontrollable and irreducible factor of measurement-interaction in the context of quantum mechanics is seen as inherent feature in nature itself. If there is any element of correctness in the kind of account of observation I am here aiming at, a reappraisal of the physicist's approach to the structure and function of physical theory becomes necessary, especially in view of those conceptual confusions which bedevil this particular field. Once these confusions are cleared, everything will fall into a pattern, even Einstein's deep dissatisfaction with the dominant view of his time that, unlike classical physics, the basic framework of quantum mechanics is deeply probabilistic in nature. The basic reason for his dissatisfaction can be traced, at least in my view, to the fact that quantum mechanics, like any other physical theory, proceeds from theoretical identifications it itself makes. As to probabilism, it enters here at the level of the experimental laws of interaction involving the experimenter and the quantum objects being experimented with – i.e. at the level of predictions of the future course of events in its domain. There is nothing in its basic assumptions, or in the field of their experimental testing, nor in the problems correlated with them, that could set final limits to the resolving power of physical theory in its own domain. The uncertainty principle states, of course, a law of measurement-interaction in this domain. But it does this by virtue of the basic theoretical identifications of the theory-problem interactive system that makes the corpus of knowledge in quantum mechanics at present. One might, therefore, wonder whether its customary interpretation is not based on the empiricist dogma that observation is the source and foundation of all our knowledge. Only if we give up this dogma in favour of the idea that observation, as a species of human interaction, invariably entails irreducible disturbance in the systems one (=the observer/knowing subject) interacts with, is it possible to argue for the view that *it* presupposes knowledge as an orientating variable which assumes problems and theories as its values. The growth of knowledge could in this context be taken to mean a progressive proliferation of theories with an ever increasing T_{EP} and T_{RP}. Thus, the limit the uncertainty principle is supposed to draw to the range of the measurement-interactions with the sub-atomic systems in quantum mechanics can be traced back to the theoretical background and T_{RP} of quantum mechanics itself. The situation here would, therefore, in no way be different from that in astrophysics. At present, as at any time, our *theories* in the latter domain permit significant interactions with the *large* only within certain limits set by their T_{RP}. Thus all our interactions at the levels of the small and the large are in equal measure T_{RP}-determined.

In so far as quantum mechanics is founded on the assumption that all measurement-interactions, including those between the classically describable experimental apparatus and the non-classical quantum objects being experimented with, are vastly complex, it is beyond its power to draw limits to knowledge in its own domain. On the contrary, what it does is to legislate in favour of those patterns of *prediction* which are capable of capturing the maximal, if not all, complexity of the processes of physical interaction at the micro-physical level: *prediction of the behaviour of the individual particle must give way to one of the whole ensembles of identically prepared particles*. Any attempt not to do so would be in violation of the very assumption of micro-physical complexity.

It does not look at all surprising that quantum mechanics should have been known more for the patterns of prediction it favours in its own field than for other aspects. No doubt, physics seeks to *describe* the microphysical domain of the small just as it does describe the macrophysical domain of the very large. The patterns of prediction of the outcome of any experimental set up will, however, differ at the two levels according to the laws of force and levels of complexity of measurement-interactions that may be involved. The inherent complexity of the microphysical domain is such – with probability and measurement entering here irreducibly as it were – that it affects all our interactions in fundamental ways ruling out certain patterns of prediction, and methodological rules thereof, as inadvisable and inappropriate. Thus given the uncertainty principle, or Bohr's complimentarity principle, a scientist must here resort to statistical probabilistic laws of whole ensembles of particles, treating particles as wave-packets, instead of the classical patterns of prediction which are useful elsewhere. If we cannot predict the momentum and the position of an electron simultaneously, we must here give up classical patterns of prediction in favour of those of a statistical nature. But how far can we remain silent about the individual particles themselves in order to describe, or predict, the behaviour of whole ensembles?[9] It was Einstein (1949: 671) who said that "the ψ-function is to be understood as the description not of a single system but of an ensemble of systems".

2.4. THE PROBLEM OF HUMAN INTERACTION

It is, I think, true of every science that it explores not only non-human interactions in and around us in the universe but human interactions in their physical, biological and other dimensions. But how is it possible to distinguish one kind of interaction from the other? Non-human interactions are, in their turn, of two types: those which involve living beings other than human beings on the one hand and those not involving any such living beings on the other. Human interactions, however, are those which take place between human beings and the environmental variables in the physical universe, or between them and the rest of the biological world (e.g. living creatures such as animals) or between human beings themselves, or even between human beings and their culturally generated *worlds*.

It is quite reasonable to suppose that there must be more to it if it is to serve as a basic distinction. As modern biology teaches us, a living organism's relationship with its physical environment, or with any other living organism, is always one of interaction resulting in the formation of all kinds of interactive systems in nature. Similarly, the relationship between human beings and the world is one of intimate interaction which generates, in its turn, interactive systems of its own unique order of complexity. But what makes human interaction, then, the kind of interaction that it is, so unique and different from all kinds of non-human interaction that neither of them can be assimilated to the other?

There may not be just one single factor but a number of factors with which we can understand their difference. It may be possible, therefore, to answer this question only in terms of a detailed study and analysis of the nature of human interactions themselves. However, I shall here discuss a possible answer which is suggested by the unique complexity and plasticity of human interaction and by the recurrent properties of its orientation and reorientation.

Like non-human interactions, human interactions too are essentially multi-variable interactions. In the case of the former, it is possible to determine to a lesser or a greater degree the values of the variables of interaction at some time in the future from their values at present. But it is not possible to do so in the case of the latter, at least with regard to all the variables of interaction which may be involved. How can we explain this essential unpredictability of the values of the variables of human interaction at any time? It is quite reasonable to assume that all non-human interactions follow a more or less fixed, rigid and closed pattern. Human interactions cannot be brought under this assumption. They do not follow any fixed or predictable pattern, since they are more complex and uniquely plastic in nature. The source of their plasticity is not to be traced to the environmental variables (= E-variables) rooted in the outside physical or biological world. It is traceable, on the contrary, to what might be appropriately called the *orientating variable* of human interaction, itself rooted in human beings as *knowing subjects*. Thus, it is quite reasonable as a first approximation, to be followed by more accurate formulations in § 2.9, to distinguish human interactions from non-human interactions in terms of *human knowledge*, conceived as their orientating variable of interaction. The claim I am here making is that most of the relations of human interaction, if not all, are informed, shaped and structured by *knowledge* as their orientating variable. Without this being so, these relations would all still be there. But they would be of a totally different character and would present a totally different – though not irreducible – order of complexity.

Now, what is the nature of this orientating variable itself, popularly called *knowledge*? Obviously, we must appeal to epistemology, the theory of human knowledge, if this important question is to be answered at all. But by bringing epistemology back in, one naturally invites all the troubles associated with the concept of human knowledge. How is human knowledge to be understood? This is one central question of epistemology which we must deal with, even if very briefly.

2.5. ON UNDERSTANDING THE CONCEPT OF HUMAN KNOWLEDGE

There are, I think, two ways to understand the concept of human knowledge, a subjectivist way and an objectivist way. Indeed, it is not only of immense philosophical significance but also methodologically necessary to distinguish between 'knowledge' in a subjective sense of the term and 'knowledge' in an objective sense of it. It is possible to draw such an important distinction today mainly because of certain important developments which have shaken the traditional foundations of philosophy as well as science. These developments enable us to see clearly that the concept of human knowledge calls for a critical and an objective analysis within the framework of the reasonable assumption that there are at least two chief dimensions to it: one *subjective* and the other *objective*. Such an analysis must assume a deeper philosophical significance once it is realized that there is an intimate relationship between human knowledge on the one hand and man's relationship of interaction with the world on the other. The kind of role the former plays in informing and structuring the latter is, I think, a unique human phenomenon. Thus it is necessary to distinguish the following two problems from each other: (1) the traditional philosophical problem of the nature of human knowledge and (2) the problem of the nature of man's relationship with the world. Although much of traditional epistemology is founded on a confusion between (1) and (2), the two problems are not identical. If they were, the physical, biological and social sciences – in so far as these are concerned with different fields of human interaction and hence with different aspects of man's relationship with the world – would be reducible just to different branches of epistemology and *vice versa*. On the contrary, any assumption to that effect will lead us to nothing but absurd consequences.

Traditional epistemology assumes quite dogmatically that subjectivity is all that there is to human knowledge, thereby making itself centre around the *knowing subject* along with the various kinds of cognitive states as its *frame of reference*. Thus traditionally and subjectivistically conceived, knowledge is conceptually inseperable from the idea of a knowing subject. This reduces the problem of identifying the true nature of human knowledge to one of identifying the most privileged kind of cognitive state of the human mind/consciousness and its possible relations with other less privileged states such as those of belief, opinion, etc. This early subjective orientation of traditional epistemology is, no doubt, responsible for a late objectivistic orientation of it. It will become clear as we proceed how important it is to distinguish the subjective concept of human knowledge, as having to do with the cognitive states of a knowing subject, from its objective counterpart, the objective concept of knowledge.

No doubt, the distinctions which traditional epistemology has always drawn between knowledge, belief and perception can be redrawn and shown to be of a much greater philosophical significance than ever before. But what is noteworthy is this: these distinctions were drawn then for wrong reasons and in a wrong manner. After all why is it that the problem of knowledge in traditional epistemology is posed as a problem of conferring a privileged status on various candidates by means of the labels 'knowledge', 'belief', 'opinion', 'perception' and the like? Is the concept of knowledge so simple a concept and are the candidates so obvious that the definition of knowledge could simply be a matter of selection of one of them for the privileged status of 'knowledge'? Questions like these raise serious doubts as to the validity of the basic philosophical framework of assumptions of traditional epistemology. One of the characteristic features of this framework is, I think, the confusion between the subjective and the objective dimensions of human knowledge. While another is its confusion of the problem of the nature of human knowledge with the problem of the nature of man's relationship with the world. These remarks along with the questions above will assume considerable significance in the light of the thesis concerning human knowledge that I would like to propose as follows.

Let it be our basic assumption that man's relationship with the physical, biological and social world is essentially one of interaction. This assumption has an interesting consequence, viz. that man, like other living beings of varying orders of complexity and organization, is a problem-solving living being, although, unlike these other living beings, man possesses further unique properties such as the traditionally recognized property of *rationality*. This consequence of our assumption in its turn leads to another which is empirically significant and testable, viz.: with all its *biologically innate* (additional) properties such as those of rationality, creativity, etc., man remains one of the different kinds of hierarchically organized problem-solving living beings which belong to a common universe. The significance of the assumption made above will become more clear if we compare it with the kind of assumption which is empty owing to its lack of an empirically significant and testable consequence. Consider, e.g., the special creation theory, which says that man is a specially created being distinct from and superior to the rest of the world of living beings – that man does not *belong* to *this* world of living beings and hence cannot *interact* with it. Such a theory obviously assumes that besides *this* world there is another – the world of 'God' etc. – 'with which any connection is impossible in principle'. Clearly, such an assumption leads to no empirically significant or testable consequence as regards the problem of the nature of man's relationship with the world.

Objectivistically speaking, it would be quite reasonable to view knowledge as an orientating variable of human interaction, that ranges over scientific theories/hypotheses and problems as its possible values. In its subjective dimension too knowledge is definable as an orientating variable of all significant human interaction. However, the difference lies in this: from a subjectivistic point of view, the orientating variable ranges over typical states of consciousness/human mind, where these states are conceivable as states of internal organization of man the knowing subject. Such states are distinguishable from other types of states of man's internal organization in the broader interactive context by their unique orientative potential, their capacity to orientate man in his complex and problematic relations of interaction with the world, physical as well as social – i.e. with the E-variables which are rooted in the world. To restate my thesis, human knowledge in its subjective dimension might be defined as an orientating variable which ranges over all those states of internal organization in man as its values which orientate and structure his interactions with the world. I think that without knowledge as such a state of internal organization, man would not interact with the world with necessary orientation in a manner different from non-human interactions. No doubt, as a rule, part of the interaction process or the result thereof acts as an input to knowledge in its present sense, as a result of which knowledge may vary from one distinguishable stage of interaction to another and also from one person to another. Indeed, we might as well lay it down as a principle of an asymmetry of internal organization that at any given time, for any two or more persons in their capacity as knowing subjects, knowledge as an orientating variable may not assume exactly one and the same value.

Our thesis above – which will be made more precise in § 2.9 – has the following implications in terms of which it might be further developed and examined. First, taking knowledge as an orientating variable of human interaction entails the impossibility of the latter without the former as a presupposition. That is, no significant human interaction can take place without the antecedent orientating function of knowledge. This suggests that epistemology's rationale of inquiring into the nature of *knowledge* is deeply rooted in the question: How is human interaction possible? In so far as it inquires into the former it must also inquire into the latter. Secondly, since knowledge is what makes human interaction possible, knowledge is neither identical with nor a form/species of the human interaction-relation/process. If the assumption (c) in § 2.2 above is correct then, on default of an infinite regress, knowledge cannot itself constitute a species of human interaction. This holds true even if it is true of all human interaction that, as a rule, a part of it returns as input to knowledge as an orientating variable. Thirdly, as a further consequence, all perception, observation, memory, etc. are not proper forms of knowledge; for each designates a form of relation of human interaction presupposing knowledge as an orientating variable. Fourthly, the idea of knowledge as an orientating variable entails that knowledge must be *conjectural* in character – a thesis most powerfully argued for by Karl Popper. For only a conjecture can answer to the characterization of knowledge as an orientating variable of human interaction. In other words, only conjectures, which are refutable, in somewhat Popper's sense and hence which are highly variable, together with problems, can constitute the legitimate range of values of knowledge as an orientating variable of human interaction.

Again, the present characterization of knowledge has far-reaching consequences for traditional epistemology, especially its problem of the foundations of human knowledge. Understood as a problem of rationality, traditional epistemology seeks to define the rationality of our K-states in terms of their relationship with the foundations of knowledge. Thus, in order to be adjudged as rational a K-state has to be firmly entrenched in or founded upon 'experience' (empirical certainties) in case of empiricism or intuitive certainties in case of rationalism. The 'given' foundations are thus conceived of not only as a corpus of basic truths but also as the source of certainty and rationality of the rest of the superstructure of human knowledge. Now, under the present (conjectural) view of knowledge as an orientating variable of human interaction, the problem of rationality can no longer arise in its traditional garb, say as the problem of the foundations or as the problem of *distance* between the foundations and the rest of our K-states. On the contrary, the whole problem is so transformed as to concern the *growth* of human knowledge, the very dynamics of its underlying patterns. Our idea of knowledge as an orientating variable of human interaction together with the idea of experience/perception as a form of interaction and as a constant source of input to knowledge implies that it should be possible for our knowledge to grow constantly both at the subjective and the objective level. And in fact our knowledge, instead of stagnating, keeps growing at both these two levels. The problem of rationality is thus the problem of investigating or reconstructing the possible underlying rational (but not predictable) patterns of the growth of human knowledge.

Lastly, as a corollary of the third consequence above, it follows further that to be a perceiving organism is not necessarily to be a knowing/understanding subject. Like we humans, many other kinds of perceiving organisms inhabit the living world, possessing varying powers of discrimination, recognition, memory, etc. But their membership of the world of perceiving organisms does not make them the same knowing subjects that we are. This consequence is, I think, especially of far-reaching philosophical significance bringing part of Plato's and even Descartes' epistemology back in. Let us consider this consequence more closely by examining the distinction between the knowing subject and the perceiving organism, which is traceable to Plato.

2.6. THE KNOWING SUBJECT AND THE PERCEIVING ORGANISM: THE STRUCTURAL AND FUNCTIONAL ASYMMETRY BETWEEN KNOWLEDGE AND PERCEPTION

Why it is very often the case that humans believe in what experience contradicts, that the human mind discerns and discloses what the eyes cannot see, the ears cannot hear and the other senses cannot feel? Is there a fundamental asymmetry of function and structure between the knowing subject and the perceiving organism and, by implication, between knowledge and perception? There is no simple answer to this complex question, since not only the history of religions and philosophy but that of science itself is characteristically full of this perpetual tension between belief and experience, between theory and perception.

There can hardly be any doubt that our relationship with the world is one of interaction most often informed by knowledge as its orientating variable. That is, it is only the living beings as complex as the humans that are capable of interacting with the world, with knowledge serving as an orientating variable of interaction. If this is granted, then *knowledge* is not itself a form of human interaction. On the contrary, it is that which distinguishes human interaction from non-human interaction. Now, in so far as human perception or sense-experience is essentially a species of human interaction, it must, in order to be possible as well as significant, presuppose knowledge in the same way that all significant human interaction does. This itself implies a sharp distinction between knowledge and perception which must in its turn lead us further to the corresponding distinction between the knowing subject and the perceiving organism – a distinction which is also suggested by what Plato and later on Descartes had to say concerning knowledge.

I think that any philosophical doctrine which identifies *knowledge* with perception or, more generally, with *experience* has the strange and absurd consequence that our relationship with the surrounding world is one of intimate knowledge: that a man's or a woman's relationship with the world around him or her for most of the time, if not for all the time, is a relationship of knowledge. Although here the precise sense of 'knowledge' is determined by how precisely the term 'experience' is defined, this is a consequence, I think, which invariably follows from all empiricist theories of knowledge including Berkeley's theory of *esse est percipi*: To be is to be perceived/experienced. Since for Berkeley 'perceived' carries the epistemological sense of 'known', his theory, in effect, says: To be (real) is to be known. Consequences of such a theory for the problem of the nature of man's relationship with the world are clear: The world, whether external or mental, will continue to be real and relevant for man in so far as he bears a relationship of knowledge/experience with it. And, in the event of this direct relationship of his being interrupted or absent, it is God's eternal relationship of knowledge with the world that guarantees the ontological/metaphysical status of the world for man. Now this theory is clearly a theory of man's relationship with the world, depicting this relationship as one of a *continuum* of knowledge interrupted only by the intervals of sleep and dreaming. But such a theory proceeds from a conflation of the problem of knowledge with the problem of the nature of man's relationship with the world, which shows that it cannot deal with the kind of questions concerning human interaction posed above.

If the distinction between the knowing subject and the perceiving organism is a valid one, the question which I wish to consider is this: Is Plato, the founder of systematic epistemology, refuting a *stronger* or a *weaker* thesis concerning the nature of knowledge when he criticizes the received theories of knowledge in order to propound his own?

(a) *The stronger thesis*: Knowledge is perception.
(b) *The weaker thesis*: Perception is (a species of) knowledge.

The two theses (a) and (b) are essentially incompatible with each other. This can be seen by drawing their respective consequences. Thus, e.g., the denial of the thesis (a) implies a distinction between the knowing subject and the perceiving organism, which is quite permissible under the thesis (b). That is to say, if we deny the thesis (a) then, in order to avoid confusion and other difficulties, we must draw a sharp conceptual distinction between the knowing subject and the perceiving organism as constituting different ontological categories altogether. I think that this is what is implied by Plato's rejection of the thesis (a) in favor of his own theory of knowledge. Again, if we accept the thesis (a) then one cannot draw such a distinction. On the contrary, one must admit then the consequence of the knowing subject being ontologically identical with the perceiving organism.

The distinction between the knowing subject (KS) and the perceiving organism (PO) thus helps us in reconstructing in more clear terms the main consequences of Plato's rejection of the theory of knowledge *as* perception in favor of his own as follows:
The stronger thesis:

(a_1) [Knowledge *is* perception] $\Rightarrow$ KS $\equiv$ PO
(a_2) $\sim$ [Knowledge is perception] $\Rightarrow$ KS $\not\equiv$ PO or, equivalently
(a_1) If knowledge is perception, then the knowing subject is identical with the perceiving organism, necessarily.
(a_2) If it is not the case that knowledge is perception, then it is not the case that the knowing subject is identical with the perceiving organism, necessarily.

The weaker thesis:

(b_1 [Perception is a *species* of knowledge] $\Rightarrow$ KS $\not\equiv$ PO
(b_2) $\sim$ [Perception is a species of knowledge] $\Rightarrow$ KS $\equiv$ PO or, equivalently
(b_1) If perception is a species of knowledge, then it is not the case that the knowing subject is identical with the perceiving organism, necessarily.

(b_2) If it is not the case that perception is a species of knowledge, then the knowing subject is identical with the perceiving organism, necessarily.

The incompatibility between the theses (a) and (b) above is thus clearly brought out by the relations of compatibility between the theses (a_2) and (b_1), with the latter implying the former; and between the theses (a_1) and (b_2), with the former implying the latter. This analysis shows that Plato is not concerned just with a mere rejection of the weaker thesis (b) above. For if he were, he could not have thereby ruled out the doctrine of the knowing subject as being essentially identical with the perceiving organism. In other words, any such rejection would leave the stronger thesis (a_1) along with its consequences essentially unscathed. Hence, in so far as it is quite reasonable to presume that what Plato is deeply concerned with is to distinguish the knowing subject sharply from the perceiving organism, the only way to do so is by refuting the stronger thesis that knowledge is perception. But there is a real difficulty with (b_2) above, if taken by itself individually. From the fact that (b_1) implies non-identity between the knowing subject and the perceiving organism ($= KS \not\equiv PO$), it does not follow that its denial, (b_2), implies identity between them ($= KS \equiv PO$). It seems to me that (b_2), being a denial of the weaker thesis (b_1), taken individually by itself is still weaker leaving various possibilities open as to its real ontological significance. This means that (b_2) would hold provided it is taken to *presume* another thesis, say, b_3 as follows:

(b_3) If perception is not a species of knowledge, then perception is identical with knowledge, necessarily.

But with (b_3)[10] as its suppressed premise, as it were, (b_2) really turns out to be the sort of thesis which is far stronger than the weaker thesis (b_1), which it denies. Perhaps it would no longer be correct to consider it a weaker thesis. And it is only by virtue of its strength secured in this manner that it would imply that the corresponding identity between the relevant ontological categories holds ($= KS \equiv PO$). Thus left to itself, without the backing of (b_3), (b_2) should look like b_4 as follows:

(b_4) $\sim$ [Perception is a species of knowledge] $\Rightarrow$ [$(KS \equiv PO) \vee (KS \not\equiv PO)$]

And this shows how weak (b_4) really is to bear any non-tautologous significance for the ontological categories of the knowing subject and the perceiving organism.

2.7. THE ESSENTIAL UNPREDICTABILITY OF THE GROWTH OF KNOWLEDGE AND THE PROBLEM OF THE PREDICTABILITY OF INTERACTIONS

Our main argument above for the thesis of our knowledge (of the external world) as an orientating variable of human interaction derives from the distinction we have drawn between the problem of the nature of the relationship between man and the world *and* the problem of the nature of human knowledge itself. This distinction is, I think, significant in yet another methodological aspect which we must consider now. Since it could be argued that without human knowledge (whatever that may mean) the nature of man's *relationship* with the world would be qualitatively altogether different from what it actually is, it is methodologically quite legitimate to subject an epistemological theory to the external, objective test of evaluating its consequences for the problem of the nature of this relationship itself. I think that the TE-theories, particularly those in the empiricist tradition, will not survive this test, since they reduce knowledge to the relations of human interaction such as those of perception, sense-experience, etc., thereby obliterating the distinction between the two problems above altogether. I shall now put forward an additional independent argument for my thesis above in terms of the idea of an essential unpredictability of human knowledge and its growth. Let us begin by a consideration of Marxism which commits itself to the view (a) that knowledge designates one particular kind or order of human-interaction relations among others; and (b) that the pattern of knowledge as a pattern of interaction is not only predictable but actually determined by human-interaction relations of a strictly economic order. For according to Marxism taken as a metaphysical doctrine the following holds true of every human society:[11]

> The economic features of a society determine all its other features. In other words, the organization of the means of production in a society determines its other features. Any change in the former will lead to a definite change in the latter, but not *vice versa*.

This doctrine can be split up into the following set of theses:

(i) Every society grows according to certain definite 'laws' of historical development.

(ii) These laws are the laws of the predictable patterns of the societal development.

(iii) These laws are always of a causal form; they are of the form of causal equations describing causal relations between the economic variables as the independent variables on the one hand and the non-economic variables as the dependent variables on the other.

Now, according to any deterministic model of society, like Marxism, the prediction – of course purely in economic terms – of its future course is not only possible but must crucially affect this course in an essentially self-confirming manner. Any social determinist will of course insist on the predictability of the future developmental course of society. Will he deny the possible crucial effect of actual predictions on that course? Perhaps he need not. For he might argue that all predictions will crucially affect the course of development of a society in a strictly self-confirming manner. What he cannot say however is that these predictions can also affect that course in a self-refuting manner. For in that situation he would have to admit interference of our predictions or theories with the societal developmental course as being crucial and at the same time as defeating the very purpose of the deterministic model. That is to say, a determinist must hold that a prediction must *necessarily* come out to be true.

Now, the main point I wish to make is this: As a piece of theoretical knowledge, a prediction of its future course must necessarily interfere with the growth of a society. Can a determinist deny this? Unless a prediction has a use, why bring it in at all? I think that any deterministic model of a society must be committed to the view that our knowledge in the form of a prediction affects the growth of that society, of its future course of development in crucial ways, even if it be essentially in a self-confirming manner. If this is so, isn't the deterministic model essentially imperfect and incomplete in the sense that it must leave at least one undoubtedly crucial kind of factor or phenomenon outside the bounds of the model? For otherwise any attempt to conceive of a higher order deterministic model of the phenomenon of knowledge itself would only lead to an infinite regress and hence be futile. The implication is clear: knowledge in the form of a new theory and what the theory predicts and hence the growth of knowledge itself cannot be predicted. I think that the very predictability of its growth would have to presuppose the availability of unpredicted knowledge in the form of a new theory, if only to constitute its prediction.

To restate our argument: In order to predict something about a society we need a theory T to constitute the prediction. But the question whether T itself is predictable or unpredictable is itself an important question. If we say 'Yes it must be predictable', it will lead us to the consequence that all theories with which we might go on predicting the course of a society are themselves predictable. And this shows an essential incompleteness as well as the paradoxical nature of a deterministic model of society. For, without ever completing the statement of perfect determinism, one must go on to say *ad infinitum*:

> All theories with which we might predict the developmental course of a society must themselves be predictable with the help of or in terms of theories of a higher order which must themselves be predictable with the help of still higher-order theories which must themselves . . .

This argument is, I think, decisive enough to show that any deterministic model of a society must fall short of perfect determinism and suffer from essential incompleteness. For it must admit the essential unpredictability of our knowledge, in the form of (our *problems* and) *theories*, which according to it, (a) provide our actual predictions concerning the socio-economic growth of our society and (b) interfere with this societal growth in an essentially self-confirming manner. The essential unpredictability of *human knowledge* shows that it falls into a category which is altogether different from all the different *kinds* of human interaction, quite often making them possible and informing them in significant ways. But I think that a far more serious consequence follows from the idea of the essential unpredictability of our knowledge, in particular from an objectivistic point of view of our knowledge of the external world.

2.8. FROM AN OBJECTIVISTIC POINT OF VIEW

One of Popper's severest criticisms of the Marxist-type deterministic and historicist model of society, its development, etc., appeals, I think, of course from his evolutionary epistemological point of view, to this very idea of the essential unpredictability of the growth of human knowledge. Thus Popper rightly argues that (a) the growth of knowledge crucially affects or determines the development of a society, especially in respect of its economic and other cultural features;[12] (b) the growth of knowledge is unpredictable – it follows an essentially unpredictable pattern;[13] and hence (c) the developmental course of a society, in so far as it is crucially affected by the growth of our knowledge, is essentially unpredictable.[14]

Apart from its damaging consequences for a Marxist-type deterministic model of a society, Popper's argument has, I think, a deeper philosophical significance. I find it quite interesting that the distinction between human knowledge and the economic features of a society is common both to Marxism and its adversary, Popper. However, I think that the differences between them relate to the nature of the relationship between the two. While Marxism singles out the economic features as the crucial determinants of the other features of a society including human knowledge, Popper singles out human knowledge as of crucial importance in determining the growth of a society including its economic growth. But I think that the Marxist model is based on the same confusion as is characteristic of the traditional empiricist theories of knowledge. It confuses human knowledge with the relations of human interaction whose pattern is allegedly predictable according to the economic principles of human interaction-relations themselves. This confusion leads once again to a complete obliteration of the basic distinction between the two distinct problems of human knowledge on the one hand and of the nature of man's relationship with the world including the social world itself on the other.[15] On the contrary, I think that the deeper philosophical significance of Popper's argument consists in the assumption, made implicitly, that *knowledge* is not identical with any relation of interaction between man and the world or between man and man. It is rather that which qualifies all or most of the *significant* human interaction-relations; it is that which makes *human interaction*, as distinct from non-human interaction, possible. Thus it determines all those social structures and organizations that arise as the *end-result* of human interactions. It seems to me that by singling out human knowledge as the crucial determinant of the growth of a society, its structures and organizations, Popper is implicitly assuming the basic framework of nearly the same distinction that I have drawn between the problem of the nature of knowledge and the problem of the nature of the relationship of man with the world, both physical and social. The latter problem is, I think, essentially the problem of the nature of human interactions which cannot be dealt with adequately without an appeal to epistemology conceived as an enterprise concerned with the former problem.

2.9. ORGANIZATION, INFORMATION AND KNOWLEDGE: INSTRUCTED INTERACTION VERSUS CREATIVE INTERACTION

It is, I think, true of every science that it seeks to understand the complexities of the natural processes in the world in terms of the subject-specific laws of interaction at various levels of complexity. Fundamental universal constraints on the form and the scope of these laws in the different sciences form, in my opinion, the proper subject-matter of a general interaction theory in the sense outlined in Pandit (1983).

A physical theory, e.g., is, I believe, inconceivable without the kind of fundamental assumptions comprising what I call a general interaction theory. In no case would it be correct to expect a physical theory itself to provide for such a general theory directly or indirectly. On the contrary, I believe that a physical theory must always be premised on a general interaction theory with at least two important jobs to perform:

(i) to *individuate the physical systems* at the ontological level so as to generate a physical ordering of nature *as* an interactive order with these systems appearing as the basic elements of the interaction-relations; and
(ii) to generate/impose a necessary *space-time-specific ordering* or restriction (e.g. an ordering of a relevant causal kind) on interactions between the elements thus individuated, at various levels.

The former is concerned with the *possibility* of the physical interactive ordering of nature itself, and the latter with the *necessary restrictions* on the interactions thus generated. It cannot be denied, I think, that, in ultimate analysis, our considerations concerning the nature of a general interaction theory and those concerning the nature of a physical theory are bound to influence each other deeply. But, one might ask, how far can the actual developmental course of physical theory itself affect our basic intuitions concerning a general interaction theory? This is, I think, a question which assumes particular importance in the context of the confusing methodological scene triggered by the contemporary debates about the interpretation of quantum mechanics. But I shall refrain from discussing it here.

A general interaction theory in the sense just considered must, however, address itself to other problems as well: e.g. (1) how are interactions in the living world generally different from those in the non-living world? And (2) how are human interactions generally different from non-human interactions (in the living world)? These questions assume considerable significance in the context of the problem of distinguishing knowledge (objective and subjective) from human interaction. For it is necessary to go into these questions, if only as part of an exercise in the general interaction theory, if my thesis that knowledge is an orientating variable of human interaction as against non-human interaction is to survive further scrutiny in terms of its consequences for other kinds of distinctions that we might wish to make. Let us therefore consider how significant or far-reaching the implications of this thesis are while we attempt to answer the questions raised above.

Assuming the framework of a general interaction theory, conceived as broadly as in Pandit (1983), it is pertinent to ask: Are all types of entities/active elements[16] of interaction in the living world so organized, in structure and function, as would warrant their characterization as *information-constrained* elements in some scientifically relevant and intelligible sense of the term? Is it a unique feature of the living world that these elements, from the most simple to the most complex, are not just *unitary* in character – just *units of organization* of some kind – but complex organizations, each *constrained*, as it were, by an orientating field of *information* or *instruction*, embodying for that element alternative possibilities of interaction with other elements?

It should not be surprising that this question employs the concept of information or instruction, besides a more general concept of *organization*. The concept of organization could perhaps be employed as a self-sufficient concept so long as a general interaction theory addresses itself to biologically neutral aspects of interaction across the worlds of living and non-living entities/processes, besides those of an abstract nature. Thus, in a biologically neutral sense, significant interaction is definable as interaction between the active elements of different orders of organization.[17] At the highest level of generality, therefore, we might speak of an *organization-theoretic* approach as regards the problem of characterizing the significant interactions across the living and non-living nature.

The concept of information-constrained interaction seems to me, on the other hand, most relevant in distinguishing the significant interactions in the living nature from those in the non-living nature. From an information-theoretic perspective, every active element in the living nature can be seen interacting with other such elements in its capacity as a non-unitary complex structure or system. Even an element relatively as simple as a *gene* in the nucleus of the cell[18] is in this sense a complex structure with a physical basis embodying or accompanied by information/instruction as to how exactly to interact with other elements in its environment. The genetic code or the DNA molecule is indeed theoretically identified as spiral-like double helices wound around each other. Its very ability to trigger self-replication (transmission) and cellular control by a process of transcription/translation – the gene structure transforming itself into a structure of protein – points to its character as an *'innately' informed system capable of informing other systems*, if only as part of an endless developmental process. Although, the concept of a *biological field*[19] *of information* has yet to be explored as to its possible applications at the cellular and sub-cellular levels and also in the case of the animal and human interactions, it is, I think, worthwhile to make the following assumptions concerning active elements in the living nature, from the relatively most simple to the most complex:

(a) Every active element is a non-unitary structure/system accompanied by coded information/instruction as to how it might interact with other systems in a stable and regular manner.
(b) Its non-unitary character has essentially to do with the type-distinct constraints of an organizational (structural) and informative (instructional) kind.
(c) And every living being – including the cellular elements – has its own field of interaction determined more or less rigidly, as it were, according to its innate (naturally given) field of information/instruction or according to the set of instructions it has acquired over a certain period of time.

These assumptions seem to me essentially to restate the fundamental distinction, drawn in Pandit (1983), between the *S-H* systems and *C-H* systems.[20] Thus one might say that the active elements in the sense of the assumptions (a)–(c) are *C-H* systems, with active elements in the non-living nature being restricted to *S-H* systems only. Consider , e.g. the prey-predator interactions: the predator and its prey interact according to their respective fields of information/instruction, with each being a *C-H* system in its own right. A newborn human infant is another example of a *C-H* system which possesses an innate field of information/instruction *constraining* its interaction with its environment in significant ways. The field of information can be seen at work here in the *awareness* in the baby of its own 'problems' which it can yet neither articulate nor solve all by itself: e.g. it cries when hungry. It is only at a later stage that it is able to articulate and solve its problems: that is, when, from a *non-verbal* infant, it transforms itself into a *fluent speaker* of the native language; and later on from a mere *perceiving organism* into a *knowing subject*. As compared to *C-H* systems in the sense of these examples, *S-H* systems turn out to be far less complex. For example, any number of large masses held together by their respective gravitational fields of attraction can be seen to form an *S-H* system. Our solar system is a good example. An atom, a molecule, a crystal and a solid, as these are known in physics, provide yet further examples of *S-H* systems.

In a sense, I think, broadly relevant not to communication engineering or statistical theory of information but to a theory of interaction according to instruction, stored or acquired, *information* might be so conceived as to imply the following:

(A) There are *systems*, from the simplest to the most complex, which are sufficiently rich to have *built into* them a whole *set of instructions* determining *rigidly* the system's field of activity and interaction with the surrounding systems.

(B) At any given time, a system can be in any one of the possible states of organization, where these states (the values of the state-variables associated with the system) are a function of the *instruction-set* of the system.

(C) At any given time, the relation between such systems might be described as one of *instruction-transfer* in the sense of one system instructing another or getting instructed by another.

(D) The concept of instruction-transfer in the sense of (C) implies the existence of systems which are capable not only of storing instruction but also of acquiring more and more of it so as to be distinguishable from *less complex* and *more rigidly informed* systems (e.g. genes, the multi-cellular organisms, the human brain, etc.).

What the theses (A-D) imply is what might be called a general systems theoretic construal of *information* as a set of instructions either stored in or acquired by a system in the course of interactions with other systems. And it follows from this that a system can acquire more and more information only by way of modification of its innately stored set of instructions. That is, only systems embodying some information can acquire more and more of it. I think that this has implications for our understanding of the systems capable of *learning* as against those not capable of it. *Learning* might accordingly be defined as an interaction between the innately stored information/instruction-set and acquired information in a system. This also means that *learning* is impossible in the absence of two or more systems interacting with one another.

I think that we have yet to understand properly the processes, structures and growth of objective knowledge, since we often tend to misunderstand these, owing mainly to the vast complexities in the interactions taking place across the living and non-living nature, some of which are yet to be recognized. Thus *knowledge*, both subjective and objective, continues to be confused with *information* which appears in *all* living systems of a natural order in the shape of embodied *instruction* for selective interaction. But I think that it can be argued that all significant human interaction reduces, on the one hand, to *instructed interaction* according to information either innately stored (such as in the genetic code) or acquired (such as in language-learning) *and* to what I propose to call *creative interaction* according to knowledge conceived as an orientating variable, on the other. That there are deeper reasons to distinguish, instead of confusing with each other, (a) the concept of knowledge from that of information and (b) the concept of a knowing subject from that of an information processing or informed system might be argued along the following lines.

Interaction in the realm of the living systems is, I believe, always *instructed interaction*, there being at any given time built into each such system, from the simplest to the most complex, a set of instructions defining or specifying the state-variables of the system's internal organization and determining the alternative possibilities of its interaction with the type-distinct systems in its environment. I think that the prey-predator interactions are a good example.

In the systems as complex as man and animal, instructed interaction must, however, sooner or later lead to a gradual process of learning in which the sets of biologically innate instructions are slowly modified or replaced by those acquired by a process of negative feedback from the interactions with a variable environment. Learning, at this level, might be seen as a form of interaction between the *innate* and *acquired* information/instruction. Learning in this sense, when reinforced by a process of negative feedback control (in the customary narrow sense of the term)[21] will generally lead to organization-variance[22], thereby equipping the learner with a stronger, and more comprehensive, set of instructions to enter in *newer*, but *safer*, fields of selective interaction.

All instructed interaction in this sense might, however, be described in game-theoretic terms as a given system's set of strategies of playing safe against other systems in its familiar and unfamiliar environment – its rivals or Darwinian competitors included. It could thus be seen as involving a Darwinian process of survival or steady-state, a process of playing safe against nature and against poorly and better instructed individuals/creatures (as in prey-predator interactions).

It follows from this that all interactions between the living and non-living matter are unilaterally *information-field-constrained* interactions. However, this is not true of all human interactions – of interactions between human beings and the physical world and of those between human beings themselves. But why is this so? I think that unlike the interactions involving animals, human interactions are not only much more strongly instructed interactions, but are characterized by what might be called *information-field-variances* in the sense that whenever there occurs rule-governed creativity or negative feedback, one information-field tends to be replaced by another. That is, the instruction-sets or information-fields are highly *variable* in the case of human beings, and much less so in the rest of the living nature.

What, therefore, characterizes the living organisms generally are, I think, what might be called the variable fields of interaction: fields of interaction activated by information/instruction, which is either innate or acquired. And most of the process of learning in the living world takes place as a process of modification of innate instruction by acquired instruction. But it seems to me that only human beings are capable of transcending the strongly instructed, even if variable, fields of interaction: only they are capable of risking everything in the quest for knowledge as distinct from information or instruction.[23] But I think that this is not surprising. For we are all knowing subjects, besides being *informed* subjects or systems. *Human interaction, when activated by knowledge, leads to creative interaction according to a growing network of theories and problems, which is what instructed interaction lacks*. Thus in yet another way it becomes clear, I think, that to be a knowing subject is not to be just an information-carrying subject or a perceiving subject, nor just a subject capable of being informed in certain ways. Further parameters are essentially involved here.

It is thus knowledge, both objective and subjective, rather than the variable information fields themselves, which makes a difference not of degree but of kind between the human and non-human interactions. The world of knowledge is neither a world of information/instruction nor one of instructed interaction, the traditional confused construals of it not-withstanding. Structures of knowledge arise only when all instruction/information, innate and acquired, is transcended and a non-Darwinian process sets in. That is, these arise essentially as an end-result of our attempts to overcome our *recurrent entanglements* in the strongly instructed fields of interaction. In this whole process language plays, I think, both a conservative and a non-conservative role. On the one hand, it is essentially our language in which our entanglements in the fields of strongly instructed interactions are realized. Without language – indeed a whole hierarchy of languages – we would just be left without any means and ability to monitor how entangled we are in these fields at any given time. On the other hand, however, it is essentially in (the language of) *science* that we can overcome such recurrent entanglements by moving on from such fields to those of creative interaction. Here knowledge is made possible as a complex structure of the methodologically constrained *problems* and *theories*, enabling us everytime to move on from the fields of *strongly instructed interactions* to those of *creative interaction*. But what is the significance of all this? Does it follow from this that scientific discoveries are possible only in the fields of creative interaction, and not in those of instructed interaction? I think that the answer is in the affirmative. Thus, on this view, scientific discoveries may be characterized as follows. They open up before us newer fields of creative interaction. But they would themselves not be possible without first inheriting *knowledge* as an orientating variable while we attempt to overcome our entanglements in the strongly instructed fields of interaction. Much of the significant human interaction, if not all, can thus be seen as creative interaction in this sense, with animal interactions generally turning out to be just those that fall in the continuum of the strongly or rigidly instructed interactions in the living nature.

It is an important consequence of this view, I think, that the essence of science as a search for knowledge lies in its creativity. Thus it is admirable for its unlimited capacity to lift the human knowing subject up from the recurrent human entanglements in the fields of strongly instructed interaction and subjective belief (in the sense of ideological commitment instead of rational belief). The processes, structures and the functions of science and scientific knowing cannot, therefore, be assimilated to those that are simply identifiable as being Darwinian in character. On the other hand, along with the Darwinian processes of competition, inter-species interaction, trial and error and natural selection, human activities such as religion, political ideology, games, witchcraft and so forth all fall into a single pattern of the strongly instructed interactions. Unlike these, however, science is unique in being able to lead men and women from fields of strongly instructed interaction to those of creative interaction.

If this is so, how are we to understand the nature of scientific knowledge systems in their structure and growth? Can their structural growth be modelled after a simpler Darwinian process, as is strongly suggested by Popper's view that science is a game of conjectures and refutations? As will become clear from my arguments in Part Two, the answer to this question is in the negative. For any attempt to do so has the serious consequence of reducing this growth just to a function of rule-governed creativity. On the contrary, I have argued in Pandit (1983) and shall again do so presently, that the structural growth of scientific knowledge is a far more complex process capable of being formalized as a function of both rule-governed creativity and rule-enriching creativity. And it is in this sense that it follows a rational, if a tortuous path: *backwards from problems to theories and forwards from theories to problems.*[24]

Ordinary games and those in the sense of game theory are perhaps paradigmatic examples of how human beings move on from fields of interaction activated or dominated by the biologically innate instructions to those dominated by the acquired ones. The essentially Darwinian character of all human interaction dominated by instruction, innate and acquired, thus lies in its underlying network of strategies of survival, or of trial and error. It is these which the better instructed/informed members of a group/population employ against the less informed/organized members of another, as, e.g. in prey-predator interactions, or one *group* of human beings employs against another in a social conflict-situation of minimizing one's loss or maximizing one's gain. It would not be unreasonable, therefore, to lay down the following requirement: that all such human interaction must be viewed as forming an appropriate subject-matter for the formalizing models of game theory. If this is a correct approach, we can conclude that if there is anything which is beyond the purview of a game-theoretic approach it is the creativity of language and science. Let me clarify further what I mean with the help of a schematic representation of some of the main theses formulated and distinctions drawn in the course of the discussion in this Chapter (see Figure 2.1.).

It follows from our discussion above that there is something common between cybernetics and the mathematical game theory. Their common meeting point lies, I think, in the fact that both aim at construction of formalizing models of the significant interactions between different systems – interactions that are information-constrained. That is, both of them aim at finding out the correct control algorithm (rule or calculus) of maintaining any two or more interacting systems in such target-states of internal organization[25] as to ensure a certain rule-governed creativity for each such system or a certain optimal strategy[26] for it to be able constantly to reduce its tendency to entropy.

The concept of negative feedback control[27] is, one might say, central to our conception of all cybernetic processes involving instructed interactions that guarantee a systems's steady-state or its target-states of internal organization against all types of forces that tend to increase its entropy. The same idea, I believe, underlies our conception of the situations of conflict that are of game-theoretic interest to the mathematical theory of games. In these situations, opposing players/groups with opposing interests interact *within* a framework of definite rules. For different types of games, which the mathematical theory of games distinguishes one from the other on an increasing or a decreasing scale of complexity, it seeks to provide formalizing models essentially as calculi or algorithms of optimal strategies. A calculus of an optimal strategy in this sense, which ensures each player's rational behaviour in a game (or a situation of conflict), clearly approximates to a *control algorithm* – a concept which is central to cybernetics.

But where do these considerations lead us? Both cybernetics and the mathematical theory of games as they are commonly understood and practiced are, I think, inherently incapable of coping with problems relating to creative interaction, considered above. The concept of negative feedback in the customary sense of a control algorithm of maintain-

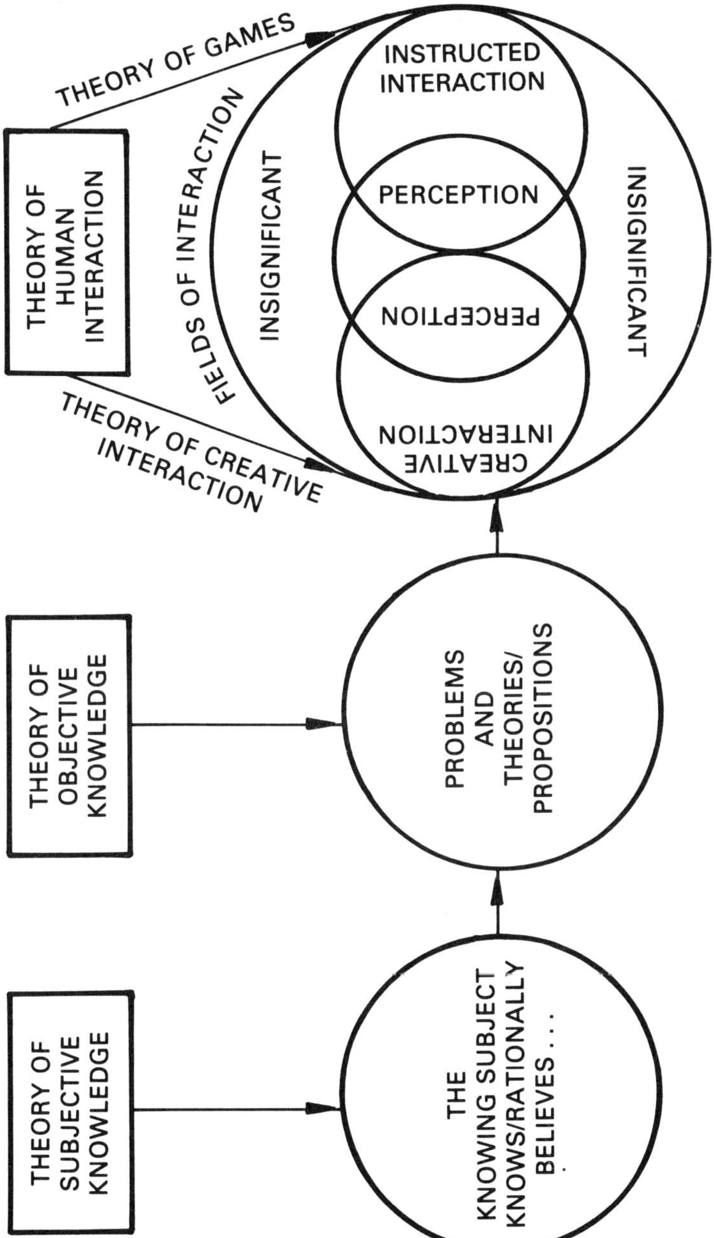

Fig. 2.1. A formalizing model distinguishing (subjective and objective) knowledge from (instructed and creative) human interaction

ing a given system in its target-states of internal organization and interaction at best suffices to take care of the rule-governed creativity, whether in living nature, in language or in science. The same is, I think, true of its game-theoretic counterpart, viz. the concept of an optimal strategy of a player for behaving rationally in a situation of conflict involving other players as well. The most important problems that lie outside the scope of these disciplines have, I think, to do with the creativity of language and science – i.e. with the rule-enriching creativity on the one hand and with structures of *knowledge* as an orientating variable of the creative interactions on the other. The concept of rule-enriching creativity is of course central here. Hence my proposal elsewhere[28], extending the concept of a negative feedback control system to cover more complex and richer systems than are covered by cybernetics. Recognition of such systems does not, however, necessarily entail a revision in the scope of cybernetics or an extension of its basic concepts as is implied or suggested by me there. I say this because one might develop a whole theory of such systems outside the basic framework of cybernetics, without, however, dispensing with the latter as a prior discipline. To avoid confusion, it is perhaps desirable to develop a theory, as part of a general interaction theory, that deals with systems that are not only capable of maintaining themselves in the rigidly fixed target-states of interaction and internal-organization but capable of entanglement in their own rules – i.e., systems of the kind that are capable of enriching/changing their own rules and target-states associated with them as and when they land in the problem-situations of rule-entanglement.[29] Such a theory would tell us, among other things, when does a situation of *rule-government* in science turn into one of *rule-entanglement*.

Since cybernetics is, I believe, very general in scope and deals with essentially *subject-neutral aspects* of different forms of instructed interaction, control and organization across the living and non-living nature – a feature which the mathematical theory of games lacks – it is worthwhile to explore the theory of games as a theory within the general systems-theoretic framework of cybernetics itself.

CHAPTER 3

INDETERMINACY OF TRANSLATION: A NON-QUINEAN FUNCTION OF CONTENT-INDETERMINACY

3.1. SCIENCE AND LANGUAGE: PROBLEMS OF THEORY-CHOICE AND TRANSLATIONAL INDETERMINACY

At what point, one might ask, do the considerations about *language* assume crucial relevance to our approach to the philosophical problems of science? Or are the linguistic and non-linguistic (= mathematical) aspects of science so disparate and wide apart that only one of them, but not both, carries methodological relevance in the context of the philosophy of science? An answer in the affirmative to the latter question is actually implied by the Sneed-Stegmüller type set-theoretic structuralist approach to the problems of the structure and growth of scientific knowledge.[1] Scientific theories must be, on this view, treated as mathematical entities (= set-theoretic structures) of a certain kind before one attacks the philosophical problems concerning them, notably those of inter-theory relations, theory-reduction/replacement and theory-choice. Only on an approach as radical as this, so is it claimed, can we deal with the Kuhn-Feyerabend type problem of incommensurability. But how? This problem, we are told, is simply dissolved if we adopt this new radical approach. That is, different scientific theories cannot be cognitively compared, one with the other, so long as they are treated as structures consisting of *true* or *false* statements. On this view, then, there is no question of our considerations about language having a bearing on those about science and *vice versa*. It is tempting to think that this might be after all the most easy way out of those very problems and difficulties (consider, e.g. the incommensurability problem itself) that would not have raised their head if there were no deeper aspects to the relationship between science and language, or to the parallelism between the problems of theory-choice and inter-theory comparison on the one hand and the more general problems of translational determinacy or translational synonymy on the other.

Any two or more languages could be made to interact through their differences so that each of them is enriched in a significant way; and interaction between any two or more theories in the same manner is bound to result in the growth of scientific knowledge. That is, if growth is predicated of science, so must it be predicated of all the languages. However, I think that the patterns of growth in the two cases must differ and hence these should be sharply distinguished from each other. As I look at it, there would be no rational pattern to the growth of scientific knowledge if there were none to the comparison of one particular theory with its rival, of one particular theory-problem interactive system with another. Indeed, as I have been arguing, the possibility of there being such a pattern rests on a number of factors, methodological variance with the fundamental principles of invariance in the background being quite notable among them. And this has, I think, essentially to do with the character of science as an enterprise comprising what I designate, in (1983), as *NFCPS* systems. In a sense it is true that once the growth of science according to a clearly discernible rational pattern is recognized, there is no escape from the recognition of the growth of *language* according to a rational pattern, the former being always subject to alternative formulations in the latter. But can we ever speak of the growth of language(s) without implying the possibility of significant comparisons/interactions across different languages from the simplest to the more complex? I think that the answer to this question is in the negative. That is, the parallelism between science and language extends up to a point only, namely to significant comparisons and interactions between (a) the candidate-theories on the one hand and (b) the candidate-languages on the other. This restricted parallelism could be understood as a simple consequence of my characterization of language, in (1983), as comprising *C H* systems, quite distinct from the *NFCPS* systems and yet presupposed by the latter as their essential embedding systems. If we accept this then we must boldly face two of its important consequences. First, no argument, whatever, that is designed to undermine the methodological commensurability and rationality in the context of science and that appeals, in the process of doing so, to translational indeterminacy in the context of language, is possible without confusing science with language. Secondly, there must be fundamental principles – something like the Chomskyan language universals and a universal grammar – by means of which the individual natural languages can be investigated, compared and *translated*, one into another. Can we then expect *translational determinacy*, when translating what the native speaker says/asks in his language into another, so that we might make a rational choice between rival *translation-claims* concerning the two languages? The standard answer to this question in the negative is already provided by Quine's (1960) thesis of the indeterminacy of radical translation. In what follows I propose to examine Quine's thesis afresh, since it is to this that Kuhn (1971) makes a fervent appeal, if only by way of a rejoinder to Lakatos: "Anyone who supposes that the points at which Feyerabend and I have aimed in introducing 'incommensurability' into our discussions of theory-choice are either trivial or obviously mistaken must simultaneously brush aside much of the contemporary literature on radical translation. I cannot think that that should be lightly done."[2] But before doing so, let us have a closer look at *science* and (natural) language in so far as one might distinguish one from the other so that a scientific theory T is clearly seen as consisting of sentences/statements which do not follow from the adoption of T's language-framework L alone. Although this makes it necessary to go into those aspects of the structure of a *language* that might provide for not just analytic and synthetic statements but contradictions as well, I shall simply proceed on the assumption that, among other things, not just the statements of T but its negation (= ~T) as well must be capable of formulation in L.

A language could be characterized by specifying a number of parameters that give it its characteristic stability and rigidity, on the one hand, and its rule-governed creativity (recursiveness) on the other as follows:

(i) a set of grammatical/syntactical rules that determine the well-formed formulae (sentences) of the language;
(ii) a set of logical and descriptive vocabulary (some of which might mean the same as certain combinations of the others);
(iii) a set of *basic* interaction-relations, on which other more complex interaction-relations keep building up in an essentially indeterminate manner, that relate (a) the members of its embedding language-community with one another and (b) certain members of the elements of the vocabulary – not necessarily all – with selected aspects of the physical world, in highly complex problem-situations.

In so far as *languagehood*, in this sense, consists essentially in the network of interactions that determine a language-community uniquely, a language cannot function in ways that are non-conservative and non-informative with respect to its embedding language-community itself. Thus, if this is so, it is inherent in the individual natural languages to function in ways that tell *more* about (the members of) their respective embedding language-communities than about the external physical world, even when their users don't intend it. As they tend to be more and more used it is their respective embedding language-communities that we know more and more about. It is to this that I would trace the reason why the problem-situations of rule-entanglement in my sense (1983) are unknown to our ordinary languages; and why these give way to science where it is not just the information about but knowledge of the external physical world which matters most. This need not be surprising, since we should not expect from a natural language what it cannot deliver. But an element of surprise is natural as this may not always be as clear as one would wish. Wherein lies, then, the most important distinction between language(s) and science? This question might be more sharply formulated by asking: How are information-carrying systems on our ordinary natural language scale different from systems of objective scientific knowledge? A question such as this would be, I think, uninteresting in absence of interesting similarities or parallelisms and interrelationships between the two types of systems. Without our languages, and hence without ourselves as languaged-animals, science would be just impossible. This has to do essentially with all those *possibilities* that are *latent* in these languages but beyond realization in terms of them. I have in mind the possibilities of abstract epistemic structures that operate, once these come into being, on a scale different from that of our natural languages, such that the following conditions hold true of them:

(i) They are abstract structures of an epistemological-ontological order, but pragmatically neutral in so far as they are indifferent to any language-communities rigidly predetermined by our natural languages.

(ii) More and more they are used, explored, evaluated and developed, more and more of information and knowledge will they yield about an external world as the subject of our objective knowledge, but less and less will be the information yielded about their original authors or builders.

(iii) While they resemble the conceptual structures in the natural languages closely in the matter of rule-governed creativity, their most remarkable and unique modality of development and growth has to do with the permanent possibility of their entanglement in their own rules, a process that finally gives way to development by rule-enriching creativity.[3]

All these conditions, which distinguish science from language in the ordinary sense of the term, are latent in the most abstract *functions* of our language that have no corresponding structural strategies to match. Consider, e.g. the following functional abilities of our otherwise pragmatically non-neutral languages, that express themselves most distinctly at two different but closely related levels:

(1) The conceptual function of formulating most abstract *what*-questions and *why*-questions that presuppose a considerable argumentative setting; and
(2) the abstract meta-linguistic function that expresses itself in the latent ability of our languages to monitor their own uses, no matter how complicated or abstract these may be.

Possibility of science then rests on the possibility of transfering generalized functions of language such as these to *specialized systems* or *structures* that are relatively more abstract. That is, the possibility of science has a great deal to do with the structural realization of certain important functions that are latent in all our languages. But obviously that is not sufficient. Their differences and their relationship can be therefore best expressed, I think, in terms of the distinction between *C-H* systems and *NFCPS* systems in my sense (1983). All *NFCPS* systems are at the same time *C-H* systems, or embedded in them, but no *C-H* system by itself is an *NFCPS* system. Only *NFCPS* systems are capable of entanglement in their own rules and hence capable of rule-enriching creativity. Our languages are just *C-H* systems. By virtue of being *C-H* systems they are mutually comparable in interesting ways. On the other hand, science comprises *NFCPS* systems, that are ever embedded in language-frameworks – *C-H* systems – that they presuppose. The presupposed language-frameworks may at any time involve large chunks from both natural and invented languages (such as mathematical systems).

Let us develop this distinction between *language* (in the ordinary natural sense of this term) and science a little further. The distinction might be better understood, to begin with, as one between *types* of language of *two different* epistemological-ontological orders, one of these being genetically but not logically presupposed by the other because it is the type of order which is *genetically primitive* or primordial. Languages of the second type might be properly called *counterpart languages* (Gegenteilsprachen) in relation to those of the genetically primitive type. Let us designate these as L and L′ (i.e. ordinary natural languages and scientific languages, respectively). The genetic presupposition relation between L (= L_1, L_2, . . . L_n) and L′ (= L'_1, L'_2, . . . L'_n) is such that without the kind of framework represented by L, and the abstract questions arising in L, there would not be any causal dynamics necessary and sufficient to cause the generation of L′. In so far as L′ really develops as a counterpart to L, the relationship between them is very complex. But once L′ comes into existence as a different range of differently ordered frameworks to formulate more precisely the questions askable but unanswerable in L, and to seek exact answers to them, it gradually sets up interaction-relations within itself creating an epistemological-ontological order of its own, quite different from that of L. One might ask: how different epistemological-ontological order does L′ really bring about in the wake of its own generation and development? So different, I think, that the epistemological-ontological significance of the *uses* of L′ – of the specific theory-problem interactive systems developed in L′ – is always strongly *pragmatically neutral* in the sense that these *uses* yield no information whatever about their authors or users. That is, the more and more L′ is used and developed the less and less will it let us know as to the nature of organization of the communities of its users. And this is so because these uses tell us more and more, and enable us to ask more and more questions, about the objective worlds that are carved out with the generation of L′ (= L'_1, L'_2, . . . L'_n) and that L′ is used to explore.

Quite the opposite is the case with L (= L_1, L_2, . . . L_n), with all its various kinds of uses, including its metaphysical uses. They are generally strongly pragmatically non-neutral, and everytime in different ways, in so far as they tend to be more and more informative as to the language-communities themselves in which they are embedded. Thus more and more L is used, for purposes ranging from most ordinary to most metaphysical, less and less is the information these uses yield about an external physical world that is generally taken for granted and referred to from time to time. This is, I believe, correlated with the fact of their ever increasing pragmatic non-neutrality in the sense that every enrichment in the use of L entails an enrichment in information about the whole language-communities themselves.

Apparently then, from a synchronic point of view, the epistemological-ontological *asymmetry* between the information-carrying systems in L and the systems of objective knowledge in L′ has to do with the pragmatic non-neutrality of the uses of the former and pragmatic neutrality of the uses of the latter. And, consequently, from a diachronic point of view, it has to do with the *opposite directions* in which the *increase* in knowledge or information, with time, takes place in the two cases. Generally, then, in the case of L, any increase in information generated by its uses at a time $t = t_2$ is significant only in so far as these tell us more about the community of its users than is possible with its uses at a time $t = t_1$. But in the case of L′, any increase in knowledge generated in the course of its uses at a time $t = t_2$ is significant only in so far as these uses tell us more, and enable us to ask more questions, about the objective world(s) themselves than is possible with its uses at a time $t = t_1$.

Since it is, I believe, a unique characteristic of all languages of type L to function in ways that must always return to pragmatically complex contexts of use rigidly determining and rigidly determined by the specific language-communities themselves, it is plausible to develop its implication further by saying that all such languages are what might be called *hard-core* languages. That is, L ranges over languages, with each language being identifiable in terms of a really stable hard-core conceptual structure or hard-core interaction-relations. Conceptual hard-cores, if there be any, have in recent philosophy of science been, I think, wrongly sought in the center of every scientific theory (scientific research programmes of Lakatos), i.e., in L′ (= L'_1, L'_2, . . . L'_n). In a way Lakatos was so right in suggesting that (conceptual) hard-cores must exist. In itself, this is a suggestion of considerable epistemological-ontological significance. But I think that they really lie elsewhere. Lakatos was also mistaken in so far as he tried to look for them in the wrong place. For, as we have seen, they have to do with those and only those *uses* of L (= L_1, L_2, . . . L_n) that tell, intendedly and unintendedly, different ‘stories’ of their respective embedding language-communities themselves. The conceptual hard-core of a natural language results in those uses of it – those interactions – that uniquely pick out a class of users as its embedding language-community. No surprise, then, that without a rigidly determined embedding language-community

– without a hard-core to sustain its network of interactions – a natural language would lose its very identity and have its very existence threatened seriously. And no surprise that the possession of different hard-cores by different natural languages should make interactions between them always interesting and deeply significant.

Consider an indefinitely larger number of old (primitive) caves as if dug deep into a mountaneous landmass from all sides. Consider these on a model somewhat different from Plato's allegory of the cave. Let us imagine ourselves as beings capable of digging such caves and also as cave-dwellers. There are then different caves with different groups of cave-dwellers living in them. We can thus imagine that each cave is, geometrically and accoustically, different from the other but in a manner that does not affect the essential structural and functional symmetry between them in their capacity as caves. We suppose further, that every cave-dweller is confined to his cave until he discovers that there might be more caves like his own in close or remote environment. As one's curiosity to know more about the surroundings develops, one starts thinking of strategies to interact with other cave-dwellers (other civilizations). One thing is beyond doubt, that it is difficult to know and understand other cave-dwellers without exploring the depths – the core-regions – of their caves from deep inside. Cave dwellers from one cave accordingly plan trips to other caves. As such interactions are accelerated, the expert visitors come to the following general conclusions:

(a) Every cave is (geometrically, accoustically and in many other ways) so different from the other that one *group* of cave-dwellers is easily and rigidly distinguishable from the other.
(b) Every cave is equally good to take a look at the outside world which one can see and think about just by sitting in the cave – all caves are equally good for the purpose of describing the outside world, for the purpose of short glimpses of it.

Our ordinary natural languages are, I claim, like these aforeimagined caves. Even after travelling long distances a natural language must return back to its cave-like innermost hard-core. Although analogies can be misleading, there is one important lesson we might draw from this analogy: *Growth* of each of these languages, if there be such growth, depends in interesting ways on translation of one into another. But what are the languages of science then? Well, they are comparable to telescopic windows that can open and extend outward in a flexible manner from any arbitrarily selected cave-top, as it were. One might imagine them so constructed as to be anchored to the network of the caves not in a one-to-one correspondence fashion but in an arbitrary manner.

Against this view, it might be argued, and quite correctly I think, how erroneously it just assumes a linguistic conception of science, and of its historical development, as a language among other languages, ruling out possible deeper aspects to the whole problem of bringing out its hidden complexities. No doubt, at every step science presupposes language, because it is embedded in its framework according to the principle of all *NFCPS* systems being systems of the type embedded in *C-H* systems.

But there exists a fundamental, though unrecognized, asymmetry, a fundamental discontinuity, between language and science. Its proper conceptualization is, I believe, a possible source of constraints on a theory of scientific change, particularly of the growth of scientific knowledge. Thus it becomes quite relevant, in the context of such a theory, to ask the following question: Why are the kinds of *possibilities* (of rational communication), either inherent in language or strongly correlated with the rigidities of its actual use, so different in nature from those of scientific practice itself? The languages themselves, no doubt, lie on a *continuum* constrained rigidly by their syntax such that there are at any time generated in their *users* indefinitely large number of possibilities of *following*, but not violating, their rules. The reason why this is so lies, I think, in the nature of language itself. To all its users it offers all its resources, once they have mastered it, only as permanent possibilities of their getting *entangled* in the network of its rules. Thus to be a language-user (= a rational communicator of meanings) is to be one caught permanently in a state of rule-entanglement, all the differences of language-use being only variations on the same theme. A distinguishing feature of language, as that of language-use, may then have to do with the manner in which it realizes, for its users, states of rule-entanglement from out of its states of rule-government. But in doing so does *it* distinguish between the two kinds of states: viz., the state of being a master of its rules and the state of being caught in their network just as a slave? No, it does not do so. There is no way *in* it to tell the difference between a fluent speaker's mastery of his language and his perfect slavery of the rigid framework of its rules. Language as a system of rational communication is *symmetric* with respect to these two kinds of fundamental states. But this raises then another question: With which conceivable way are we left then to overcome, if not to avoid, its states of rule-entanglement? Is it possible to over-step them in some transcendental manner by stepping into something new and different from language? The only possibility of doing so would have to be some kind of strategy to break through the master-pretension in mankind to master everything through the mastery of language. But that is the kind of possibility, one can rightly argue, which can never arise *in* language itself, only *outside* it. For, as we have just noticed, it is a recurrent feature of language to land its users in perfect situations or states of rule-entanglement. It is in these situations, then, that the master-pretension in man – the languaged rational communicator – finds its first and best outlet: That with his mastery of language he can capture those highest metaphysical truths about the universe and himself which are liberating for one and all. Different ideologies of religion all over the world are its special manifestations. Before one may lay one's claims to upholding any one of them, one is already caught in its network without even realizing how this is so. This reminds me of Kuhn, Feyerabend and others whose failure to distinguish language from science assumes most glaring proportions at this point where it becomes necessary to ask:

> What is the nature of that original need of mankind which gave rise to *science*?

By way of a conjecture, we can answer this as follows: The original need for it must have arisen as a need to *break* the symmetry of language between the twin-states of rule-government and rule-entanglement. This was something impossible to fulfil so long as one stayed *inside* (one's) language. What it demanded, therefore, was a creative leap forward taking us beyond language. The result of such a creative leap, in the form of modern science, is exactly what neither Kuhn nor Feyerabend has been able to appreciate fully. We are not for ever caught in an *hermeneutic* circle of linguistic preunderstandings.

The thesis which I wish here to put forth may now be stated more simply as follows: What distinguishes *science* from language, among other presuppositions, is this: Given the kind of possibilities which are strongly correlated with language-use, the possibilities of science pull in an altogether opposite direction. Thus, they are the possibilities of breaking through *its* states of rule-entanglement by a process of violation of those rules which may characterize or define its movement at any stage of historical development. The very direction of its movement is, therefore, always a more complex type of movement from states of rule-government through those of rule-entanglement to those of rule-enrichment. It is its rule-entanglement states which are here of particular interest to us. For it is in these states themselves that a science tends to function just as a *language*, just as a system which is at once self-complete and self-sufficient. This happens when the scientists take their task to be one of rational communication, becoming victims to the master-pretension in mankind that with the mastery of language man can ensure his mastery of the profoundest metaphysical truths about the whole universe. For example, is not the current problem-situation in quantum mechanics, like the situation which prevailed for quite some time in the classical physics of the last century, a good approximation to the rule-entanglement states of the kind I am here referring to? I think that the answer is in the affirmative.

But sooner or later such states of a science, where there is no way to distinguish the master from the slave or the slave from the master, will give way to *methodological variance*, as I like to call the scientific processes of rule-change and rule-enrichment. Methodological variance in science may come about either from the side of experimentation or from chance discoveries of the newer complexities which take the scientists by surprise. (One might boldly ask: Did science have a beginning in poetry? One could possibly say yes. Because the first violation of language – by over-stepping its rules and breaking through the primordial hermeneutical circle – takes place here).

Does not this entail that the fate of every scientific theory is its essential epistemological incompleteness? Before one has articulated and then mastered its language, one must prepare for learning an altogether different theory whether already in sight or not. One is reminded here of what Newton said: Like a child all that he accomplished was to collect pebbles on the sea shore. The child of science must stay, to use the metaphor of Newton, as a child for ever. But what happens then to the scientist's frequent claims of a theory's epistemological completeness? Such claims were put forward in the 18th and 19th centuries in the case of classical Newtonian physical theory. Its tremendous success together with the philosophical misconceptions about the connection between theory and experience – that the latter can generate the former in a non-creative, algorithmic fashion – must have prevented the many scientists then from recognizing clearly how scientific creativity and theory-construction or theory-change are correlated with each other. Again, the completeness debate triggered by the creation of quantum mechanics in the late 1920s is a case in point.

Let us not confuse the aim of science, if it has any, with its method. Let us not ask, first of all, what does it aim at and, then, when is it rational to accept one theory in place of another. What it cannot aim at is truth (= one and only one true theory) because of the false presupposition that all sciences aim at just one particular thing. The question of the aim of science can also be understood as a question of *what* would complete the states of scientific change it goes through, with its corresponding question of rationality as one of *how* to complete them. On the contrary, if we allow each science to ask itself what *kind* of theory – with what mathematical structure – should it aim at and, in the light of an answer to that, what kind of method of theory-choice, we can understand these as twin-questions of how to bring about scientific change and how to accelerate the same. What science cannot, therefore, aim at is the kind of theory which is truly complete in the sense that its laws reduce all known problems to their solutions without posing any new problems. But methodologically speaking, it can progress only from one particular state of rule-entanglement to another by a process of theory-change from one complete theory to another such theory. In other words, every science passes through those *linguistic* phases where it *tends* to become just another language at once complete and self-sufficient. It is these which, I think, Kuhn has predominantly in his mind when he speaks of normal science as quite distinct from revolutionary science. His whole idea of the stablizing phases of normal science could be best understood, therefore, in terms of my own thesis that every science passes through linguistic phases of rule-entanglement such that one particular group of scientists, overcome by *a false sense of language* where there is really none, may just refuse to listen to or talk to another similarly organized group. (Imagine a native speaker of a language who is full of aversion to talking to a foreigner in the latter's language.)

On the other hand, it is the reduction-theorists, interested in rational reconstructions of scientific theories, who assume, more than anyone else, a highly over-simple linguistic picture of science and scientific change. Thus a scientific theory is recognized as good by the degree of its formalizability using some well-known or well-defined language-system with an underlying logic of its own. Reducibility of its theories to a language among the already existing languages is the mark of all good science. It never occurs to the reduction-theorist how this takes away from Gödel's theorem all the relevance it has for science. Today, after a prolonged debate begun by Einstein and Bohr, we are being told that quantum mechanics is a theory of indefinitely many *possibilities* with our observational/measurement interactions playing a decisive role in determining which one of them is actually realized on given initial conditions. The underlying idea may be alternatively expressed by declaring that quantum mechanics is nothing but a language of many possibilities, *all* of them being calculable, from their respective initial conditions, as to their probability on satisfying certain conditions, predictive rules and operations. Sometimes this is expressed in a different way by showing how all the difficulties of understanding and using quantum mechanics reduce to those of a fundamental shift in *logic* the transition from classical mechanics to quantum mechanics itself demands. These are thus taken to reduce to difficulties of communication, as it were, between the communities of physicists brought up to divergent, even incompatible, linguistic expectations. For example, consider how Niels Bohr added a linguistic dimension, where he could have left room for deeper physical considerations, to his entire debate on this issue with Einstein. Thus his principle of complimentarity could be better understood, I think, as a result of the linguistic strategy he chose against all possible misgivings against quantum mechanics. And I think that it is this kind of strategy, already flourishing in the Kuhnian and Feyerabendian variations on it, which calls for a closer analysis.

3.2. QUINE'S THESIS OF THE INDETERMINACY OF RADICAL TRANSLATION

According to Quine, translation of what the native *says* in his language into another language is essentially *indeterminate*[4] except where the native's sentence happens to be an observation/occasion sentence.[5] More precisely, it is his set of theses that:

(A) In respect of being under-determined by all possible data, translational synonymy and theoretical physics are indeed alike. The totality of possible observations of nature, made and unmade, is compatible with physical theories that are incompatible with one another, correspondingly the totality of possible observations of verbal behaviour, made and unmade, is compatible with systems of analytical hypotheses of translation that are incompatible with one another.[6]

(B) Though linguistics is of course a part of the theory of nature, the indeterminacy of translation is not just inherited as a special case of the under-determination of our theory of nature. It is parallel but additional. Thus, adopt for now my fully realistic

> attitude toward electrons and muons and curved space-time, thus falling in line with the current theory of the world despite knowing that it is in principle methodologically under-determined. Consider, from this realistic point of view, the totality of truths of nature, known and unknown, observable and unobservable, past and future. The point about indeterminacy of translation is that it withstands even all this truth, the whole truth about nature. This is what I mean by saying that, where indeterminacy of translation applies, there is no real question of right choice; there is no fact of the matter even to *within* the acknowledged under-determination of a theory of nature.[7]

Quine argues his way to the thesis (A) above about translation essentially by arguing first for his thesis of the indeterminacy of *radical* translation in his (1960).[8] Translation of a hitherto untouched/unknown and hence unfamiliar language into a familiar one is designated as 'radical' in the sense that an enterprising linguist has no antecedently given manual of translation (a general scheme of translation) to fall back upon if he undertakes such translation. He has, on the contrary, a formidable task of laying down a system of what Quine calls analytical hypotheses of translation with due regard to the native's observable verbal behaviour in common perceptual situations such as those in which both the linguist and the native happen to be exposed to the stimulation of a rabbit scurrying by in their visual fields.

Quine's thesis (B) above that, "where indeterminacy of translation applies, there is no real question of right choice" is not to be looked at merely as a corollary of his thesis (A) of the methodologically under-determined rival systems of analytical hypotheses all compatible with the totality of the observations of the native's observable verbal behaviour. For what the former asserts is the relativity of all translation to the systems of analytical hypotheses, thereby rejecting the idea of an absolutely correct translation as a myth sustainable only under a strong belief in propositions as linguistically neutral/invariant meanings. Perhaps, this needs to be asserted additionally in so far as one is likely to miss it if one merely asserted the thesis (A): That in translating what is said by the native in his language into another language alternative systems of analytical hypotheses can be set up which are equally methodologically under-determined by 'experience' in the sense that, while being incompatible with one another, they are compatible with the totality of the verbal behaviour of the native, observed and unobserved.

Given the two theses (A) and (B) above, what Quine expects as a consequence is "a change in prevalent attitude toward meaning, idea, proposition".[9] Conversely, it is "because of the uncritical persistence of old notions of meaning, idea, proposition"[10] that what these theses assert escapes recognition.[11] In other words, it is Quine's view that the reason "why one thinks that one's question 'what did the native say?' has a right answer independent of choices among mutually incompatible manuals of translation"[12] is traceable to "the stubborn notion that we can tell intuitively which idea some one's sentence expresses, our sentence any way, even when the intuition is irreducible to behavioral criteria".[13]

3.3. TRANSLATIONAL DETERMINACY: QUINE'S BEHAVIORAL CRITERIA

The root of the entire problem of translation as Quine raises it, I think, is not the traditional doctrine of propositions as linguistically neutral/invariant meanings. It lies, I believe, in Quine's own highly over-simplifying assumptions which lead him to impose highly reductionistic behavioral criteria on language and translation.[14] According to Quine, wherever these behavioral criteria are satisfied language can be made sense of in terms of stimulus meanings[15] without indeterminacy threatening it in any way. That is, *translatability and meaning without indeterminacy* coincide for Quine with the domain of observational language – the language of occasion sentences/highly observational sentences. The rest of our language is 'translatable' and 'intelligible' strictly within the limits of indeterminacy.

Since, according to Quine, translation and intelligibility in this latter sense are relative to one's system of analytical hypotheses, without involving any absolute question of correct/incorrect translation, one is here placed, I think, in a situation of anarchism, a situation where the only possible universal rule should turn out to be a negative Feyerabendian one: *anything goes*.[16] Anarchism in the present context means the failure of the Quinean behavioral criteria when it is strictly a matter of our understanding and translating a language in aspects other than those of the Quinean observation/occasion sentences. Such a failure implies a failure in choice purely on the basis of these criteria. And this is only another way of describing the kind of situation designated as the methodological under-determination of rival systems of analytical hypotheses of translation of what the native says in his language (in aspects other than those of its observation sentences). In ultimate analysis, however, given Quinean behavioral standards of translation and intelligibility *without indeterminacy*, to translate or not to translate a language in aspects other than those of its observation sentences should be, I think, one and the same thing, rival translations being after all 'equally right' relative to their respective systems of analytical hypotheses.[17]

3.3.1. *Language Viewed as a System of Dispositions to Verbal Behavior*

For a behaviorist like Quine indeterminacy must, I think, reign supreme in all translation where the behavioral criteria laid down by him are not fulfilled. At least from the point of view of the problem of intelligibility, translation and even learning, indeterminacy is bad if only because the non-fulfillment of behavioural criteria is bad. Obviously, in the case of their fulfillment, there would be no indeterminacy in Quine's sense and hence no cause for concern. In such a supposed situation, all language, not just a small portion of it, indeed all that the native *said* in his language would be available and intelligible for the foreigner-linguist. But all this, again, not in any absolute sense whatever; only in a relative sense – relative to Quinean behavioral criteria. And, in this way, one must come back to the basic starting point of Quine's entire exercise behind this thesis of the indeterminacy of radical translation, the very basis of his philosophy of language:

> Let us then recognize that the semantical study of language is worth pursuing with all the scruples of the natural scientist. We must study language as a system of dispositions to verbal behaviour, and not just surface listlessly to the Sargasso sea of mentalism.[18]

Quine's persistent attack on propositions as *linguistically invariant* meanings shows how strongly inclined he is to view as *mentalistic*[19] any philosophy of language which is not premised on the behavioristic assumption that language is a system of dispositions to verbal behavior.[20] This not only exposes the dogmatic aspects of Quine's philosophy of language; it raises a number of questions that have remained, I believe, essentially unasked:

(1) How would Quine as a behaviorist accommodate the *rules* of language in his conception of language as a system of dispositions to verbal behaviour? Is a dispositional account of rules possible?
(2) Given Quine's thesis of indeterminacy of radical translation, how would he account for the fact that any number of foreigners can gradually so learn the native's language as to express in it all that they can express in their own language? In other words, Quine's

thesis does not seem to threaten in any way the learnability relation between different languages. And, hence, indeterminacy of radical translation in Quine's sense is compatible with learnability without Quinean indeterminacy. But how can Quine reconcile with or account for it?

(3) Is all philosophy of language, which is premised on the assumption of propositions as linguistically neutral/invariant meanings, necessarily *mentalistic*, as is assumed by Quine?

3.3.2. *Methodological Indeterminacy of Propositions Conceived Non-Mentalistically*

I shall consider here only question (3) above with a view to examining Quine's thesis of indeterminacy of radical translation and his attitude to propositions more closely. Quine's approach to what is and what is not mentalistic is, I think, dogmatic, being pre-determined by his commitment to behaviorism, particularly the *twin* doctrines that mind is a system of dispositions to behavior and that language is a system of dispositions to verbal behavior. And I think that it is this dogmatism which prevents him from recognizing the possibility of a mentalistic as well as a non-mentalistic construal of propositions.

Conceived mentalistically, propositions/meanings at once come under the requirement that they must be determinate, as determinate as any other mentalistically conceived entity. Without a determinate character, propositions so conceived would as well cease to be of any ontological significance whatever. A mentalistic construal of propositions must, however, be ruled out if only for the reason that it would make propositions as elusive as the mental world itself taken as a whole. In any case, it would *reduce*, I think, all thought and knowledge just to subjective states of the mind. That is, it would reduce the world of objective thought and knowledge – the Popperian[21] world$_3$ – to the mental world. And this is not what the generations of philosophers – from the earliest times to our own – committed to propositions as linguistically neutral meanings must have ever intended.

Conceived non-mentalistically, as they ought to be, as belonging to the Popperian world$_3$, various interesting aspects of propositions come to light as follows:

(a) Propositions are not only abstract, linguistically invariant meanings corresponding to different languages; they are related in intimate ways to equally linguistically neutral *problems*.

(b) They together with these problems are the *developmental structures*[22] of objective knowledge. That is, they are revisable and replaceable according to a *rational pattern*, making growth of knowledge possible.
(c) They are inherently indeterminate in and by themselves, their indeterminacy being a function of their intimate relationship with an indefinitely large number of other propositions/problems, formulated and unformulated, known and unknown.

The kind of indeterminacy I have here in mind is long known to reign supreme in empirical science, particularly from P. Duhem onwards. In Duhem it finds expression in his holistic approach to the problem of theory-evaluation in science: it is always a *system* of theories/hypotheses as a whole, and never a single hypothesis in its individual capacity, which serves as the unit of appraisal and which is confronted with favourable/unfavourable observational/experimental results. And I think that it is this which led Quine to say in his essay 'Two Dogmas of Empiricism'[23] that empirical science as a whole must be taken as the unit of empirical content. When Quine said this he was far from being a dogmatic behaviorist. It is interesting to note how it is only much later that he turns into a dogmatic behaviorist paying merely a lip service to this original Duhemian insight:

> I do not know how, in general, in terms of behavioural dispositions, to approximate to the notion of understanding at all, when the sentences understood are standing sentences. Perhaps it cannot be done, taking standing sentences one by one.[24]

Similarly, at another place Quine says:

> Observation sentences peel nicely; their meanings, stimulus meanings, emerge absolute and free of residual verbal taint. Similarly for occasion sentences more generally, since the linguist can go native. Theoretical sentences such as 'Neutrinos lack mass', or the law of entropy, or the constancy of the speed of light, are at the other extreme. It is of such sentences above all that Wittgenstein's dictum holds true: "Understanding a sentence, means understanding a language". Such sentences, and countless ones that lie indeterminate between the two extremes, lack linguistically neutral meanings.[25]

3.4. THE IDEOLOGICALLY NEUTRAL PROBLEM OF THE CONDITIONS OF TRANSLATIONAL DETERMINACY

It follows from these considerations, I think, that for translation to be determinate it is not enough either (a) that the Quinean behavioral criteria are fulfilled uniquely or (b) that there be propositions as linguistically invariant meanings. But Quine thinks otherwise. According to him the failure or the non-fulfillment of the behavioral criteria entails the absence of propositions. Thus, for him, if there were anything approximating to propositions corresponding to the non-observational sentences of our languages, they would show themselves up in terms of the behavioral criteria and then there would be no indeterminacy. Since no such thing shows itself up in terms of the Quinean behavioral criteria, we cannot admit propositions as the linguistically neutral meanings of the non-observational sentences of our languages. Hence the indeterminacy of translation in the sense of Quine.

What Quine fails to take note of is, I believe, the possibility that propositions, along with *problems* in science, might be conceived non-mentalistically as belonging to the Popperian world$_3$ in the precise sense indicated above; and they might themselves be so inherently indeterminate as to affect all translation in fundamental ways. From an objective world$_3$ point of view, one might argue that for translation from one language into another to be determinate it suffices (i) that there be propositions in the sense of (a)–(b) above and (ii) that these propositions themselves be quite determinate in the following sense:

(d) A proposition is determinate if it can be taken individually in isolation from other propositions/problems and subjected to an appraisal without reference to an indefinitely large number of other propositions or appraisals thereof.

But (d) clearly clashes with our very characterization (a)–(c) of a proposition above that makes it inherently indeterminate. And indeterminacy of propositions/problems, in our sense, by itself suffices, I think, to entail the indeterminacy of translation in a sense which must be distinguished clearly from the Quinean sense.

Obviously, taken in this sense, the determinacy/indeterminacy of meaning is methodological in character. And it is indeterminacy in this sense, and not in any other sense, which seems relevant from the point of view of the indeterminacy of translation. That is, for translation to be determinate/indeterminate what is both *sufficient and necessary* is, I think, this: (a) that there be propositions as linguistically neutral meanings; and (b) that these propositions be themselves determinate/indeterminate in a methodological and not in a semantical sense. That this is so will be clear only if we are able to distinguish the semantical from the methodological concept of meaning and define what might be called a fundamental asymmetry between them.[26] The asymmetry I have in mind has remained unsuspected in absence of any attempt to distinguish the two concepts from each other. It is thus all the more necessary in the present context to ask: what kind of determinacy/indeterminacy conditions can we lay down so as to decide in terms of them whether or not a given set of sentences in a given language or a given scientific theory is determinate/indeterminate in respect of its content either semantically or methodologically or in both the aspects? Or, how do we look at the semantical and methodological concepts of sentence-meaning? I think that if the Quinean behavioristic onslaughts on meaning/propositions are to be met adequately we must not only draw a clear distinction between the two concepts of meaning but show how the methodological concept relates to content-indeterminacy in a sense relevant to a non-Quinean indeterminacy of translation.[27]

3.4.1. *The Criterial Character of Quinean Indeterminacy*

Our reasoning up to this point shows that it is possible to ground a non-Quinean indeterminacy of translation in a fundamental indeterminacy of propositions/problems in the Popperian world$_3$. And this shows once again how under Quinean dogmatic behaviorism a very important question remains unasked, namely: what is the determinacy/indeterminacy of translation, if any, a function of?

For Quine this question, formulated as it is in an ideologically – and, hence, behavioristically – neutral manner, does not arise at all. As a behaviorist, he assumes determinacy of translation as what must be guaranteed by the fulfillment of the appropriate behavioral criteria for translation, intelligibility and language-learning. Thus, for him the only question worth asking is this: what *behavioral* criteria must be invoked such that their fulfillment guarantees translation without indeterminacy and their non-fulfillment indeterminacy of translation? Obviously, this question assumes too much under the burden of its behaviorist ideology and makes little or no sense to one who may not be a behaviorist in the sense of Quine.

Quinean indeterminacy of translation, it thus turns out, is at best *criterial* indeterminacy in the following sense:[28] wherever, while translating from one language into another, the Quinean behavioral criteria for determinacy breakdown there the indeterminacy must reign supreme. But what I am arguing for is the stronger thesis that there is indeterminacy in a more fundamental sense, more fundamental and more pervasive than the Quinean criterial indeterminacy. And, therefore, the question which Quine obliterates and fails to ask is this: Is there indeterminacy/determinacy of a kind which does not depend upon the failure/fulfillment of criteria, behavioral or non-behavioral in character? I think that the reason for this lies deep in his scepticism[29] about propositions and in his assumption that it is possible to explain the determinacy/indeterminacy of translation without commitment to propositions as linguistically neutral meanings. But, one might ask, isn't this like cutting the very branch of a tree on which one is sitting?

3.5. INDETERMINACY OF TRANSLATION: A NON-QUINEAN FUNCTION OF CONTENT-INDETERMINACY

Now the real problem before us is, I think, one of deciding between the two ways of posing the problem of indeterminacy, the Quinean and the non-Quinean way, namely

(1) *The Quinean formulation:*
What behavioral criteria must be invoked such that their fulfillment guarantees translation without indeterminacy and their non-fulfillment indeterminacy of translation?

(2) *The non-Quinean formulation:*
Given the reasonable assumption that in translation one tries to understand in one's own language the propositions/problems expressed/stated in another language, what condition(s) must these propositions themselves fulfil in order to guarantee translation without indeterminacy? Or, simply: what is the determinacy/indeterminacy of *translation* in the sense indicated a function of?

I think that only a strong ideological commitment to a Quinean behaviorism could prevent one from recognizing that, unlike question (1), question (2) is ideologically neutral between mentalism and behaviorism and hence genuinely philosophical in character. The same ideological neutrality is, I believe, discernible in the remarks made above which already contain an answer to this question. In any case, a closer look at these remarks will show how fundamental the indeterminacy of translation is, taken not in Quine's but in a non-Quinean sense. What these remarks imply is the view that propositions/problems as linguistically invariant meanings in the Popperian $world_3$ are inherently indeterminate in a methodological sense. And it is only because of their inherent indeterminacy that translation is always indeterminate: there is always more than one way of translating what the native *says* in his language. Non-Quinean indeterminacy of translation is thus a function of the deeper indeterminacy of the propositions/problems themselves. Whereas, the Quinean indeterminacy of translation allegedly results from a failure of the Quinean behavioral criteria of intelligibility. Again, the non-Quinean indeterminacy reigns supreme in language as a *whole*, there being no privileged sentences not affected by it. Quinean indeterminacy is, on the other hand, restricted largely to non-observational or non-occasion sentences[30] (what Quine calls standing sentences) – a consequence, I think, quite natural under behaviorism in Quine's sense. In this way we might conclude that if indeterminacy reigns supreme in science and in language it is essentially non-Quinean in character.

3.6. EPISTEMIC STRUCTURALISM: PROBLEMS AND PROPOSITIONS IN RETROSPECT

What I have been aiming at through my main arguments above is this among other things: for translation to be determinate it is not enough that there be linguistically neutral propositions/meanings. What is necessary and sufficient is that propositions/meanings themselves be determinate in the sense defined above. In this way what I have shown further is this: the determinacy/indeterminacy conditions for translation and meaning can and must be formulated in a manner which is ideologically neutral between rivals such as behaviorism and mentalism. I think that, as a consequence of all this, I have exposed the absurdity inherent in Quine's doctrine that makes the indeterminacy of translation (in his sense) a function of non-fulfillment of the behavioral criteria. Its absurdity becomes now quite obvious: while indeterminacy of translation might be admitted as reigning supreme in the domain of languages in the sense explained above, it would make no sense at all to require or to suggest that the behavioral criteria too reign supreme in the sense of Quine.

I think that one chief reason why Quine rejects the idea of linguistically neutral propositions/meanings has to do with his strong commitment to criteria of identity steming from a deep-rooted traditional statement-view of the epistemic structures in science. This is understandable in so far as epistemology has always rested on one doctrine of epistemic structure or the other, almost invariably presupposing a statement-view of the epistemic structure admitted as relevant. I think that any sound criteria of identity for propositions are bound to fail so long as propositions are looked at or sought to be hunted out as 'abstract entities' capable of being considered either in their individual capacity or as whole groups. The hollowness of this approach is revealed, I think, by epistemic structuralism (see Chapter 8, § 8.3 below), which identifies the epistemic structures in our language generally and in science particularly as the developmental structures of objective knowledge. On this principle, then, the 'identity' of propositions is inseparably linked with that of problems and *vice-versa*, each *in its turn* either presupposing or leading to the other in the interactive pattern of the growth of objective knowledge (see Figure 7.1). Any philosophical approach to the present problem which fails to take note of this is bound to land itself sooner or later in scepticism of a variety already advocated by Quine.

Ontological scepticism concerning propositions – on Quinean lines particularly – cannot however survive critical scrutiny from the point of view of the epistemological ontology I adopted in (1983); nor from the point of view of the very logic and dynamics of type-distinct questions formulating type-distinct linguistically neutral theoretical problems. For what the Quinean scepticism assumes, if only as a target of attack, is the customary dogmatic identification of the epistemic structures in language, and in science, with statements that are capable of being true. But the whole dogma, along with the scepticism raised on it, is undermined once the relevant epistemic structures are properly re-identified as type-distinct developmental structures of objective knowledge – such as *theories* and *problems* in science – with interactions between them invariably leading to growing theory-problem interactive systems as the proper subjects of epistemic appraisal in science (see Figure 7.1).

I think that epistemic structuralism offers yet another way to combat indeterminacy of translation in Quine's sense. Consider the specific case of theory-problem interactive systems in physics, if only to be followed by a consideration of propositions and problems in general across different languages. In so far as it is correct to say that the classical (Newtonian) physical problems concerning motion and force and Newtonian mechanics taken together with the law of universal gravitation can be seen as pragmatically[31] *simpler* and special cases of pragmatically more complex problems and theories, viz. Einstein's special and general theories of relativity, it is also correct to think that the former *can* be 'translated' into the latter. Newtonian mechanics and the law of universal gravitation are, in other words, *characterized* by and based on subject-specific rules of problem-formulation and theory-construction that are much simpler pragmatically as compared with those which one can discern in the special and the general theories of relativity. Since the former are derivable as a special case from the latter, they could be shown to rest on over-simplifications, among other things. To whatever degree this is true of these two types of theory-problem interactive systems, to the same degree it will be correct to say of the pragmatically simpler system S_1 that it is translatable into – because it is in some appropriate sense derivable from – the pragmatically more complex one S_2. Consider in this context the concept of *mass* which appears as a scalar concept in Newtonian mechanics, always remaining a *constant*, and hence is much simpler compared to the concept of *mass* as it appears in Einstein's special theory of relativity. Here it makes its appearance as a function of velocity – as a variable – and hence as a more complex physical magnitude. Yet, under the assumption that in certain simpler types of situation the velocity with which a material body is moving is a negligible fraction of c – the velocity of light – the Newtonian concept of mass is applicable as a special case of the concept of mass as introduced by the special theory of relativity. The main point underlying these generalizations can be formulated now more generally as follows:

> The growth of scientific knowledge follows an interactive pattern in which theory-problem interactive systems can be seen moving from the pragmatically simpler frameworks of subject-specific rules of problem-formulation and theory-construction to more complex ones, with the simpler ones being *always* derivable from, and hence *translatable* into, the more complex ones as special cases.

In a sense this is applicable in case of propositions and problems generally as these are identifiable or expressible across different languages. Consider the language – L_1 – of a primitive tribe with a much simpler structure and function or applicability than that of our own language L_2. But notice that, for Quine if different enterprising linguists undertake to translate whatever the native says in L_1 into L_2, they will come out with widely incompatible sets of analytical hypotheses leading to such different translations that there is no question of any one of them being right to the exclusion of the rest of them. This will be particularly true of the native's non-occasion sentences that Quine calls standing sentences. What is here most interesting is Quine's distinction between these two types of sentences, thereby implying, perhaps *unintendedly, that L_1 is essentially not different from L_2*. For the distinction is *assumed* by Quine to be valid across all languages from the most primitive, hitherto untouched jungle language to the more complex like our own. Indeed, without an assumption as important as this, there will be, I think, hardly anything left in different languages for Quine to argue for his thesis of the indeterminacy of radical translation.

I think that if we follow the clue which this particular feature of Quine's thesis provides, there is no escape from the following assumption which any enterprising linguist seeking, however unaided, to translate L_1 into L_2 must make to begin with:

> L_1 is *essentially* similar to L_2: that is, in structure and function both, the two are essentially similar, even if L_1 is pragmatically far simpler than L_2. The implication of this being, among other things, that the linguistic practices associated with L_1 and L_2 are more or less equally rule-governed practices developed around commonly agreed upon paradigms – paradigms in terms of which concepts are developed, taught/learned and extended in the day-to-day use of language. In other words, this means: *that L_1 and L_2 and similarly other languages show similar patterns of development, learning and use.*

If one proceeds on this reasonable assumption, one could then fill many details such as the one discernible, I think, in Quine himself: since any translation of what the native says in L_1 into L_2 will be indeterminate, in his sense, with regard to standing sentences, if not with regard to occasion sentences, what must be common to L_1 and L_2 is precisely the applicability of the fundamental distinction between standing sentences and occasion sentences themselves. And it will then follow as a further consequence in the light of our argument concerning physical theory-problem interactive systems above that L_1, being admittedly pragmatically simpler than L_2, must be derivable from and hence translatable into the much more widely applicable L_2, if only as a special case. In this way, it might be concluded that Quine's thesis of the indeterminacy of translation does not hold true at all where the direction in which translation is undertaken is one from a pragmatically simpler language to a more comprehensive or complex one, but where the two candidate-languages must be, nevertheless, essentially similar in the sense of Quine's own fundamental distinctions.

But what about translation following the opposite direction: viz., from a more complex and highly developed language to a simpler one? Quinean indeterminacy, I believe, does not hold true even here. What holds true here, on the contrary, is content-indeterminacy in the sense discussed above. This may now be restated as follows:

> Translation of what is *said* and *asked* in a pragmatically more complex language into a simpler one perhaps as a generalized case of a translation of a pragmatically more complex theory-problem interactive system governed by more complex rules into a simpler one governed by simpler rules must suffer from content-indeterminacy resulting in maximal loss of those very components that are essential to the candidate-language under translation.

I think that this is to argue, in other words, for translatability, clearly within the limits of content-indeterminacy but only in one direction, namely from what is pragmatically simpler to what is pragmatically more complex. And this is an important consequence which it is difficult to obtain under Quine's own approach.

CHAPTER 4

ON THE IMPOSSIBILITY OF ANY ENTERPRISE CONCERNING SELF-KNOWLEDGE WITHIN TRADITIONAL EPISTEMOLOGY

4.1. THE TRADITIONAL DOCTRINE OF SELF-KNOWLEDGE AND THE CONCEPT OF A PERSON

As an abstract concept, the concept of a knowing subject is as old as epistemology and might be regarded as a lasting contribution of this philosophical enterprise. There is not a single theory of knowledge – whether in the East or in the West – which is not woven around it, or which does not employ it as a central concept. Thus, throughout the different epochs of philosophical thinking concerning the nature of *human knowledge* in different fields of learning, all attempts at its definition/ characterization invariably fall back upon this concept. This is as much true of Plato as it is of Rene Descartes, the two great system-builders in traditional epistemology (TE). Yet, while all this is true, the concept of a knowing subject remains the least analyzed concept in TE. Is it due to this concept's fundamental role in TE? Or is there some other reason for the indifference to its analysis?

For the purpose of the present discussion it is necessary, I think, to state at the very outset two of my theses which I employ as my chief assumptions but leave unargued for the present, having argued for them elsewhere.[1] These are: (a) the thesis of a parallelism between a subjectivistic epistemology (SE) and an objectivistic epistemology (OE), where the latter is conceived as an abstract enterprise concerning the structure and growth of (objective) scientific knowledge; and (b) the thesis that to be a person is to be a hierarchically organized growing problem-solving *knowing subject*. Among other things, the thesis (a) enables us to admit on the one hand the reality of the world of human knowledge in a subjective sense by admitting TE as a species of SE and that of the world of objective knowledge on the other. As a consequence, TE as a species of SE is seen as involving a perfectly legitimate enterprise concerning those and only those knowledge-states (K-states) which presuppose a knowing subject as their invariable epistemic subject of predication. Hence TE is incapable of dealing with the types of K-states that do not involve any such presupposition. On the contrary, only an enterprise as abstract and objectivistic as OE can take care of such K-states. The thesis of SE-OE parallelism however leaves the possibility of their interaction open. Coming now to the thesis (b) above, among other things, it enables us to say that the concept of a person entails that of a knowing subject precisely in the sense in which TE assumes it. Proceeding within these assumptions, what I propose to make as an immediate target of my criticism are the following philosophical doctrines: (1) *self-knowledge* in the only legitimate sense of one's knowledge of one's own self – and hence as distinct from one's knowledge of other 'selves' – exists as beyond doubt; and (2) it is primarily because of the *existence* and *nature* of self-knowledge – i.e. of the way in which first-person statements are *made* and known to be true – that there is a special set of philosophical problems about the nature of persons (selves) and personal identity.[2]

4.2. THE IMPOSSIBILITY OF ANY ENTERPRISE CONCERNING SELF-KNOWLEDGE WITHIN TRADITIONAL EPISTEMOLOGY

In what sense, if any, are persons knowable? In so far as the concept of a person entails that of a knowing subject, let me pose the question: In what sense, if any, are knowing subjects knowable? Traditionally, it is assumed without any question that self-knowledge exists: there can be no doubt as to the existence of the type of knowledge that either one has or is capable of having about his/her own self. And what is questionable, on the contrary, is one's knowledge of *other* selves/minds. The underlying reasoning here has been somewhat like this: one's knowledge of one's own self is infallible and hence the only knowledge in the true sense of the term; while one's 'knowledge' of other persons/minds – being fallible – cannot be called knowledge in the true sense of the term. Now the thesis which I wish here to argue for can be formulated in terms of my thesis of SE-OE parallelism. As was stated above, SE centres around the knowing subject as its very invariant frame of reference in the sense that it deals with *subjective knowledge* conceived as a special kind of mental state (of infallible belief). Within SE, therefore, a K-state *presupposes* a knowing subject – i.e., knowledge in the subjective sense is inconceivable without the assumption of the knowing subject as the invariant frame of reference or as the invariant epistemic subject of predication. It follows from this that, in the context of SE, any idea of knowledge without a knowing subject will be a contradiction in terms. Now if this is so, it is not possible at all to speak of one's knowledge of one's own self without committing a contradiction or generating a vicious circle. Self-knowledge in the sense of one's knowledge of one's own self in the capacity of a knowing subject is inconceivable.

In other words, the knowing subject *cannot* be conceived as an object of knowledge within SE, it being its very limit, as it were. For it will be true of any subjective K-state concerning one's own self as a knowing subject *that* it presupposes an unknown knowing subject and hence a knowing subject that must remain unknowable from the point of view of SE. From this we might conclude that self-knowledge in its traditionally conceived sense is impossible.

From the viewpoint of SE, every K-state presupposes a knowing subject to whom such states must be, in ultimate analysis, ascribable. Only this may not be as obvious in the context of the so-called 'self-knowledge' as it is in that of our knowledge of the external world or of other persons. Now if to be a person is to be, among other things, a problem-solving knowing subject that transcends all anchorage (and hence that does not belong among the inmates of any one of the Popperian $world_1$, $world_2$, $world_3$)[3] the question that needs answering is this: Is it meaningful/intelligible at all to speak of 'self-knowledge' in the alleged sense of one's knowledge of one's own identity as a person/problem-solving knowing subject? Is not 'self-knowledge' in this sense something that is logically impossible? For every specific self-knowledge-state (SK-state), like any K-state, *presupposes*, I think, the unknown knowing subject as a subject to whom such a state must be ascribed. That which is so obviously presupposed invariably in all the contexts of K-states cannot itself be reduced to an object of knowledge without a *vicious circle*. In SE, it is the knowing subject, if anything, which forms the limit of the knowable, like the eye that can see everything except itself. But, one might ask, how can we then even speak of a knowing subject as the invariant epistemic subject of predication in SE? What is the nature of our *presuppositional* knowledge to the effect that something invariant – the knowing subject – is presupposed as the epistemic subject of predication in SE? How might our talk of the knowing subject as a fundamental presupposition in SE be legitimized without leading to any *vicious circle*?

My main argument for my thesis of the impossibility of any enterprise concerning 'self-knowledge' within TE could, I think, be given a more decisive form as follows:

(I) Within TE as a species of SE all K-states must satisfy the following general conditions:
  (1) Given a knowing subject S and his successive/intervalled K-states S (K_1), S (K_2), S (K_3), . . . , the following epistemic relations must correspondingly hold true:
    (a) S $knows_1$ that p
    (b) S $knows_2$ that q
    (c) S $knows_3$ that r
  (2) A K-state S (K) that is generative of the epistemic relation, viz., 'S $knows_2$ that *q*' must be, in order to be genuine, definable as a Belief-state (B-state) such that it is true to say: 'S believes/accepts that *q*, and S has ground that makes his belief that *q* 'rational', where the pragmatic predicate 'believes' occurs as a primitive.

(II) Conditions (1)–(2) above have the following consequences:
  (3) All K-states in the sense of conditions (1)–(2) presuppose the knowing subject as the very frame of reference or as the fixed subject of epistemic predication.
  (4) All SK-states in the sense of TE are a species of K-states as characterized in (3).
  (5) All SK-states, therefore, presuppose the knowing subject as the subject of epistemic predication.
  (6) All K-states other than SK-states are states concerning subjects other than the knowing subject. That is, the only major difference between the class of K-states and the sub-class of SK-states is that all SK-states concern nothing but the knowing subject.
  (7) All K-states other than SK-states are, therefore, the only kind of K-states that are possible.
  (8) All SK-states are, therefore, impossible; so is any enterprise concerning them conceived within the framework of TE.

Now, what is puzzling is, I think, the philosophical doctrine that says that it is the existence of self-knowledge and questions such as 'What am I?' which primarily give rise to problems about the nature of persons/personal identity.[4] My counter-question is this: Is it legitimate to regard the first-person psychological statements as statements about persons and 'What am I?' type questions as raising a problem about the nature of persons? How is it that the question 'What am I?' can be said to raise the question 'What is a person?' What the contemporary philosophers dealing with these problems do not realize is, I think, that it is in ultimate analysis impossible to ask 'What am I?' type questions. And if that is so, any question of a transition from such a question and its alleged answers to the question 'What is a person?' does not arise. For consider what the uncritical asking of the question 'What am I?' involves. It is, I think, part of this question's anatomy and presuppositional setting (which includes language) that only a knowing subject can ask it and that everytime it is asked it must be addressed to the very 'same' knowing subject who is asking it. It is thus a distinguishing feature of this question that it entangles us in the paradoxical situation of a knowing subject seeking to capture himself as an *object* of his (subjective) knowledge. But this very presupposition of a knowing subject capturing himself as the object of his knowledge is full of paradox or contradiction and hence the question 'What am I?', which rests on this presupposition, is an unaskable one. The philosophical implication here is clear: Self-knowledge, as traditionally construed, is impossible and hence a pseudo-concept. The far-reaching implication is, I think, equally clear: those who in the recent developments in the field of epistemology and ontology look to the unquestioned existence of 'self-knowledge' as the source as well as the rationale for raising questions of personal identity and nature of persons are mistaken – the entire enterprise (Shoemaker's included) is misconceived.

4.3. FIRST-PERSON PSYCHOLOGICAL SENTENCES: SELF-CONSCIOUSNESS, OBJECTIVE KNOWLEDGE AND HUMAN INTERACTION

It might be argued that my rejection of the traditional problem of 'self-knowledge' as a pseudo-problem is too hasty and sweeping to do justice to a significant reformulation which the problem has undergone in contemporary philosophy. For, with the traditional theory of the 'self' as a type-distinct substance already dismantled, the traditional problem so shifts itself to the linguistic scene of the first-person psychological/quasi-psychological sentences as to reduce to the problem of how the statements expressed by such sentences come to be *asserted* and then *validated* by us: the whole problem of the criteria for making and knowing such statements to be true. That it is precisely in this manner that the problem of self-knowledge in the sense of TE is (dogmatically) believed to shift itself to the relatively firmer grounds in the linguistic scene of (chiefly three types of) first-person psychological/quasi-psychological sentences is what characterizes Shoemaker's enterprise[5] and those closely similar to his in contemporary analytic philosophy.[6] The sentences in question are typified as follows:[7]

(a) 'I am in pain'; 'I have a toothache' and the like, that express what are called experience statements (E-statements);
(b) 'I see that there is a rabbit scurrying by'; 'I see that there is a Japanese cedar in the garden' and the like, that express what are called perceptual statements (P-statements);
(c) 'I remember that the Chinese junipers were planted a year ago' and the like, that express what are called memory statements (M-statements).

This contemporary philosophical strategy of preventing the problem of self-knowledge from meeting the fate of its traditional presuppositions of the self as a 'substance' must however fail. It is, I think, open to a number of decisive criticisms over and above the one made already above, that can be extended to it. First of all, I think that the only way in which the problem of self-knowledge in its traditional (Cartesian) sense can be said to shift itself to the linguistic scene relating to E-statements, P-statements and M-statements is to assume the basic framework of TE which dictates *that self-consciousness implies self-knowledge*. If we add to this the further assumption – which, I believe, operates implicitly most of the time in the contemporary analytical philosopher's construal of the problem of self-knowledge – that the rich and the variable scene of self-consciousness is most easily captured or becomes most clearly accessible only in the rich and the variable linguistic scene of sentences expressing E-statements, P-statements and M-statements respectively, the contemporary linguistic construal of the problem of self-knowledge will follow as a logical consequence. Now it is, I think, not only naive and simplistic but fundamentally erroneous to take self-consciousness as implying self-knowledge. It is simplistic in so far as any such view must assume the following simple but highly objectionable model for all knowledge:

> To know a thing is either to be *directly* aware of it or to show reflective awareness/consciousness of that thing with what-

> ever observable properties it has; that is, our knowledge of objects in the external world reduces, in ultimate analysis, either to our direct or to our reflective awareness of these objects, their observable properties and relations; and that, similarly, it is only in states of self-consciousness that one can be in states of self-knowledge with one's own self as the object of knowledge.

As is the case with traditional epistemological thinking in the West, much of the epistemological thinking in the ancient systems of Indian philosophy is, I believe, bedevilled by this model. That this is so can be easily gauged from what are usually cited as the examples of knowledge, the types of examples being generally common to all the systems. The model might, however, be disposed of along the following lines of argument. Let us suppose that our cognitive relationship with our environment or the objects therein is describable in terms of what might be called a *continuum of states of awareness* which is subject-dependent and hence inconceivable without a human subject. We can speak of such a continuum since each of us shows – continually from the childhood learning to the adulthood interactions – either 'direct' or 'reflective' awareness of different kinds of objects in his/her environment. We might be said to do so most of the time in our very manner of *dealing with* our environment. But we might also be said to do it in our very *thinking about* it. Once an awareness continuum in this rough sense is admitted, it becomes possible to conceive of our 'experiences' relating to the external objects/events as what could be plotted on this continuum. This is precisely what we should do if, e.g., we hear a person, who is either watching or just thinking of the night sky, remark: The moon has moved further away from its position an hour earlier. That is, in this case and the similar ones, both the speaker and the understanding hearer may be said to show (reflective) awareness of the moon in apparent motion around the earth – leaving aside any question of what this implies as to one's awareness of the structure of the solar system. Now, will it be a case of (reflective) awareness in the same sense if we suppose a different situation in which a person – uninitiated in enterprises like modern science – is occupied with the thought of 'the sun moving around the earth' simply because that is what his 'experience' discloses to him every time he considers their relative positions? I think that nothing short of a constant scientific thinking can prevent one from plotting a thought like that on one's awareness continuum. It is well-known that in earlier times – prior to the Copernican revolution, to be precise – it was the 'experience' of sun's *apparent* motion which very much dominated human thinking about and human awareness of the universe: that is, *human interaction then was characterized by a different kind of cognitive orientation and a different order of complexity.* But what is it now which prevents one from plotting a specific thought/belief like that on one's awareness continuum? I think that it is quite correct to maintain that there is something in our current systems of knowledge which would knock it out if one tried to do so. More generally then, one might argue, it is the context of our (fallible) knowledge – the context of our problems and theories – which provides the principle of selection, thereby determining the orientation and the order of complexity of our interactions with our environment including ourselves. It follows from this that nothing short of a radically different context of problems and theories – the whole geocentric world-view in short – could have, at an earlier time, made it possible for one to so orientate and order his interactions with an external world as to plot the 'experience' of 'the sun moving around the earth' on his awareness continuum. That is, only to someone seriously concerned to define the structure of reality around him and willing to subscribe to the geocentric world-view as a solution would such an 'experience' be forthcoming.

I think that if we introduce at this point the concept of knowledge as a kind of *subject-independent continuum of problems and theories* – subject-independent in the sense of belonging to the Popperian $world_3$ – the considerations above can be seen to lead to the following significant consequences:

(A) At a *primitive level*, a completely *unplotted* subject-independent continuum of knowledge – this would be the case if there were no human language in which alone is it possible to formulate this knowledge – is quite compatible with a heavily plotted subject-dependent awareness-continuum.

(B) This compatibility holds irrespective of the manner in which one may plot one's experiences on the awareness continuum.

(C) It is the actual plotting done on the continuum of *knowledge* which *can* significantly, but need not, affect the plotting done on the awareness continuum, without the converse being true.

I think that all the three consequences are intimately interrelated and are perhaps ideally borne out in the case of children and highly intelligent animals. This is particularly true of the consequence (A). This consequence is essentially one of compatibility between one's 'ignorance' about one's environment – strictly in the sense of one's inability to clearly conceive of the *relevant* context(s) of problems and theories and hence one's inability to enter in properly oriented significant interactions – and one's otherwise heavily plotted awareness continuum. (A more detailed discussion of this important point must, however, await another occasion.)

Now, all the three consequences, (A)–(C), tentatively formulated as they are, when taken together may look quite puzzling and far from offering any decisive argument against the doctrine that awareness – in the sense considered above – implies knowledge. But what is interesting is, I think, this: they will cease to be puzzling if we drop the very assumption that awareness implies knowledge. And we can do this precisely for the following reason, which is also the reason why the three consequences above hold: Awareness, like self-awareness, implies not knowledge but interaction. More precisely, any specific state of awareness/self-awareness must be viewed, I suggest, as a complex state of interaction with various properties which are worth exploring in detail. And, I believe that one important respect in which *human* interaction at various levels of complexity differs from non-human interaction is this: it is the former and not the latter which *presupposes* knowledge or allows itself to be so *informed* by knowledge – where knowledge is conceived as a continuum of problems and theories whether explicitly formulated or not – as to acquire a unique order of complexity and organization. It follows from this, I think, that it is knowledge in the present sense of this term which makes human awareness possible at all levels of complexity whatever, instead of awareness leading to knowledge.[8]

It is thus erroneous to hold that self-consciousness implies self-knowledge in any sense whatever. On the contrary, what it does imply is, I think, a certain uniqueness that permeats most if not all *human interaction*. That is, it implies a certain order of complexity in human interaction. That this is so might be seen more clearly once we realize how erroneous it is – the error being traceable to TE and the specific theories developed thereunder – to take E-statements, P-statements and M-statements as statements of what we 'know' or as representing knowledge-states of any kind. That is, in my view at least, their very construal as statements expressing knowledge-states is fundamentally mistaken. On the contrary, I think, such types of statements, and those which they may entail, could be taken as the end-results of interactions taking place at various levels of complexity and organization; and not as implying any knowledge-states whatever. In a nutshell, such statements codify the end-results of interactions in which we ourselves are involved deeply; they do not just report 'what we know about ourselves'.

Secondly, while all P-statements and M-statements are essentially elaborations of the corresponding statements made by means of sentences in which neither the personal pronoun nor any psychological expression occurs, the relationship between them may be very complex. Thus when one says simply that "there is a rabbit over there" one might as well say what one wants to say more appropriately, and without risking any charge of faulty perception, by saying that "I see that there is a rabbit over there". That is, the P-statements and M-statements may be regarded as most appropriate for saying what one might simply and elliptically say by means of sentences in which neither the personal pronoun nor any psychological expression occurs. In any case, there is an aspect of their relationship which warrants the thesis – to which analytic philosophers such as Shoemaker subscribe – that all P-statements and M-statements entail such (apparantly) psychologically neutral physical-object statements or other types of such statements as relevantly correspond to their (non-personal) sentential components.[9] Now, if this is so, the problem of self-knowledge, in the sense of its contemporary construal, cannot simply relate to first-person psychological/quasi-psychological statements; it must also, and perhaps more importantly in the present context, relate to all sorts of physical-object statements. For P-statements and M-statements, as Shoemaker himself notes, entail such statements invariably. But isn't it paradoxical that what begins as an enterprise concerning the problem of self-knowledge should turn out to be one which relates to the kinds of sentences which we normally employ to make non-psychological, physical-object statements about the external world?[10] The answer to this question is, I think, in the affirmative in so far as we have come face-to-face with a *reductio-ad-absurdum* type consequence of the contemporary construal of the traditional problem of self-knowledge.

It follows from this that the problem of self-knowledge – both in its classical Cartesian and contemporary linguistic formulation – is a pseudo-problem in TE. For all that its linguistic formulation does is to shift the problem from its classical entrenchment in the 'self' as the kind of substance allegedly designated by the personal pronoun 'I' to one in the varied linguistic scene of the first-person psychological/quasi-psychological sentences of different categories. It is now very clear how in this process the problem is only shifted from one strong-hold (in metaphysics) to another (in language) but always *within* TE. Whereas what is warranted by our discussion above is its complete dissolution within TE or its complete shifting from TE to OE, implying, among other things, that the only residual problem left concerns the nature of persons/human beings conceived as comprising a subject of objective knowledge in their own right.

4.4. PERSONS AS A SUBJECT OF OBJECTIVE KNOWLEDGE

But we might still ask: Is there any other sense in which it would be possible to speak of the knowing subjects as a possible subject of objective knowledge or of our knowledge of the knowing subjects/persons? An answer in the affirmative is already anticipated by my thesis of SE-OE parallelism, where by OE we understand an abstract enterprise that centres around the epistemic structures of science, thereby dispensing with the knowing subject altogether. By the epistemic structures of *science* I understand such developmental structures of knowledge as *theories* and *problems* that not only codetermine its structural identity but interact to form the most interesting interactive systems of objective knowledge. Now, there are, I think, no grounds *a priori* to warrant our ruling the knowing subjects or persons out as a possible subject of objective knowledge within the framework of OE. Notice that OE is free from any presuppositions that could entangle it in subjectivism or compel it to fall back on the knowing subject as a frame of reference. In a nutshell, this means that the knowing subjects, like any other kind of ontic categories, are knowable from the point of view of OE. And if knowing subjects and hence persons are knowable in this sense, any philosophical enterprise that aims at a philosophical characterization of the very concept of a knowing subject/person would be quite legitimate, without however any such enterprise concerning 'self-knowledge' being possible.

Since there is no doubt about the existence/reality of human beings in their capacity as persons/knowing subjects – the fundamental presupposition of TE – the world of persons must be admitted as yet another domain of the subjects of objective knowledge. All of us are knowing subjects. The job of a knowing subject is, among other things, to know, first of all, the knowing subjects other than himself; and only then, with those other knowing subjects as a frame of reference, as it were, to know himself or herself *if only obliquely*. While we can fallibly know other persons, we can keep conjecturing and asking questions about ourselves but only with other persons/knowing subjects as a constant frame of reference. It is this, I think, which makes persons an appropriate subject of objective (scientific) knowledge.

It is clear from the preceding discussion that there is a sense in which persons exist and are knowable, although all that we can know in this context is bound to share the fallibility of all objective scientific knowledge. If what I said implies the impossibility of 'self-knowledge' as traditionally conceived, it implies at the same time the definability within OE of what must invariably be assumed as indefinable in SE.

4.5. THE PHILOSOPHICAL SIGNIFICANCE OF SCEPTICISM IN HUME AND WITTGENSTEIN

Let us now consider whether in the light of my theses above a better understanding of the sceptical views of David Hume and Wittgenstein, among others, is possible, since the essential points in their views seem to have suffered considerable distortions at the hands of the commentators.[11] It might be quite plausible to say, as an interesting consequence of my main thesis concerning 'self-knowledge' here, that it saves us from the kind of philosophical embarrasment to which David Hume was driven by TE quite inescapably, as it were, and which he himself summed up excellently as follows: "For my part, when I enter most intimately into what I call *myself*, I always stumble on some particular perception or other, of heat, or cold, light or shade, love or hatred, pain or pleasure. I never can catch *myself* at any time without a perception, and never can observe anything but the perception . . .".[12] I think that when Hume said this and concluded that a human being/person is "nothing but a bundle or collection of different perceptions, which succeed each other with an inconceivable rapidity, and are in a perceptual flux and movement"[13] what he was proposing, in effect, though quite implicitly and perhaps without himself realizing it, was to shift the problem of 'self-knowledge' from *one's own 'self'* not to the linguistic scene of the complex first-person psychological sentences that one may utter from time to time but to what might be described as a scene of our *hypotheses* about the ontological order of human beings/persons/minds. If my interpretation here is not basically erroneous I would consider it as yet another reason why there is a certain irony in the contemporary analytic philosophers' commonly shared construal of the problem of 'self-knowledge' as one which relates to the linguistic scene of our first-person psychological/quasi-psychological sentences.

Let us now turn to Wittgenstein's *Philosophical Investigations* (*PI*). The philosophical import of Wittgenstein's remarks, in (*PI*: §§ 243–304, 404–410), concerning the nature and role of the first-person psychological sentences in our language is not as clear as one would wish. Nevertheless, what these remarks can be shown to imply is not that there are no such things as *persons* or subjects of experiences etc. but that 'self-knowledge' is impossible – i.e. the thesis that the sentences in question cannot be accorded a descriptive and hence an epistemic status in the traditional fashion. In other words, Wittgenstein's scepticism concerning such sentences is, I believe, two-fold and implies that the classical problem of 'self-knowledge', having lost its metaphysical ground of the Cartesian 'self', cannot be shifted to the linguistic scene of our first-person psychological sentences. Thus, consider the following remarks (conclusions) – *PI*: § 246:

> It can't be said of me at all (except perhaps as a joke) that I *know* I am in pain. What is it supposed to mean – except perhaps that I *am* in Pain?
>
> Other people cannot be said to learn of my sensations *only* from my behaviour, – for *I* cannot be said to learn of them. I *have* them.
>
> The truth is: it makes sense to say about other people that they doubt whether I am in pain: but not to say it about myself.

Taking these as implying the impossibility of self-knowledge, and not as denying the existence of the subject of experiences, one reason why Wittgenstein arrives at such a sceptical conclusion is already available in (*PI*: § 244). Here Wittgenstein invokes an argument from language-learning against the first-person psychological sentences being accorded any descriptive status – a scepticism about their very logical grammar being traditionally modelled along those of the 'object and designation' model on the one hand and the 'descriptivist' model on the other.[14]

The underlying pattern of the argument here is as follows: For any sentence to be able to express or represent a knowledge-state, whether in the infallibilist or the fallibilist sense of the term, it is necessary for it to have the logical grammar of a descriptive sentence. But the first-person psychological sentences do not possess such a logical grammar, their appearance to the contrary not-withstanding. Hence such sentences do not express any knowledge-states whatever.

Now, one may view 'description' (of a thing) and 'ascription' (of properties) in this context as typical notions so closely interlinked that one may at times use them interchangeably. More generally, we might say that description of a given thing in a given context implies a kind of ascription of a set of properties to that thing. Taken in this sense, description presupposes, then, not only an identification of the given thing as a thing of a certain relevant category but an identification of those properties that one might *ascribe* to it. If it could be shown how, in the context of the first-person psychological sentences, one might legitimately invoke and employ relevant criteria of identity in this sense, a ground would thereby be prepared for rejecting Wittgenstein's two-fold scepticism above concerning logical grammar and 'self-knowledge'. But that this is not possible is precisely what Wittgenstein is concerned with in (*PI*: §§ 288, 404–408).

Wittgenstein's views concerning logical grammar, 'self-knowledge' and personal identity are, I believe, essentially being missed when we are told by his commentators that he along with others is a no-subject theorist.[15] How could, one must ask, Wittgenstein be seen as simply denying the subject of experiences and the like when his scepticism sketched above does not extend to sentences other than the first-person psychological sentences; and when, on the contrary, in (*PI*: §§ 281–283, 404), he shows considerable concern with the problem of personal identity as relating to or arising out of the third-person psychological sentences. He, therefore, concludes that

> . . . There is a great variety of criteria for personal identity.
>
> Now which one of them determines my saying that 'I' am in pain? None. (*PI*: § 404).
>
> . . . Only of a living human being and what resembles (behaves like) a living human being can one say: it has sensations; it sees; is blind; hears; is deaf; is conscious or unconscious.[16]

Over and above the various arguments which Wittgenstein offers for his views, the following additional argument deserves, I think, a separate consideration:

> "I" is not the name of a person, nor "here" of a place, and "this" is not a name. But they are connected with names. Names are explained by means of them. It is also true that it is characteristic of physics not to use these words.[17]

That is, 'I' is neither an ordinary nor a logically proper name. It is a demonstrative expression like 'this' or 'here'. And, it is quite true that demonstratives such as these play a unique role in explaining names, even proper names, – explaining their uses. It follows in yet another way that 'I am in pain' type sentences do not describe anything. But what about 'He' in the 'He is in pain' type sentence, one might ask? Is 'He' a name? If not, how can the sentence 'He is in pain' be said to describe anything at all?

4.6. THE 'PRIVATE LANGUAGE' VERSION OF THE PROBLEM OF SELF-KNOWLEDGE

I would like to conclude the discussion above by showing how one of the decisive ways of arguing for the very idea/existence of 'self-knowledge' (in TE) as beyond doubt is to argue for the possibility of a 'private language' (Lp) satisfying the following conditions, among others:

(1) The logical grammar of a 'private language' is a species of the logical grammar of language (L) in general: $G(Lp) \in G(L)$.
(2) The class of users (U) of a 'private language' is logically restricted to a unit class: $U(Lp) \equiv$ Unit class.
(3) The sentence set S(Lp) of a 'private language' is a set of first-person sentences.

What (1) and (2) entail *respectively* are the following propositions:

(1′) Lp is learnable.
(2′) Lp is not learnable.

But (1') and (2') arc clearly a contradictory consequence of what one must assume minimally if a 'private language' is to be a logical possibility. Among other things, then, one must assume learnability as an essential condition on any language that is sufficiently rich to allow the formation of sentences that are *descriptive* of things/processes of a particular ontological category whatever. Before a language-user/languaged animal such as man is able to *describe* things/processes already suitably designated in his language, he must learn the rules thereof with whatever structure the language may initially have.

Learnability in this or in a similar sense implies, I think, interaction-relations of various kinds ranging over language-users as ordered pairs of speakers and hearers.[18] Thus, with the language-users as the values of the individual variables of interaction, and interaction-relations themselves as the values of the predicate variables, learnability implies asymmetrically ordered pairs of speakers and hearers in mutual interaction as follows:

$$I\,(S_1, H_1), I(S_2, H_2) \ldots I(S_n, H_n)$$

where I, S and H respectively abbreviate 'interaction relation', 'speaker' and 'hearer'.

It is clear from this that the non-learnability of a 'language' in this standard sense would imply the idea of a 'language' with *isolated* (as against *ordered*) Leibnizian monads as its 'users'. And this is not only inconceivable by itself but contradicts all that the concept of a (learnable) language implies.

It follows from this yet in another significant way that, if one of the decisive ways of formulating the idea of privileged 'self-knowledge' is to argue for the possibility of a 'private language', 'self-knowledge' in the sense of TE is impossible and so is any enterprise concerning it that is sought to be conceived within TE.

The consequence of non-learnability as implied by the idea of a 'private language' is itself quite serious. Consider, e.g., the type of situation it would create for the linguist or for the linguistic theory itself. Consider any number of 'private languages': $Lp = Lp_1, Lp_2, \ldots Lp_n$. In no case would these or our assumption about their existence provide any subject-matter in terms of generalizations/hypotheses, existential and universal, which a linguist could investigate. Whatever may be said to be true or false of such a 'language' is, on the contrary, the sole concern and prerogative of its monad-user who cannot *interact* with other monad-users of other 'private languages'. By definition, then, a 'private language' is allegedly of such structure and use to its monad-user as is beyond quantification – indeed beyond the limits of expressibility or formalization by means of an appropriate logical machinery. Thus, a 'private language' does not and *cannot* exist as a language for the linguist: the very idea of such a language makes no difference to the enterprise of linguistic theory. Nor *can* it exist for the full-blooded interacting human languaged, rational animal.

PART TWO

Gewiß ist ein Gedankenexperiment eine Abstraktion. Aber diese Abstraktion ist dem Physiker, und zwar sowohl dem Experimentator wie dem Theoretiker, bei seiner Forschungsarbeit ebenso unentbehrlich wie diejenige, daß es eine reale Außenwelt gibt. Denn ebenso wie wir bei jedem Vorgang, den wir in der Natur beobachten, etwas voraussetzen müssen, was unabhängig von uns verläuft, müssen wir auf der andern Seite danach trachten, uns von den Mängeln unserer Sinne und unserer Messungsmethoden möglichst zu befreien, und von einer höheren Warte aus die Einzelheiten des Vorganges zu durchschauen. . . . Gerade die großen Geister und Bahnbrecher der Physik, die Kepler, Newton, Leibniz, Faraday, wurden getrieben von ihrem Glauben einerseits an die Realität der Außenwelt, andererseits an das Walten einer höheren Vernunft in oder über ihr.

Max Planck (1949: 294–295)

Every discoverer of a new physical principle makes an important contribution to philosophy, even though he may not discuss it in philosophic terms.

Henry Margenau (1949)

CHAPTER 5

METHODOLOGICAL ESSENTIALISM IN SCIENCE AND IN PHILOSOPHY

> The world we live in has a structure,
> and even though there may be several
> different ways of looking at it, each
> providing some insights, they would
> all have to respect that structure,
> which is as "real" as the person whose
> portrait a painter is doing.
>
> C. H. Waddington (1977)

5.1. METHODOLOGICAL CONVENTIONALISM IN SCIENCE AND IN PHILOSOPHY

The successive ideal-type models of language, of its essential structure and function, which one finds in Russell's logical atomism, early Wittgenstein's *Tractatus Logico-Philosophicus* (*TLP*)[1] and Carnap's earlier formulations of the principle of logical empiricism all proceed from what might be called the methodological framework of essentialism. For the problem they address themselves to is the problem of identifying/ defining the one and only one essential structure of our language (especially the languages of the sciences). And they do so with a view to effecting a clear and sharp demarcation between the cognitively/ factually meaningful languages and the 'language' which is not meaningful. This commonly shared philosophical concern with the essential logico-semantical structure – the logically necessary properties – of all languages is of course traceable to their deeper commitments to the equally essentialistic enterprise of traditional epistemology – i.e. the traditional problem of defining knowledge, its essential structure and its foundations. In a way, it might be a useful strategy to tackle the problem of *the* structure of knowledge by tackling the problem of the structure of our language. For it is only in language that all knowledge-states can find adequate and objective expression. And from this one might conclude that any constraints on meaningful language, once these are identified, will entail principles concerning the structure and limits of knowledge. But how is one to assess essentialism in the sense just described from a historical perspective? Can one say, for example, that the nominalist onslaughts such as those of the eighteenth century British empiricist philosopher George Berkeley pose no threat to its basic principles? This question assumes all the more significance when considered in the context of the brand of nominalism which is much more vigorously being advocated today. A clarification of essentialism in its present sense is in order before we attend to this question more closely.

I take essentialism strictly in the sense of a highly abstract, *subject-neutral* methodological principle that restricts the logical forms of the fundamental subject-specific problems in philosophy and in the individual sciences to those of the '*what*-questions' and the explanation-seeking '*why*-questions', where the latter type of questions presuppose those of the former type. Essentialism so construed is, in ultimate analysis, traceable to the ancient philosophical systems of Plato and Aristotle. Ever since they made available their respective metaphysical versions of the doctrine of universals/essences, essentialism has assumed different forms and enjoyed an ever-increasing dominance.

In its capacity as a subject-neutral methodological principle in the above sense, essentialism may operate simultaneously at several levels of analysis/inquiry. Any assessment of its adequacy or relevance will therefore be possible only relatively to the specific levels involved. Thus its appraisal at the level of the individual sciences should be, as far as possible, independent of any appraisal of it at the level of philosophical analysis of concepts whatever. This *rule* assumes all the more significance in so far as it is possible to distinguish different levels of inquiry/analysis not only within each individual science but also within philosophy. Thus of every science it is true to say that the *level* of inquiry that it is engaged in at any given time is specified by the type of assumptions it makes and the type of questions that it asks: *what*-questions, *why*-questions or some other types.[2] Similarly, in philosophy one might quite legitimately determine, in terms of the type of questions being asked and the assumptions being made, the precise level of philosophical analysis that may be involved at any given time: e.g. synchronic, diachronic or some other level.

Essentialism – more precisely, the versions associated with the names of Plato and Aristotle – seems to have come a long way ever since Berkeley launched his nominalistic onslaughts on it in the 18th century. But it has again come under fire in the works of Karl R. Popper and later Wittgenstein. It is, however, no less characteristic of these more recent onslaughts than of those by Berkeley that they are insensitive to subtle differences between the levels of analysis/inquiry just considered. This is one reason why these thinkers show little awareness of the important fact that whenever essentialism – as a methodological principle – is suppressed at a particular level of analysis, it raises its head at another level. And it is this which deserves serious attention in any critical appraisal of Popper's and later Wittgenstein's radical scepticism with regard to essentialism.

5.1.1. *On the Subjects of Objective Knowledge and Carnap's Methodological Conventionalism*

Before I discuss Popper's and Wittgenstein's rejection of essentialism, a few remarks about Carnap are in order. Like young Wittgenstein, Carnap too began as a methodological essentialist. Thus on the fundamental problem of *what* structure – the one and only one essential structure – our language must have if it is to describe the world of facts meaningfully and precisely, his position was essentially the same as that of young Wittgenstein. I think that it is his essay 'Testability and Meaning' which marked a radical change in his whole approach. How he changed his position is best recounted in his own words:[3]

> Previously, influenced also by Mach and Russell, I too accepted a molecular language. According to the positivistic principle of testability in its most radical form, I restricted the atomic sentences to sentences about actual experiences . . . But I was wrong in thinking that the language I dealt with was *the* language, i.e., the only legitimate language, – as Wittgenstein, Schlick and Lewis likewise seem to think concerning the language-forms accepted by them. Consequently I made the mistake of formulating my epistemological view in the form of an assertion – as most philosophers do – instead of in the form of a suggestion concerning the form of language. At present I think that the whole question is a matter of choice, of convention; and further, that a molecular language can be chosen as the language of science, but that a non-molecular, generalized one is much more suitable and, in addition, closer to the actual practice of science.

At a more fundamental level of the problem of the range of the meaningful sentences – the problem of what kind of structure our language as a *whole* can possess without inviting non-sense constructions in it – Carnap again gives up essentialism in favor of a conventionalist position. Thus to require that analytic-synthetic dichotomy be built into the very framework of our language is no longer to be seen as part of an essentialist move to define language once and for all, i.e., as part of an ideal-type model of it. On the contrary, it is rightly seen as part of a conventionalist strategy. In Carnap's own words:

He (Schlick) includes in the range of meaningful sentences only synthetic and analytic sentences but not contradictory ones . . . In my view . . . this question is not a theoretical question of truth but a practical question of decision concerning the form of the language-system, and especially the formative rules . . . if we decide to include in the range of (meaningful) sentences of our language only analytic and synthetic sentences (or even only synthetic sentences), then the formative rules of our language become indefinite. That means that in this case we have no fixed finite method of distinguishing between the meaningful and the meaningless, i.e., between sentences and expressions which are not sentences. And this would be a serious disadvantage.[4]

Thus, Carnap consistently believes that all philosophical problems, including those concerning science, are about the structure of language.[5] As regards the fundamental problem of what to include and what not to include in this structure, there is a major shift in his position from methodological essentialism to methodological conventionalism. This is quite evident from the passages just quoted above. But it is also evident from his approach to questions of method and reality which a scientific realist might address to the kind of theories the individual sciences generally aim at:

My own view . . . is essentially this. I believe that the question should not be discussed in the form: "Are theoretical entities real?" but rather in the form: "Shall we prefer a language of physics (and of science in general) that contains theoretical terms, or a language without such terms?" From this point of view the question becomes one of preference and practical decision.[6]

This particular move by Carnap can be safely interpreted as favoring methodological conventionalism on questions about the methodology of science, such as the following: What is the nature of the subject-specific methodological rules of problem-determination and theory-construction in the individual sciences?[7] It is at this point that Carnap's methodological conventionalism comes very close to that of Popper, for whom these rules are just a matter of decision or convention within a particular scientific community. But Carnap's methodological conventionalism is evidently based on his linguistic conception of science. Thus, according to him, science is nothing but another *form* of language. Its implication for scientific change, its nature and rationality, is clear. Thus, one may look at scientific change as a kind of language change or conceptual change. Variations on this conception have become a commonplace in the recent philosophies of science of Thomas Kuhn and Paul Feyerabend. But this is precisely the kind of approach which I reject, in Pandit, (1989), if only because it is full of confusion between language and science.

As a step in appraisal, it is important now to ask a very unconventional question, in terms of Popper's ontological categories of $world_1$, $world_2$ and $world_3$, as follows: Under which conditions of transformation do the $world_1$ and $world_2$ objects/processes *become* the subjects of objective knowledge? Certainly, as one might answer, not when a child *learns* by what name/label to call an object; not when one looks or stares at it as an already labelled object; not even when one claims to stumble on it, to perceive it or to recognize it. The correct answer is, I think, equally unconventional: *an object becomes a subject of our objective knowledge only when it is brought under the network of problems and theories as in science*. Anything not even remotely connected with such a network cannot participate in interactions between the $world_1$, $world_2$ and $world_3$. That is, *it is at the point of interaction between* $world_1$ *and* $world_3$ *on the one hand and* $world_2$ *and* $world_3$ *on the other that an object is transformed into a subject of objective knowledge*. It follows from this that the general attitude to *object-domains* in science is not only confused but deep-rooted in traditional epistemology. The attitude is characteristic of philosophers and scientists alike, in their talk of entities as being observable or unobservable, interpreting the latter either realistically or on instrumentalist lines. Instrumentalist approach to scientific theories and theoretical entities is generally associated with various forms of empiricism, phenomenalism of Ernst Mach and logical empiricism of Carnap and others being quite notable among them. The philosophical attitude underlying these takes inspiration from Occam's razor and treats the object-domains in science just as domains of *objects* subject to the requirement that they must not be multiplied beyond necessity, where 'necessity' is pre-conceived in terms of the doctrine of one and only one true description of an object-domain. It is in this and similar ways that requirements as reductionistic as that of verifiability or verification are brought in to cut scientific description to a 'necessary' size and to eliminate unobservable objects/entities that appear as superfluous elements of description. If anything important is lost sight of under this attitude and in this process, it is this. In doing science and in philosophically understanding it what matters is not whether an object/entity is observable or unobservable – or whether statements thereof are verifiable or not verifiable – but whether it is a (potential) subject of objective knowledge in the sense defined above. From our definition above it follows logically that in order to be a subject of objective knowledge, an entity/process in the $world_1$ and $world_2$ does not have to be necessarily an observable one; being observable or unobservable is irrelevant. What is relevant is whether our theory-problem interactive systems tend to catch it in their growing network as a *subject* for theoretical investigation.

This reminds me of the insight underlying later Carnap's distinction between internal and external questions concerning the existence or reality of entities.[8] The philosophical importance of this distinction consists in bringing out hitherto unrecognized complexities in questions of ontological significance. More than scientists, it is philosophers themselves who raise questions of existence/reality of this or that type of entities as simple questions subject to a purely philosophical solution. Phenomenalism, or the instrumentalist interpretations of scientific theories on the one hand and Quinean scepticism concerning propositions/meanings on the other illustrate this approach.

> If someone wishes to speak in his language [says Carnap] about a new kind of entities, he has to introduce a system of new ways of speaking, subject to new rules; we shall call this procedure the construction of a linguistic *framework* for the new entities in question.[9]

Carnap's internal questions relate to "the existence of certain entities of the new kind *within the framework*";[10] and external questions to "the existence or reality *of the system of entities as a whole*".[11] What is here noteworthy is the implication that the very asking and answering of internal questions presupposes what Carnap calls a linguistic framework and our very acceptance of it for certain purposes. Such questions are accordingly either *theoretical* (in the case of the ordinary natural thing language and the theoretical frameworks in the sciences) or *logical* in the case of purely formal sciences. Now, to take an example, what sort of a question is it to ask:

Do black holes exist? Are they real?

Such questions are, of course, deeply theoretical in nature, presupposing a whole theoretical framework in physical theory, astronomy and cosmology. In order to answer these questions, it is necessary to remember, with Carnap, that "to be real in the scientific sense means to be an element of the system".[12] I think that this can be interpreted as follows. To be an object in the $world_1$ or the $world_2$ is to be a subject of objective knowledge – i.e., a subject of scientific discovery and investigation in the sense defined above. Thus, Carnap's insight boils down to this. Internal questions relate to the existence of entities or objects *within* the $world_1$, $world_2$, $world_3$, and so forth. But external questions, which too have the appearance of being ontological questions of existence, relate to each of these worlds as a *whole*. This explains why Carnap is of the view that the latter questions are philosophical questions not of a theoretical but practical nature.[13] Now, Carnap seems quite right to the extent that questions and assertions concerning a given framework cannot be exactly similar in nature to those about certain entities within that framework. However, one might disagree with his position in respect of the characterization of external questions as practical questions of decision concerning the structure of our language.[14] For built into this is methodological conventionalism in the sense under discussion, which finds expression in Carnap's remark:

> we are free to choose to continue using the thing language or not; in the latter case we could restrict ourselves to a language of sense-data and other 'phenomenal' entities, or construct an alternative to the customary thing language with another structure, or, finally, we could refrain from speaking.[15]

On Carnap's view, followed to its logical conclusion, all that a philosopher can do is to work out his ontological commitments to this *system/world* of entities or that, to this theoretical framework or that, strictly within the framework of methodological conventionalism (in the present sense of this term). And all this while the scientists themselves might be busy raising theoretical/internal questions presupposing a certain minimal commitment to (or belief in) those very systems or frameworks. The problem which arises out of such a situation and which Carnap never considers is this. What is there in Carnap's position which could prevent methodological conventionalism, already operating from outside, penetrating deeper into science itself? As far as I can see, there is nothing. On the contrary, and if seen in the light of our discussion in Chapter 6, Carnap's methodological conventionalism is, I believe, damaging for his otherwise useful distinction between external and internal questions.

In any case, the failure of methodological conventionalism, whether in science or outside it, lies, I believe, in its inability to account for the following aspect of creativity in science: *all scientific discoveries are discoveries of subjects of objective knowledge*. If a scientist discovers a new type of entity in a given space-time region, it is discovered, I think, not just as an item of information among other such items but as a subject of objective knowledge. The acceptance or rejection of the very framework of such entities cannot be just a matter of practical decision regarding the structure of language.

5.1.2. *Science and Popper's Methodological Conventionalism*

Popper's attempt to identify and criticize what he calls methodological essentialism[16] is explicitly addressed to the sciences rather than philosophy itself. "The school of thinkers whom I propose to call *methodological essentialists*", says Popper,

> "was founded by Aristotle, who taught that scientific research must penetrate to the essence of things in order to explain them. Methodological essentialists are inclined to formulate scientific questions in such terms as 'what is matter?' or 'what is force?' or 'what is justice?' and they believe that a penetrating answer to such questions, revealing the real or essential meaning of these terms and thereby the real or true nature of the essences denoted by them, is at least a necessary prerequisite of scientific research, if not its main task".[17]

According to Popper, methodological essentialism is opposed by methodological nominalism which requires a different way of formulating scientific problems: e.g., 'how does this piece of matter behave?' or 'how does it move in the presence of other bodies?'.[18] Methodological nominalism as a framework for science, enabling it to ask 'non-essentialistic' 'how-questions', is hailed by Popper as what is 'victorious in the natural sciences'.[19] Citing the examples of physics and biology, Popper says that the former

> does not inquire . . . into the essence of atoms or of light, but it uses these terms with great freedom to explain and describe certain physical observations, and also as names of certain important and complicated physical structures.[20]

The same is said to be true of biology.

It can be shown, I think, how all that is in this way attributed to the actual scene in the natural sciences is not quite true. That is, it can be shown that the basic framework of these sciences remains one of methodological essentialism. But before this more or less factual question of what is actually true of the history of the sciences is settled, a fundamental philosophical question arises:

> Which one of the two rival doctrines of methodological essentialism and methodological nominalism is preferable as a framework of the sciences? Or, is it possible for the sciences to do away with the former completely?

What sort of an answer, if any, is available in Popper's discussion on essentialism? That his position is from the very outset quite radical should be clear from the fact that for him there is no question of choice between these two doctrines. Thus, for him, while the first question does not arise at all, the second question must be answered in the affirmative. The reason for this radicalism lies in Popper's commitment to the belief that while methodological essentialism would only suit the pseudo-sciences (e.g., psychoanalysis) the sciences must fall back upon methodological nominalism. I think that this is what is implied when Popper remarks:

> . . . it is one of the main differences between psycho-analysis and the wave theory of light that while the former is still largely essentialistic and metaphorical, the latter is not.[21]

Popper also believes in a fundamental unity of method between the natural and the social sciences, and advocates the extension of methodological nominalism from the former to the latter, the actual scene in the social sciences being, in his view, lamentably though, one which is dominated by methodological essentialism.[22] Popper seems to view this (proposed) extension of methodological nominalism to the social sciences scene as leading to a very important *subject-specific* methodological principle, viz., the principle of 'methodological individualism' as opposed to that of 'methodological holism'.[23] This principle is best formulated by Popper himself as follows:

> . . . the task of social theory is to construct and to analyse our sociological models carefully in descriptive or nominalistic terms, that is to say, *in terms of individuals*, of their attitudes, expectations, relations, etc.[24]

I think that one chief reason why Popper rules out methodological essentialism as conducive only to pseudo-science has to do with his mistaken belief that its very framework commits one to a programme of ultimate explanations of things in terms of their unchanging 'essences' – to "the view that in science we can aim at, and obtain, an *ultimate explanation by essences*".[25] Describing it variously as essentialism, as the Aristotelian view, as a part of the Galilean view of science, Popper subjects the doctrine that science aims at ultimate explanation – the sort of explanation which cannot be further explained – to detailed criticism.[26] But what does his criticism aim at? According to Popper, "it merely aims at showing the obscurantist character of the role played by the idea of essences in the Galilean philosophy of science . . ."[27]

In other words, [says he],

'my criticism tries to show that whether essences exist or not the belief in them does not help us in any way and indeed is likely to hamper us; so that there is no reason why the scientist should *assume* their existence'.[28]

Judging from the type of examples which Popper cites from Descartes through Newton to Faraday, what worries him most in the present context is a certain aspect of essentialism: viz., its alleged potential for raising obstacles to thought, to the growth of the sciences and indeed to the otherwise imminent scientific revolutions. Since essentialism leads to obscurantism, so might one conclude with Popper, it is not "one of those extra-scientific beliefs . . . which a scientist need accept".[29] It is interesting to consider how Popper illustrates his point by taking the Newtonian theory of gravity – or more precisely a certain attitude to this theory – as an example. After stating a certain essentialistic interpretation of Newtonian theory, which is attributed to Roger Cotes,[30] Popper says:

This essentialistic view of Newton's theory was on the whole the accepted view until the last decades of the nineteenth century. That it was obscurantist is clear: *it prevented fruitful questions from being raised*, such as 'What is the cause of gravity?' Or more fully, 'Can we perhaps explain gravity by deducing Newton's theory, or a good approximation of it, from a more general theory (which should be independently testable)?'[31]

I think that Popper's criticism fails in an important respect. The criticism is essentially muddled on a number of crucial points, invariably hitting the wrong target, viz., certain extra-scientific attitudes towards science in general and also towards the specific 'scientific' concepts and theories that have been formulated since the time of Aristotle. It fails, I believe, in its chief aim of having methodological essentialism knocked out. This is so because it moves away from its actual target – viz., a certain framework for scientific problems – to one that has, in my view, essentially to do with (a) ideological commitment *in* science and (b) one's very attitude to science as a whole.

It is interesting to see how Popper (1972b) reaffirms his adherence to his criticism of essentialism (in his *Open Society* vol. ii, chapter II, section ii, and elsewhere).[32] Here essentialism is restated as a cluster of doctrines(a)–(c), each intimately connected with the other as follows:

(a) science aims at *ultimate explanations* in terms of essences;[33]
(b) in every single thing there is an essence, an inherent nature or principle . . . which necessarily causes it to be what it is, and thus to act as it does.[34]
(c) . . . it is the essential properties inherent *in each individual or singular thing* which may be appealed to as the explanation of this thing's behavior.[35]

In particular, in connection with (b), Popper has this to say: "I reject all 'what-is' questions: questions asking what a thing is, what is its essence, or its true nature".[36]

This is followed by Popper's (1977) latest outburst against essentialism as he goes on to argue:

> . . . as I have often pointed out, 'What is' questions are never fruitful, although they have been much discussed by philosophers. They are connected with the idea of *essences* – 'what is the self essentially?' – and so with the very influencial philosophy which I have called 'essentialism' and which I regard as mistaken. 'What is' questions are liable to degenerate into verbalism – into a discussion of the meaning of words or concepts, or into a discussion of definitions. But . . . such discussions and definitions are useless.[37]

I think that, for a proper understanding and assessment of methodological essentialism, it is necessary to separate two kinds of issues from each other. Thus, on the one hand, there are those issues which relate to the very logical forms of scientific *theories* and *problems* as the epistemic structural codeterminants of objective knowledge systems in the sciences. They raise important questions of structure-property correlations in these systems, which will be a subject of detailed discussion in Chapters 7 and 8 below. But, on the other hand, there are those issues which relate to our attitude to these systems or to scientific explanation. One of the very useful ways of separating the two kinds of issues is to ask:

(1) Do the individual sciences raise subject-specific *what*-questions?
(2) If yes, do such questions, and the assumptions appropriate to them, in any way commit us to what Popper calls 'ultimate explanations by essences'? Do they in any way commit us to the view that scientific knowledge is infallible?

(3) How are such questions, if any, related to other types of important questions in the sciences – e.g., the *whether*-questions and the explanation-seeking *why*-questions?
(4) What is the nature of ideological commitment in science?

I think that it is necessary for every individual science, such as physics, to make, if only to begin with, subject-specific assumptions rich enough to raise subject-specific *what*-questions such as the following: (1) What are the physical properties of metal-objects? (2) What is the nature of motion of a force-free material object? (3) What is, in geometric terms, the nature of planetary motion/orbit? (4) What is the nature of the force by which one material body attracts another at-a-distance? (5) What is gravity? (6) What is the basic/invariant structure of all forms of life? (7) What is the structure of a cell? (8) What is the structure of a gene? (9) What is the nature of the invariant cognitive structure in the child which determines the possibilities of acquisition by him of the knowledge of his language? (10) What is the nature/structure of one's knowledge of a particular language? And so on. What is here noteworthy is not just the great variety of these and similar questions, ranging from physics to linguistics, but the fact that each of these subject-specific questions presupposes subject-specific theoretical assumptions which are essentially falsifiable in character. It can be shown that these assumptions themselves must have, in their turn, arisen as tentative solutions of the antecedently known problems of a purely theoretical nature.

By the time an answer to a subject-specific *what*-question is framed, whether in the framework of ordinary natural language or in the framework of advanced scientific theory, there will be sufficient ground in preparation for a different problem-situation prompting a *whether*-question, if only to be followed sooner or later by an explanation seeking *why*-question. Thus, it is true of every science that its *why*-questions presuppose, as their theoretical bases, those theoretical assumptions/statements which can themselves be shown to have been originally formulated as answers to its *what*-questions and *whether*-questions respectively. For example, consider the following explanation seeking *why*-question:

Why do metal-objects expand when heated?

Among other things, what it presupposes is a theoretical basis in the statement *that all metal-objects expand, with their temperatures raised, when heated*. This statement, in its turn, answers the antecedently determined theoretical problem of structure-property correlations: Whether metal-objects possess important thermodynamic properties and/or relations with other physical magnitudes? Thus, the movement of a science from its *what*-questions and *whether*-questions to *why*-questions (and *how*-questions) can be shown to follow a methodologically significant pattern. There is nothing in the structure of scientific problems formulated by means of *what*-questions and *whether*-questions which could prevent the movement of a science from such problems to those formulated by means of the explanation seeking *why*-questions, or from solutions of the latter to the former types of problems, according to a definite pattern. It follows from this that the subject-specific *what*-questions and *whether*-questions in a science do not commit the scientist, or the philosopher-methodologist, to so-called 'ultimate explanation by essences' in the sense of Popper.

For an in-depth study of the interactive growth of objective knowledge systems (= theory-problem interactive systems, as we will call them), it may be useful to begin with the framework of minimal theoretical identifications as *primitive* as is deeply embedded in ordinary natural language. For this may serve as a simpler illustration of the kind of framework which makes it possible to raise more questions – *what*-questions, *whether*-questions and *why*-questions – than it itself either presupposes or can answer, *giving way*, in the process, to theory-problem interactions at the advanced level of scientific theory. Of this framework it must be said, in any case, that it itself represents a vast range of our original theoretical responses to a vast range of *asked* and *unasked what*-questions. These include questions of the type a child is seen asking when it is still learning a language, and others are seen anticipating on his/her behalf when, for one reason or other, he/she is unable to ask.[38] Thus, on the basis of this framework, and the one which it must ultimately *give way* to in the sciences, the following schematic view of theory-problem interaction emerges:

> What is the nature of objects in the domain X as against those in the domains Y, Z, K and so forth (which may all be identifiable under various classifications as trees, plants, insects and so forth)?
>
> ↓
>
> Objects in the domain X are all metal-objects.
>
> ↓
>
> Whether metal-objects possess any thermodynamic properties

and/or relations to other physical magnitudes?

↓

All metal-objects expand, with their temperatures raised, when heated.

↓

Why do metal-objects expand when heated?

↓

The different macroscopic physical properties of different substances including metal-objects must be (*explained as*) functions of variables ranging over microscopic (and sub-microscopic) complexities of the properties and interactions of the structural elements of matter at molecular and atomic levels.

↓

What is the molecular nature of matter and of the inter-molecular forces themselves which hold the molecules of a given material body together? What is the specific heat of a substance? And so forth.

↓

. . . .

As a rule, it is true of every subject-specific *what*-question (consider, e.g., "what is the structure of the gene?") that, like every *whether*-question and *why*-question, it presupposes a theoretical basis in the corresponding subject-specific assumptions which are essentially falsifiable (in the sense of Popper) and without which such questions could not be asked. This implies, once again, that there is nothing in the structure of the subject-specific scientific problems formulated by means of *what*-questions, *whether*-questions and *why*-questions which could warrant the (unacceptable) view that scientific knowledge is infallible. Since the two doctrines of (a) ultimate explanations by essences and (b) the infallibility of scientific knowledge imply each other, they must stand or fall together. In this way, I think, it becomes clear that any Popper-type criticism of methodological essentialism cannot survive closer scrutiny.

In order to turn to the next question concerning ideological commitment, let us ask: what is it (in classical physics) which prevented fruitful questions from being asked and properly addressed to Newton's law of gravitation? And what sort of questions were these? As we have seen above, Popper singles out the following questions: What is the cause of gravity? Can we explain gravity by deducing Newton's theory from a more general theory? Newton's theory says that between any two particles, small or large, there is a force inversely proportional to the square of the distance between them. It was addressed to the question: What is that force which could be so invoked as to explain in one sweep Kepler's laws of planetary motion (with necessary correction) and Galileo's findings in terrestrial mechanics? That there is nothing either in the statement of Newton's theory of gravitation or in the structure of the problem itself, to which it was addressed as a solution, which could have prevented the questions above from being asked can be seen clearly from the progressive movement from Newtonian dynamics to Einstein's special and general theories of relativity. Thus, it is not methodological essentialism but a certain deep-rooted ideological commitment to Newtonian physical theory, method and strategy which alone could prevent fruitful questions from being asked and significant changes in the search-cum-discovery procedures from being made. I have dealt with the issues of ideological commitment in the sciences before in my (1983). It should suffice here to remark that it arises out of complex pragmatical relations of interaction between theory-problem interactive systems and their builders in the individual sciences. Thus, it follows a distinctive pattern of its own.

What further philosophical defence of methodological essentialism is possible? I think that it is a drawback in Popper's formulation and criticism of methodological essentialism that it is being projected more as an ideology than as a simple philosophical principle. As I have suggested at the very beginning of my discussion, it is better understood as an ideologically neutral abstract principle of problem-determination at various levels of scientific inquiry/analysis, including those of philosophy. However, in so far as the terms 'essence' and 'essentialism' might be said to evoke memories of deep-rooted metaphysical doctrines associated with the names of Plato, Aristotle, Descartes and others, it would be less misleading to speak of *methodological structuralism* instead of methodological essentialism. Thus, it can be argued that all 'essentialistic' *what*-questions, *whether*-questions and *why*-questions are, in final analysis, questions relating to different types of structures that are postulated at different levels of analysis. For example, consider the following subject-specific questions and their different contexts of inquiry:

(1) What is knowledge?
(2) What is language?
(3) What is cell?
(4) What is gene?
(5) What is cellular growth?
(6) What is gravitational field?

I believe that it is true of each of these questions that it relates to the particular kind of *structure-property correlations* and structural invariances which are either postulated or relevant at a particular level of analysis, whether scientific or philosophical. Thus, the correct understanding of question (1) concerning 'knowledge' is not to view it as a demand for an 'essentialistic definition' of some Platonic entity or other, but as a demand for the specification, or identification, of such correlations and invariances in systems of human knowledge, whether scientific or subjective. The same is true of the remaining questions (2)–(6). As regards the last question (6), it is interesting to note that it arises out of the following set of novel assumptions:

> Like electro-magnetism, gravity is a field-characteristic of matter. It must itself be some kind of structure governed by structure-laws where we believed in the action-at-distance Newtonian gravitational force.

This clearly implies that this is not the type of question which is permitted in a straightforward manner in the framework of (Newtonian) subject-specific methodological rules of problem-determination. On the contrary, what it illustrates is how the individual sciences can progressively move from one set of such rules to a different, even incompatible, set altogether, and correspondingly from one set of theory-problem interactive systems to another set. The question "What is gravity?" can, no doubt, be raised both in the Newtonian and the Einsteinian framework. But it can be answered differently in these two different, though related, frameworks. The same is true of concepts such as mass, velocity, space and time and the like. Radical methodological variance in the individual sciences, as and when it occurs, involves revolutionary transitions (without incommensurability) from the given frameworks of subject-specific rules of problem-determination to newer ones. This is very well illustrated by Einstein's STR and GTR when compared to Newtonian framework of rules for physical problem-determination and theory-construction. In so far as methodological variance in this sense implies variance in the very theoretical bases of subject-specific scientific problems, it will be only proper to require that these problems be, as a rule, formulated in the following manner: "What is meant by gravity in the Newtonian framework?", "What kind of magnitude is mass in Newtonian mechanics?", "What is meant by gravity in the Einsteinian framework?" and so on. All this is permissible. But it does not follow from this that the types of theories/concepts under consideration, along with their respective frameworks, are essentially incommensurable in the sense of Kuhn and Feyerabend. Radical reformulation not only of scientific problems but of their theoretical bases and solutions has to do with methodological variance, and the growth of scientific knowledge, which follows a rational pattern. If the concept of gravity is marked by a movement from our (earlier) conception of it as an action-at-a-distance Newtonian force to the newer conception of it as a structure and as a field-characteristic of matter, this is made possible by the changes at the deeper level of *what*-questions, *whether*-questions and *why*-questions, and their respective theoretical bases.

It follows yet in another way that *the logical structure* of scientific problems formulated by means of *what*-questions, *whether*-questions and *why*-questions, does not in any way come in the way of methodological variance in the sense of my (1983), or in the sense just indicated. Thus, it does not come in the way of the growth of scientific knowledge. As questions relating to different types of structures postulated at various levels of inquiry, even the seemingly familiar *what*-questions, *whether*-questions and *why*-questions can be seen to assume fresh significance as and when there are significant changes in their respective theoretical bases. Thus, at no point of the inquiry that one may be engaged in can one complain of obscurantism of a philosophy of ultimate/infallible explanations, unless one is committed to such a philosophy from the very beginning.

5.2. ESSENTIALISM IN PHILOSOPHY: POPPER'S AND WITTGENSTEIN'S CRITICISM

At the level of philosophical inquiry itself, however, essentialism is no less misunderstood and no less a target of methodological criticism. In this context, it would be interesting to consider the criticisms formulated by Karl R. Popper and later Wittgenstein, each striving in his own way to formulate and solve philosophical problems concerning science and language *outside* the methodological framework of essentialism. The differences in their procedures have mainly to do with the differences between their respective problem areas. And the consequences of the criticisms which they formulate seem quite far-reaching in character and damaging for methodological essentialism. They are as far-reaching as is indicated by Popper's conception of methodological rules in science as conventions, and of epistemology as the theory of the growth of scientific knowledge, on the one hand and by Wittgenstein's doctrines of family resemblance, language-games and forms of life on the other. However, these criticisms call for a critical scrutiny, if only because it can be argued that (i) they fail to hit their real target (ii) all attempts to formulate and solve philosophical problems outside the framework of methodological essentialism lead to the sort of awkward consequences which would not be acceptable even to its staunchest adversaries like Popper and Wittgenstein; and (iii) it is possible to so defend essentialism as to salvage it as a subject-neutral methodological principle of fundamental importance both to philosophy and to the sciences.

More precisely, a closer look at Popper's and Wittgenstein's philosophical works would reveal the following: (a) Popper's attempt to eliminate and replace methodological essentialism lands him in the kind of situation where he has no choice but to embrace *methodological conventionalism* pregnant with anarchist implications of the sort that are quite independently brought home to us by Paul Feyerabend in his (1978a; 1978b); (b) but evidently his attempt only succeeds in so *suppressing* essentialism at one particular level of analysis as to let it raise its head at another level; and (c) Wittgenstein's attempt in the same direction in his (1968) lands him with more sweepingly radical doctrines such as the doctrine of *family resemblances* as a sufficient basis for the application of our concepts/predicates – as different as 'language', 'game', 'tool', 'man' and the like – at different levels of language-use. What is here noteworthy is how Wittgenstein seeks the elimination of essentialism at the philosophical level itself: at the level of analysis of language, mind and philosophical problems generally.[39] As I shall argue subsequently, his doctrine of family resemblances could be viewed as an attempt to move away from the essentialistically conceived *subjects of predication* to those about which it is no longer possible to ask questions of the form: What is X? Thus, the doctrine facilitates, for Wittgenstein himself and for others, as I have designated it in my (1983), the philosophical strategy of so changing, in the subject-specific contexts, the very subject of predication/appraisal as to obtain the consequences desired from the very beginning, such as the alleged incommensurability of theories in the sense of Kuhn and Feyerabend.

It might be argued that the distinctive dialogue-structure of Wittgenstein's (1968: e.g. §§ 89–128) conceals many a point of continuity from *TLP* to *PI*. And, in any case, at many places (e.g. §§ 89–128) *PI* does convey an impression of its deeper commitments to essentialism; although the 'essence' is no longer conceived in the older *Tractarian* fashion as something hidden from our view capable of being dug out only by an appropriate kind of deeper analysis – an analysis of the complex into the simple.[40] It is no doubt true that in his (1968: §§ 89–92, 129) Wittgenstein advocates some sort of a doctrine of the *essences* as being conspicuous, transparent, all the time open to our view or as what is expressed by grammar and yet as something which we need to be constantly reminded of. Nevertheless, if one is to be guided by his doctrine of the family resemblances, one can only conclude that Wittgenstein, in his (1968), is no longer an essentialist in any non-trivial sense of the term. Along with philosophical tasks, then, we find here essentialism trivialized to the point of its mutilation beyond recognition.

I believe that it is possible to salvage essentialism from this disreputation into which it seems to have fallen owing mainly to the type of criticism mentioned above; and to defend it as an ideologically neutral abstract, subject-neutral principle of fundamental methodological significance to the sciences and to philosophy as well. It should be possible to do so, if an account is to be provided of an invariant feature of all analysis, viz., essentialism suppressed/avoided at a given level of analysis raises its head at the subsequent/higher levels of analysis. This is non-trivially illustrated not only in Popper's own analysis of the growth of scientific knowledge, but also in Wittgenstein's analysis of language in terms of a continuum of 'language-games' from the most simple to the most complex.[41]

5.2.1. *Philosophy of Science and Methodological Conventionalism of Popper*

Popper's anti-essentialistic approach to philosophy of science is best summed up in his own words:

> . . . what is to be called a 'science' and who is to be called a 'scientist' must always remain a matter of convention or decision.[42]

This suggests that there need not be anything invariant about the set of properties in terms of which we can distinguish science from non-science. That is, there need not be any invariant properties of science as its distinguishing properties. And, on the contrary, the way we distinguish the one from the other will depend on what properties we choose/decide upon as its defining properties. Thus depending on our decision, among other things, the set of properties we choose at one particular time may be completely replaced by a different set chosen at another time. And the whole process of understanding, and engaging in, the enterprise of science by anchoring it to one set of conventions now and to another set later on can go on *endlessly*.

But sooner or later one must ask: why should this be so? Why cannot we meaningfully ask about the essential/invariant character of science while leaving room for characterizations that may change from time to time? It won't suffice to answer just: well there simply are no essential properties of science, so inherently essential as to distinguish it from non-science. For that will only repeat what one might say somewhat in the later Wittgensteinian fashion: well there simply are no properties of language, so inherently essential as to demarcate it from all other conceivable systems of (animal) communication. At least one reason why later Wittgenstein thought like this has to do with the mistaken analogy drawn between language and games. But the alleged analogy is as pre-conceived and pre-mature as many of the doctrines in Wittgenstein's *Tractatus are*, singled out by Wittgenstein himself in his (1968) for purposes of criticism. Does Popper's thinking on the nature of science suffer from some similar difficulties? The answer is in the affirmative. Consider, e.g., what Popper says concerning science:

> . . . Just as chess might be defined by the rules proper to it, so empirical science may be defined by means of its methodological rules.[43]

Again:

> Methodological rules are here regarded as *conventions*. They might be described as the rules of the game of empirical science. They differ from the rules of pure logic rather as do the rules of chess, which few would regard as part of *pure* logic: seeing that the rules of pure logic govern transformations of linguistic formulae, the result of an inquiry into the rules of chess could perhaps be entitled 'The Logic of Chess', but hardly 'Logic' pure and simple. (Similarly, the result of an inquiry into the rules of the game of science – that is, of scientific discovery – may be entitled 'The Logic of Scientific Discovery').[44]

It would seem, given what Popper says, that it is the taken-for-granted pre-conceived analogy between science and games which heavily prejudices Popper's approach to the methodology of science. And in this he is hardly different from later Wittgenstein. One wonders whether it is at all necessary to let one's approach to science or language be so prejudiced at the very outset as to leave one's deeper assumptions not only unstated but unexamined. The hastily drawn analogies like the ones cited above seem only symptomatic of these deeper assumptions. But what could these assumptions be in this case?

A closer look at Popper's remarks above shows that they do not just imply the (weaker) thesis that science and games are in certain ways analogous. Indeed anyone who holds such a view could do so without commitment to methodological conventionalism as a stronger thesis that the methodological rules of science are nothing but conventions. What is noteworthy is the unambiguity and the specificity with which Popper compares science with the game of chess. But any such comparison would make sense if it is premised on the assumption that science is a 'game' in its own right. I think that methodological conventionalism in Popper's sense is rooted in some such deeper assumption assimilating science to games. Thus the chief reason why Popper views the methodological rules of science as *conventions* is traceable to his remark that these rules "might be described as the rules of the game of empirical science". Empirical science must be viewed as, so we are told, one among many games, each of which can be defined by means of its rules: that is, empirical science belongs to the *family* of games. And, of course, who would doubt the *conventional character* of the rules of activities as simple as games. Isn't methodological conventionalism, therefore, the simplest and the most convenient way to solving our problems in the philosophy of science, such as the problem of demarcation between science and non-science? In a way, the answer is in the affirmative. For an assimilation of the kind one commits in this way is a sure way to an oversimplification which, in any case, will appear as a species of simplification. But where will the simplification sought through an assimilation of subjects as disparate as science and games lead one ultimately? The answer is simply that it is bound to land one in a doctrine with anarchist implications. For all that this type of strategy reduces to is the strategy of so changing, in the subject-specific contexts, the very subject of predication/appraisal as to achieve the sort of consequences that are desired from the very beginning.

Popper's view of science can now be seen falling into a pattern. The methodological rules by means of which one may define science are to be viewed as conventions in exactly the sense in which the rules of the game of chess are conventions. The reason why this is so has to do, we are told, with the 'fact' that science itself is one among many such games. But, as we have seen, nothing short of the strategy already alluded to above could make science look as one among many games. That Popper does employ such a strategy should also be clear from the fact that one of the things which he wants to avoid from the very beginning are the essentialist questions such as: What is empirical science? What is its structural identity from an epistemological/methodological point of view? By telling us that there is such a thing as the *game* of empirical science which, like the game of chess, is definable by means of its conventions, Popper, I believe, leaves these essentialistic questions untouched. Far from eliminating these questions, what Popper brings about is a change in the very subject of predication/appraisal in the present context: i.e., a change from the subject of *science* to that of *games*. But having done that, can Popper avoid asking the question: What is a game? Popper simply assumes that there is no longer any essentialistic question left for him either to avoid or to tackle. Even the most important, though unrecognized, fact that the whole discipline of the mathematical theory of games would have been impossible without the essentialistic question concerning games/game-like situations does not concern him.

I believe that there could not be a better ground for methodological anarchism than is provided by Popper's methodological conventionalism. In so far as methodological conventionalism requires that "what is to be called a 'science' and who is to be called a 'scientist' must always remain a matter of convention or decision" and in so far as it is itself rooted, as shown above, in deeper assumptions assimilating science not to language but to games, it amounts to a clear invitation to methodological anarchism in the following sense: anybody – even the inductivist – is welcome to conceive of empirical science and its rules in the manner that he likes, provided one's conception permits us to play the game of science as a *useful game* of 'scientific discovery'. Does it mean that 'the logic of scientific discovery' or an inquiry into the rules of 'the game of science' is subject to no constraints whatever? To what sort of constraints could an inquiry into the rules of chess be subjected? The only constraint one can think of is this: that one's inquiry allows the interested parties to play the game of chess, as usual, as a game of strategy or as a game of winning and losing. And if we say similarly of an inquiry into the rules of 'the game of science' that it should permit the interested parties to play the game, as usual, as a game of scientific discovery, a host of questions must be answered before such a restriction becomes meaningful: What is 'the game' of scientific discovery? Which are the parties that are interested in playing this 'game'? But aren't we now moving in a circle? For we are back at the old problem of demarcation.

Of an inquiry into the rules of the game of chess one may quite legitimately require that it must leave the essential character of the game undistorted; that it must not distort but illuminate the unique features of the game. But what requirements can one legitimately impose on an inquiry into science, its rules and its structure? Can we require of such an inquiry that it illuminate the game-like properties of science, whatever? It would seem quite natural to do so only if we conceive of such an inquiry, with Popper, as one into the rules of 'the game of science'. But then it would be too much to have already assumed either that science is nothing but one among many games or that we know beforehand what its unique features as a game are and hence what restrictions an inquiry into it must be subjected to.

5.2.2. *Conventionalism and the Game-Theoretic Conception of Science*

We have seen that as a methodological conventionalist, Popper is strongly opposed to essentialism: What we call 'science' is a matter of convention/decision. And to this one might add quite consistently: at one and the same time more than one set of conventions is possible, indeed sets of conventions incompatible with one another. Given then any two rival theories T_1 and T_2 in their field, methodological conventionalism in Popper's sense implies, however unintendedly, that there can be no good reasons to prefer T_2 to T_1, since each may be a good and acceptable theory on one set of methodological conventions concerning science (and what it aims at) but unacceptable on another such set. It is interesting that I. Lakatos, among others, does not question Popper's essentially game-theoretic conception of science or its anarchistic implications for the methodology of science: he endorses and seeks to improve on it. The result is a Lakatosian version of conventionalism as distinct from the Popperian. Thus, e.g., while for Popper only *basic statements*, among the totality of scientific statements including *theories*, can be accepted as *conventions* in 'the game' of science,[45] for Lakatos even *theories* themselves can be.[46] We have, however, seen that when Popper speaks of science as what can be defined by means of its rules, he already sees it as analogous to games of chess: the rules in terms of which we *decide* to define science are themselves to be accepted as conventions, as is also true of the games like chess. Following Wittgenstein's quite influential 'language-game' view of language formulated essentially in terms of his doctrine of family resemblances, philosophers have lost little time in announcing programmes extending the analogy of the games further to the deeper levels of analysis. This is done, notably by Jaakko Hintikka, by bringing in game theory as an essential tool of semantical studies.[47] But the strategy underlying the whole game-theoretic approach to science and language can serve, in my view, neither Popper's nor later Wittgenstein's programmes seeking to eliminate essentialism. And one wonders how long we still should take to realize that even game theory[48] itself assumes the framework of what I call methodological essentialism.

Popper's basic, though unstated, assumption assimilating science to games cannot stand critical scrutiny. Whether any activity is or is not a game will be simply decidable, in ultimate analysis, by ascertaining whether or not it has game-like properties either in the ordinary or game-theoretic sense of the term. It must be pointed out here, if not argued, that we cannot say of empirical science that it is a game in either of these two senses.[49] In any case, of course, it can be argued that the activities/situations of the type called 'games' are extremely simpler in character and more rigidly rule-governed when compared to problem-situations characteristic of science. Popper's assumption is thus exposed as an application of the strategy of so changing, in the present context, the very subject of predication/appraisal – from science to games – as to achieve the consequences that are desired from the very beginning. One such consequence, we have seen, has to do with the (Popperian) elimination of essentialistic questions such as: what is the nature of empirical science? What are its unique epistemic structures, given its character as a complex theoretical problem-solving enterprise? And so forth.

5.2.3. *The Tension between Objectivism and Conventionalism*

The crucial question now is this: How much can one assume about science before one inquires into its structure and its rules with a view to demarcating it from pseudo-science? That without some minimum assumptions about its objective features no such inquiry should be possible is beyond doubt. What restrictions one may impose on such inquiry will, therefore, depend on the assumptions one makes in this context. If one follows Popper, one's assumptions will amount to methodological conventionalism: that science is one game among many other games and that its definition is ultimately a matter of decision/ convention. But methodological conventionalism in this sense lands us in methodological anarchism. Any definition and any method would do, so long as science is projected as a game, just as any definition of chess would do, so long as chess survives under it as a game. I think that it is the anarchist implications of methodological conventionalism such as this which are quite independently developed by Feyerabend in his (1978a and 1978b). But it must be admitted that there is discernible in Popper's approach and thinking a kind of tension – between his objectivism and conventionalism – which has concealed these anarchist implications most of the time. This tension comes to surface in his own conception of the rules of the 'game' of science as a hierarchical structure with the following supreme rule as its apex, serving "as a kind of norm for deciding upon the remaining rules":[50] " . . . that the other rules of scientific procedure must be designed in such a way that they do not protect any statement in science against falsification".[51] In other words, we might put it as follows. A science will employ no search-cum-discovery procedure which has not been "constructed with the aim of ensuring the applicability of our criterion of demarcation".[52] This criterion – Popper's supreme rule of falsifiability – requires an empirical scientific system of statements to be falsifiable. Thus, it requires that a science employ in its search for solutions of its problems only such statements or theories as are *capable* of refutation by experience.

But what is it which is so basic about this criterion or which makes it of such supreme importance? Is its acceptance, or rejection, also a matter of decision or convention? I think that only a detailed analysis of the structural identity of objective knowledge systems in the sciences would enable us to arrive at a satisfactory answer. Popper, however, considers any such analysis quite unnecessary to his entire methodological enterprise.[53] One of the reasons for his reluctance here may be traced to science itself. Its statements are susceptible to criticism and revision. They can be, therefore, replaced by better alternatives.[54] Anyone who recognizes scientific change and growth as distinguishing features of science cannot escape consequences thereof. Thus, an adequate criterion of demarcation must be consistent with this recognition. Moreover, any inquiry into method with a view to distinguishing science from non-science not by its logical/epistemic structure but by its methods – "by our manner of dealing with scientific systems: by what we do with them and what we do to them"[55] – must be premised on a prior inquiry into the structural identity of objective knowledge systems in the sciences themselves. In a nutshell, any methodological enterprise dealing with the growth of the objective knowledge systems must be premised on a prior inquiry into their structural identity.

Consequences such as these are quite damaging for the Popperian version of methodological conventionalism. The constraints which a sound definition of empirical science, and a sound methodological enterprise based on it, must be subjected to cannot be, as we have just seen, a matter of decision or convention. The acceptance or rejection of a proposed criterion of demarcation, and of a proposed set of methodological rules, irrespective of the detailed nature of these constraints, cannot, therefore, be just a matter of decision or convention. Defining science and laying down its methodological rules cannot be as simple as that. Even if we may choose to define science by means of the principle of falsifiability, it must be for deeper reasons. These may have to do with the structure-property correlations in the objective knowledge systems in the sciences. But it is only within the objectivistic framework of methodological essentialism, or methodological structuralism, instead of the anarchistic framework of methodological conventionalism, that we can inquire into the detailed nature of these correlations (turn to Chapter 8 below for a detailed discussion).

From the discussion above it follows clearly that there are serious gaps in the objectivistic philosophy of science of Popper, resulting in a constant tension between his objectivism and his anarchistic methodological conventionalism. He has chosen the latter in order to carry out the programme of his methodological enterprise concerning science. But when it comes to actually carrying it out, he falls back on objectivism as the chief source of constraints on his enterprise. Objectivism, which he claims as his basic ontological commitment,[56] pulls him in the direction of methodological essentialism. But his methodological conventionalism pulls him in the opposite direction of anarchism. What we can conclude from this is Popper's failure to recognize the following important fact. The reason why our acceptance, or rejection, of a proposed criterion of demarcation – such as the falsifiability criterion or the one proposed by the inductivist – cannot be just a matter of decision/convention is better explained within objectivism instead of methodological conventionalism, as the following discussions will show. The same kind of argument can be brought forward against Popper in the context of scientific change, theory-problem interaction and the recurrent entanglement of a science – or its *CSD* and the associated search-cum-discovery procedures – in difficult problem-situations. Thus, one might here ask: can the conventionalist immunization strategies, such as *ad hoc* rescue operations one may employ to save a theory (whose *potential* falsification becomes *effective* falsification in the sense of Popper), be ruled out/rejected as inadmissible or irrational by one who is himself a conventionalist with respect to the very question of demarcation? The answer is clearly in the negative, Popper's own approach to this question not-withstanding.[57] Thus, only a philosopher who is a methodological essentialist in my sense could do so. As I shall argue later on, *theories and problems* are not only the epistemic structural codeterminants of objective knowledge systems in the sciences but also their developmental structures. Any strategy of *ad hoc* rescue operations by resorting to *ad hoc* hypotheses in a science will be, therefore, seen as a strategy of rendering them as their pseudo-developmental structures.

Let me restate some of the important steps in our attempt to identify the gaps in the objectivistic philosophy of science of Popper. As regards the important question of demarcating science from non-science by its methods, Popper has adopted the framework of methodological conventionalism. But as to the question of formulating an adequate theory of the *growth* of scientific knowledge, or of the rationality of scientific change by scientific revolutions, Popper no more remains a methodological conventionalist pure and simple. None of these problems, nor its solution in terms of a formalizing model, is thought to have to do anything with conventions. On the contrary, the important fact that science can grow, or that it can aim at *truth*, is allowed by Popper, however stealthily, to operate as a fundamental constraint on his demarcation criterion and, of course, on his whole theory of scientific method.[58] As our examination of Popper's approach shows, it is not possible to carry out the programme of methodological conventionalism consistently at any level of analysis whatever. In the present context, this particular consequence is hardly surprising. As a framework for his methodological enterprise concerning science, methodological conventionalism is itself a consequence of the deeper assumptions which assimilate science to games. These assumptions can be, in their turn, understood as a consequence of the strategy of changing the subject of predication/appraisal from the sciences themselves, or their theoretical systems, to games.

Thus, it is Popper himself, more than anyone else, who initiates the strategy of so changing the subject of predication/appraisal – even if half-heartedly as we have seen – as to obtain the consequences which are desired from the very beginning. It is his aversion to methodological essentialism, if not objectivism as such, more than anything else, which dictates these consequences in his case. But when it comes to actually carrying out the pre-conceived programme of his methodological enterprise at various levels of analysis, what we find is this. Methodological essentialism, suppressed at one particular level of analysis (e.g., the problem of demarcating science from non-science), raises its head at other levels of analysis (e.g., the problems of the growth of objective knowledge systems and their rational scientific appraisal). This is, of course, true of Popper's approach to different problems in his (1968, 1972a, 1972b, 1976, 1977) where he lands himself in a constant tension between methodological conventionalism and methodological essentialism. The tension comes to surface most clearly in his conventionalist approach to the basic question of demarcation and in his somewhat essentialistic approach to the rationality of scientific change by scientific revolutions. A simplistic answer to the former question is what has been attempted by Popper through his falsifiability criterion of demarcation. But his answer to the questions of scientific change and growth of knowledge turns out to be very complex. It leads to a whole Popperian ontology[59] of $world_1$, $world_2$ and $world_3$ interactions on the one hand and to the idea that the best theory in its field is the one which is more truthlike than its rivals on the other. The resulting contrast between the two types of questions, and their answers, is quite striking. It demonstrates clearly that methodological conventionalism, inspired as it always is by certain oversimplifying assumptions, can only lead to a proliferation of such assumptions.

5.2.4. *Kuhn, Lakatos and Feyerabend*

It might be argued that methodological conventionalism as applied to (the language of) science must be extended to all *language* so as to answer afresh the familiar question: What is language? I believe that it is possible to look at Popper's strategy as applied to science as a special case of later Wittgenstein's strategy as applied to language. But before I turn to consider Wittgenstein's approach, it should be of considerable interest to consider briefly a certain (hitherto neglected) aspect of T. S. Kuhn's view of science.[60] Kuhn's view raises the same problem as was considered above, bringing his view much closer to that of Popper than I had thought earlier.[61] I believe that the assumption that science is easily and simply characterizable in terms of a set of *game-like properties* is no less at work in Kuhn's conception of science than it is in that of Popper. It is in this assumption that Popper and Kuhn can be seen to share a common approach to science, its very structure. Thus both are methodological conventionalists. For that is exactly what the common assumption that science is (like) a game dictates. That this is so will become clearer as we proceed.

Now, I take it as a clear instance of the operation of the strategy of so changing the subject of predication as to suit one's pre-conceived programme when Kuhn propounds the view that (in 'normal science') "it is the individual scientist rather than current theory which is tested".[62] At first, any such view would appear to be quite puzzling. But once we go deeper into the very assumptions that facilitate it what we find at work is the doctrine that science, or what Kuhn calls 'normal science', is in ultimate analysis a *kind* of game or, as he puts it, a "puzzle-solving" enterprise, with the current/dominant theory as its 'paradigm' – as the very source of the rules/standards of the game.[63] And it is this doctrine which determines Kuhn's view of the structure and functioning of 'normal science':

> . . . when engaged with a normal research problem, the scientist must *premise* current theory as the rules of his game. His object is to solve a puzzle, preferably one at which others have failed, and current theory is required to define that puzzle and to guarantee that, given sufficient brilliance, it can be solved.[64]

What does Kuhn understand by 'normal science' as a puzzle-solving enterprise, and, by implication, by 'normal scientist' as a puzzle-solver? He himself explains, contrasting it sharply with Popper's use of the term 'problem-solver':

> I use the term 'puzzle' in order to emphasize that the difficulties which *ordinarily* confront even the very best scientists are, like crossword puzzles or chess puzzles, challenges only to his ingenuity. *He* is in difficulty, not current theory. My point is almost the converse of Sir Karl's.[65]

I now believe that in essence this is not the converse of what Popper says concerning science. What brings Kuhn closer to Popper, appearances and Kuhn's perception to the contrary not-withstanding, is the former's view that in (normal) science each paradigm, or what Kuhn (1974) now prefers to call disciplinary matrix, carries certain standards/definitions – of scientific problems and their solutions – which are *internal* and hence *unique* to it.[66] Thus these standards are bound to change radically with every change in one's allegiance to a given paradigm in the sense of a disciplinary matrix as a whole. That is, our very conception of a problem/theory as scientific is bound to change radically as we move from paradigm to paradigm, from disciplinary matrix to disciplinary matrix. Demarcation of what is to be called science as against non-science is a function of one's ideological commitment; and scientific revolutions, therefore, are a function of ideological variance. It is then the group-ideology, and the totality of commitments that go with it, which will always determine the way science is conceived of and pursued at a particular time. Now, how is all this different from Popper's doctrine that what we call 'science' must always remain a matter of decision/convention? One could here very well argue that there is no easier way leading to conventions than one's allegiance to a given group-ideology, and no easier way to enjoy group-rationality than one's deep involvement in a disciplinary matrix.

In any case, I cannot imagine a better application of the strategy of changing the subject of appraisal through an implicit appeal to devastatingly oversimplifying assumptions than the one Kuhn's view considered above entails. Unlike Popper, however, Kuhn repeats the same strategy at the level of the analysis of scientific revolutions. For here, once again, he falls back on the oversimplification that scientific revolutions are closely analogous to political revolutions – an aspect to be discussed in Chapter 6, § 6.5.

I believe that it would be no exaggeration to conclude from our discussion above that Popper's adversaries such as T. S. Kuhn and Paul K. Feyerabend can be seen thriving on precisely the same strategy as is discernible in Popper himself. The commonly shared strategy of so changing the subject of predication/appraisal as to suit one's preconceived programme is facilitated by methodological conventionalism which assumes various forms from most moderate to most radical. Methodological conventionalism is, as I have shown, a sure way to anarchism, which ultimately leads to that non-commital doctrine of

Feyerabend (1978a, 1978b) which 'says' that 'anything goes', or that *all* methods must be explored, or that *all* methods are equally good or bad, or that no method is either good or bad in itself.[67] Now, what is it which gives rise to methodological conventionalism itself? It is, I think, the deeper sceptical assumptions to the effect that science is not characterizable in terms of any set of inherent features whatever: there is no unique set of features peculiar or *essential* to science, to its structure and development. And, on the contrary, the only way to a proper perspective here is to view science in terms of its family resemblances to, and comparisons with, *other* subjects/activities such as, e.g., religion, witchcraft and games/puzzle-solving activities (Popper and Kuhn).[68] It is with assumptions as far-reaching and devastatingly assimilating in effect as these that a process of oversimplification sets in, preparing the ground for the strategy of changing the subject of appraisal/predication to operate. And it is in this and similar ways that methodological conventionalism, illustrated in Popper's and Kuhn's view of science, with its anarchist implications brought out in Feyerabend's epistemological anarchism, arises.

Finally, a word about ideological commitment. One might ask: Is science one ideology among others? Is ideological commitment to a paradigm (= a disciplinary matrix), to an assumption or to a program, that which a science shares in common with other activities/ideologies? I believe that closely similar answers in the affirmative to this question are already discernible in the positions of Kuhn and Lakatos and also, quite explicitly, in that of Feyerabend.[69] Both Kuhn and Lakatos imply, I believe, in what they say concerning science, that ideological commitment is a *central*, and hence an important feature of empirical science, its very structure and growth. This can be seen, e.g., in the case of Lakatos, in the very concept of 'a scientific research programme' which is central to his view of what might be called the structural identity of science. Scientific research programmes are conceived by him as the *kind* of programmes that are *wedded to* or that centre around a 'hard core'. And by this Lakatos means this: ideological commitment to a set of theories at the core is the heart of the matter in a scientific research programme. That is, the *central feature*, and hence the *starting point*, in a scientific research programme is the ideological commitment to a given theory. For a Lakatosian scientific research programme is a complex whole of "a conventionally accepted (and thus by provisional decision 'irrefutable') 'hard core'" and "a 'positive heuristic' which defines problems, outlines the construction of a belt of auxiliary hypotheses, foresees anomalies and turns them victoriously into examples, all according to a pre-conceived plan".[70] There is also a third component to it: a 'negative heuristic' in the form of the principles that (a) protect the hard core from experimental refutation and (b) rule out radically different sorts of explanatory attempts. In the words of Lakatos: "All scientific research programmes may be characterized by their '*hard core*'. The negative heuristic of the programme forbids us to direct the *modus tollens* at this 'hard core'".[71] The positive heuristic, on the other hand,

> consists of a partially articulated set of suggestions or hints on how to change, develop the 'refutable variants' of the research-programme, how to modify, sophisticate, the 'refutable' protective belt.[72]

I think that this whole doctrine concerning the structural identity of science goes a long way to meet the demands of methodological conventionalism on the one hand and of analysis of science in terms of distinct phases of ideological commitment on the other. As against this, I have argued, in my (1983), that ideological commitment is only a secondary feature arising out of the primary features of science. *It is only peripheral, and not central, to science*, arising out of interactions between the epistemic structures of science and their builders. Moreover, it is shown there how pragmatical imperatives suffice to take care of ideological commitment in science.

There are perhaps other aspects to Lakatos' methodology of scientific research programmes. For example, John Worral thinks, and quite rightly, that the most "recent attempts to provide an objective and generally applicable criterion of scientific merit have started from"[73] the following two assumptions:

> A theory is scientific rather than pseudo-scientific if it is capable of receiving genuine 'support' from the 'facts'. One scientific theory is better than another rival theory if it is better supported by the facts than its rival.[74]

"But when does a fact", asks Worrall, "provide genuine support for a theory? And when do the facts support one theory better than another?"[75] Contrasting the Popperian answers to these questions with those of the methodology of scientific research programmes, he says:

> The methodology of research programmes in providing new answers to these fundamental questions *corrects* Popper's methodology in two important ways. It recognizes . . . the programmatic aspect of scientific achievements and it imports this programmatic aspect into methodological appraisals.[76]

In so far as this is any indication of the reasons for improving on Popper's methodology, the methodology of scientific research programmes must fail at least in this aspect. The programmatic aspect of science, if any, is best represented, from an epistemological/methodological point of view, by the very character of its epistemic structures as being developmental structures. That this is so will not be recognized so long as one mixes up history of science with its epistemology. And that is precisely what the methodology of scientific research programmes does. However, recognition to this effect is, even if implicitly, discernible in the tensions and the *unstated assumptions* of Popper's falsifiability criterion of demarcation. These remain unstated mainly because of Popper's strong commitment to methodological conventionalism – to his own idea of the game of science. Furthermore, in its second aspect in the sense of Worrall, the methodology of scientific research programmes only grounds methodological appraisals in those of a historical nature. The methodology singles out a *scientific research programme* as the unit of appraisal, which can be appraised only historically/heuristically. On the contrary, for epistemic appraisals proper to be possible it is necessary that the epistemic structures in science and interactions between them are identified beforehand.[77] It is also not clear how one could attribute to a research programme what I call the resolving power of a theory,[78] this being a very important aspect of the dynamics of theory-problem interaction in science.

Being deep-rooted in the tradition of the explanation-theorist's statement-view of the epistemic structure in science, falsificationists, paradigm-theorists and the research programme-theorists are, I believe, essentially unable to account for scientific revolutions. In any case, however, only oversimple views are generally available. For example, as regards the methodology of scientific research programmes, Worral has this to say:

> When do scientific revolutions occur, according to the methodology which I have been presenting? They occur when a new programme is introduced which challenges some already accepted programme and which becomes and remains more progressive than its accepted rival.[79]

The subject of scientific revolutions could not be treated, I believe, in a manner more superficial than this, which is essentially in the best traditions of the explanation-theorist. A *CSD*-theorist, as I would like to call myself, in the sense of my (1983), if only for purposes of a contrast, is able, on the other hand, to present a more adequate account outside the explanation-theorist's tradition – see Chapter 6, § 6.5 and Chapter 9, § 9.1.

I believe that we could lay down as a fundamental constraint on any sound methodology of scientific appraisal the requirement that it does not resort to the ideological strategy of changing the subject of predication/appraisal. It becomes necessary to make this condition explicit, since most of the recently formulated methodologies and doctrines seem essentially to thrive on this very sceptical strategy. Thus on the one extreme it has been suggested by the paradigm-theorists like Kuhn that the unit of appraisal is a scientific community as a community of puzzle-solvers committed to a paradigm (= a disciplinary matrix); and on the other, by thinkers like Larry Laudan (1977)[80] that it is a research tradition as a whole, or that it is a research programme itself (Lakatos). All these, including the anarchist tendencies that they lead to, speak of the poverty of their epistemological bases, their lack of a rational basis in adequate theories of epistemic structures and of the growth of these structures. Once the proposed *general condition* on the alternative methodologies of science is heeded, temptations to resort to the strategy will be, one hopes, held back.

5.2.5. *Later Wittgenstein's Philosophy of Language: Criticism*

In his (*PI*: §§ 1, 3, 31, 54, 66–67, 69–71, 75) one notices Wittgenstein invariably invoking the predicate 'game' of our ordinary language framework whenever he raises a question concerning the nature of our *language*, its structure, its rules, its unique features of vagueness, etc., its foundations and the nature of our learning of it. One might ask: why does Wittgenstein resort to this sort of approach? Is it because he assumes that what is true of the activities falling under the predicate 'game' must also be true of the activities falling under the predicate 'language'? I think that for an adequate answer we must turn to (*PI*: § 7) where Wittgenstein goes so far as to explicitly *introduce* the predicate 'language game' ('Sprachspiel') as what turns out to be a systematically ambiguous predicate. Here Wittgenstein makes it very clear that he intends to use this predicate in various senses: although it is very difficult to make out what these senses are. Thus on the one hand it is the very *process* by means of which children learn their native language – the processes of ostensive teaching and ostensive definition as considered by him in *PI*, §§ 6, 8, 9, 27–36 – which he wants to call a language game. On the other hand, however, Wittgenstein also speaks of the (complex) *whole* of "language and the actions into which it is woven" as a language game (P1: § 7). The question as to whether this refers to the adult use of language must be left open. There is also a third possible sense which Wittgenstein hints at (*PI*: § 7) when he says that he "will sometimes speak of a primitive language as a language game".

At first it would seem that it is Wittgenstein's real intention to replace the concept of language by the concept of language-game. But I think that Wittgenstein's intentions are not as radical as that. On the contrary, the concept of a language game is being introduced with a view to developing a strategy to have a fresh look at language, its structure and function – a strategy which he suggests, in (*PI*: § 5), as a way out of our misunderstanding of language and language-learning:

It disperses the fog to study the phenomena of language in primitive kinds of application in which one can command a clear view of the aim and functioning of the words.

A child uses such primitive forms of language when it learns to talk. Here the teaching of language is not explanation, but training.

Thus the following interpretation of the concept of language game as introduced by Wittgenstein suggests itself. In so far as the processes – which are identified as ostensive teaching and ostensive definition respectively – by means of which a non-verbal infant learns his native language *resemble* language (as the adults use it), we may look at these processes as (developmentally) primitive forms of language. But in so far as these processes *resemble* games in their characteristic simplicity and entertaining character, we may look at these as games. Accordingly, we may speak of these processes as language games, meaning thereby systems of communication – from the most simple to the most complex – which are at work at various levels of language-learning. That Wittgenstein intends to speak of language games in this restricted (genetic) sense only is also borne out by his following remarks:[81]

Language games are the forms of language with which a child begins to make use of words. The study of language games is the study of primitive forms of language or primitive languages.

Systems of communication as for instance (1), (2), (3), (4), (5) we shall call "language games". They are more or less akin to what in ordinary language we call games. Children are taught their native language by means of such games, and here they even have the entertaining character of games. We are not, however, regarding the language games which we describe as incomplete parts of a language, but as languages complete in themselves, as complete systems of human communication. To keep this point of view in mind, it very often is useful to imagine such a simple language to be the entire system of communication of a tribe in a primitive state of society. Think of primitive arithmetics of such tribes.

Wittgenstein's conception of a language game seems, however, much more complex at the very outset. What complicates his conception is the fact that he introduces the predicate 'language game' simultaneously in several senses, and without much warning. Thus, it is only hinted at that not only the *processes* of language-acquisition but also *what* one learns is to be called a language game. This is what he does soon after making the lengthy remark quoted above:[82]

When a boy or grown-up learns what one might call special technical languages, e.g., the use of charts and diagrams, descriptive geometry, chemical symbolism, etc., he learns more language games, (Remark: The picture we have of the language of the grown-up is that of a nebulous mass of language, his mother tongue, surrounded by discrete and more or less clear-cut language games, the technical languages.)

This confirms one's suspicion that the concept of language game as introduced by Wittgenstein is central to his strategy of characterizing language essentially in terms of what is (dogmatically) thought to be a wide-ranging analogy between it and games at various levels ranging from language-learning to the adult use of language.[83] So central indeed that our talk of language at various levels of analysis, e.g., of meaning, rules, etc. reduces to our talk of language games, a vast multiplicity of them.

At first, one may find it very difficult to believe that Wittgenstein could have ever thought of illuminating the nature of language by an analogy with games. But gradually one realizes that if the puzzle is to be knocked out of his whole conception of a language game, it must be understood as *underlining* a wide-ranging analogy between language and games at various levels, from language learning to the adult use of language. But it can be shown that one cannot view language as analogous to games without blundering on fundamentals.

Now, this whole assumption of a wide-ranging analogy between language and games raises serious problems. Isn't it a pre-condition for drawing of analogies between different concepts that they be on par with each other in the sense that they belong to a single level of discourse? I think that the answer is in the affirmative. But this is precisely what is not true of our talk of language on the one hand and of our talk of games on the other. That the two concepts of language and game are not on par with each other should be clear from the fact that our language in which we talk about the world and about ourselves contains the predicate 'game' among other predicates; but 'language' is not one of them. Strictly speaking, the predicate 'language' belongs to a metalinguistic framework which we must employ if we wish to talk about languages. The concept of game on the other hand belongs to a certain conceptual setting which is rooted in our language; but this is not so with the concept of language itself. The activities we ordinarily call language include, among other things, our employment of such predicates as 'game', 'tool', 'man', etc. in varying contexts; but they do not include any employment of the predicate 'language' itself. The reason why this is so is simple: the predicate 'language' belongs to a higher level framework, perhaps in the same way that 'means', 'true', 'false', 'valid', etc. do. Wittgenstein's decision to introduce the expression 'language game' seems therefore to commit a type-mistake.

I believe that, like many other philosophers, Wittgenstein's problem is essentially one of how best to understand our language, the complexities of its structure and function, and not one of how to understand this or that one-place predicate that belongs to its framework. Now, if this is so and if the one-place predicate 'game' *presupposes* the whole framework of our language where it belongs among other one-place predicates, how can one single out this very predicate to illuminate the character of what it presupposes – viz., the whole setting of our language?

The assumption that there is a wide-ranging analogy between games and language entails that our language has game-like properties. The only reason why a philosopher of language may proceed from this assumption can be provided by another but equally mistaken assumption to the effect that, since the properties of language are much easily discernible in the *simpler* activities called games, the best strategy to study the complexities of language at various levels of analysis is to compare it with these simpler activities themselves. With assumptions as assimilating in effect as these, it becomes necessary to ask: What game-like properties does our language have? No one would deny that, as rule-governed activities, games are far more simpler than languages. And what is more relevant for Wittgenstein is this. There is something about games as rule-following activities which the *Tractarian* conception of rules as a calculus misses completely. This may be expressed by saying that, in the case of games and also of ordinary language, rules and practices constantly merge into one another with neither assuming any rigid form whatever. However, all this may well be questionable.

Assuming, however, with Wittgenstein that games are much simpler to study as examples of rule-governed activities that are embedded in ordinary life-situations, it becomes relevant to ask: Given all the properties that are the properties of games as rule-governed activities, can the much more complex rule-governed activities called language be said to have only some or most of the game-like properties? Now, there is no doubt that Wittgenstein assumes that languages have game-like properties. But this by itself does not imply that the supposed game-like properties of language exhaust all its properties. For if we say that human beings have animal-like properties, we are not thereby implying simply that human beings are nothing but animals. What is implied is that human beings are a sub-class of animals while leaving the question of their further properties wide open. It follows that even if one proceeds on the assumption that language has game-like properties, the question of what further properties languages have is left open. But it remains to be seen whether and how this is possible under the overall position of Wittgenstein, particularly his theory of family resemblances (Familienähnlichkeiten) (*PI*: §§ 65–77).

It might be correct to state the following theses (a)–(d) about games:

(a) Games are rule-governed activities informed by the flexibilities and inexactitudes of ordinary life-situations.

(b) When we look closely at the proceedings that we call games we see nothing but "a complicated network of similarities overlapping and criss-crossing: sometimes overall similarities, sometimes similarities of detail".[84]

(c) Games form a *family* in an interesting and non-trivial sense of the term. That is, there *is* nothing common/essential to all that we may call games.

(d) If we consider, e.g. how we learn/explain what a game is, we will realize that it is not *necessary* that there be some one set of properties in order for us to be able to apply the predicate 'game' to different activities.[85]

The thesis (d) is obviously stronger than the thesis (c). For the theses (b–d) to be clear it is, however, necessary that the concept of *family* be made sufficiently clear. In other words, how is one to understand the very characterization of games as forming a family, or as forming a "complicated network of similarities overlapping and criss-crossing"? Wittgenstein's answer to this very question is heavily loaded with sociological overtones in so far as his whole thinking here is influenced by the concept of family as a social structure:

> I can think of no better expression to characterize these similarities than "family resemblances"; for the various resemblances between members of a family: build, features, colour of eyes, gait, temperament, etc. etc. overlap and criss-cross in the same way. And I shall say: 'games' form a family.[86]

There is something fundamentally wrong with this answer of Wittgenstein. There can be no doubt that games cannot be said to form a family in the same sense in which we speak of a social structure as a family. I think that if Wittgenstein's theory of games as forming a family is to serve its purpose, we must look for an alternative way of understanding the concept of family itself. This is also necessary if the family resemblance theory as an account of our concepts as they pass through various stages is to be rescued from a narrow sociological interpretation which Wittgenstein himself seems to impose on it.

One philosophically fruitful way of elucidating the concept of family in the present context is, I think, to consider one-place predicates like 'game' through as many different stages as these: (a) introduction of the predicate into the framework of our language; (b) the process of learning the use of the predicate; and (c) the extension, if any, of the predicate's use to cover newer situations. These are, in other words, the three stages of concept-formation, concept-learning, and the extended application of a given concept. It is relatively simpler to consider a predicate at the level of language-learning. I think that Wittgenstein is right when he suggests, in (*PI*: § 69), that a predicate like 'game' can be taught/explained to someone only by *describing* games to him and adding: "This and similar things are called 'games'". Wittgenstein's elaboration follows in (*PI*: § 75):

> What does it mean to know what a game is? . . . Isn't my knowledge, my concept of a game, completely expressed in the explanations that I could give? That is, in my describing examples of various kinds of games; showing how all sorts of other games can be constructed on the analogy of these; saying that I should scarcely include this or this among games; and so on.

It follows from this that it is a sort of ostensive elucidation which plays a central role in one's learning of a predicate like 'game'. For no better structure for learning through ostensive definition/elucidation is conceivable than the following: *This and similar things/activities are called 'X'*. Now, besides having an explanatory character of its own, an ostensive definition is distinguished by the following features. It must involve some appeal to what is chosen/assumed as *the paradigm* of all that one intends to cover under a given predicate; and it must employ demonstratives like 'this', 'that' and the like. The two features must go hand in hand. The explanatory character of an ostensive definition has to do with the question-answer setting that it presupposes. That is, someone must ask actually or in his mind: What is this thing/activity called? And someone must be able to reply: This is called a 'game' and so on.

Let us now attend more closely to what I said is central to one's learning of the predicate 'game', viz., a description such as the following: This and similar things are called 'games'. Here the underlying suggestion is this. A predicate like 'game' can be taught/explained by pointing to the *examples* of activities/things of the sort to which the predicate in question applies; and one's *description* of the examples cited is – necessarily as it were – restricted to their individual features and the complicated network of their similarities. And this implies that one's description of these examples is already a description projecting them as nothing but a *family*. But if this is so, what is the principle at work here which makes this possible? What makes it possible for one to take a number of *similar* things as forming a *family* and hence as coming under a single predicate. Similarities by themselves, howsoever great in number these may be, would not warrant our looking at things as forming distinct families. Nothing short of a deeper principle could warrant our conceptual movement from a world of chaotic similarities to one of distinctly ordered families/classes.

The kind of answer we are looking for is already hinted at above in the suggestion that one implicitly appeals to something as the paradigm when what one says is of the form: This and similar things/activities are called 'X'. It is what one chooses/assumes as the paradigm in the given context which determines one's *choice* of the examples and one's description of their similarities. The complicated network of similarities that one discerns and chooses as a mark of a distinct family are *paradigm-determined*. With a radical change in one's paradigm, one's conceptual activity is bound to change radically. With the principle of paradigm-determined similarities in the background, it would seem that essentialism has been knocked out of our very understanding of our language which abounds with one-place predicates like 'game'. There are no *essences* or rigid concepts corresponding to these one-place predicates. The things/activities over which we conceptualize by applying one and the same one-place predicate are not bound by any common essence whatever. And it would seem that this devastating consequence permeats all the three levels of concept-introduction, concept-teaching and concept-extension or concept-innovation. But I think that it is too early to decide whether this is really what Wittgenstein achieves. For a number of aspects of the supposed principle of paradigm-determined similarities, we have just introduced to rescue Wittgenstein's account, remain unexplored and a number of questions unasked.

Our immediate task is, however, to turn to language itself: How does Wittgenstein's whole assumption of a wide-ranging *analogy* between games and language affect his philosophical thinking on the latter? I think that it is in (*PI*: § 65) that Wittgenstein's approach to language shows itself most clearly as one which is heavily laden with his approach to games:

> Here we come up against the great question that lies behind all these considerations. – For someone might object against me: "You take the easy way out! You talk about all sorts of language-games, but have nowhere said what the essence of a language-game, and hence of language, is: What is common to all these activities, and what makes them into language or parts of language. So you let yourself off the very part of the investigation that once gave you yourself most headache, the part about the *general form of propositions* and of language".
>
> And this is true – Instead of producing something common to all that we call language, I am saying that these phenomena have no one thing in common which makes us use the same word for all, – but that they are *related* to one another in many different ways. And it is because of this relationship, or these relationships, that we call them all "language". I will try to explain this.

The explanation that Wittgenstein develops in the subsequent sections (*PI*: § 66–75) is in terms of examples relating to one-place predicates *in* our language, notably the predicate 'game'. And we have already discussed above some of the relevant details in this explanation.

Wittgenstein's approach to language is thus dictated by his picture of games chiefly in two respects. First, in respect of his scepticism to the effect that there *is* no such thing as a common *essential* character to all that we call language; that any essentialistic approach to language is misconceived and futile since there *is* nothing out there in language corresponding to the basic assumptions of such an approach. Secondly, in respect of his doctrine that it is in terms of the family resemblances themselves, characterizing our language as they do, that we can account for the very fact which essentialism is usually invoked to explain, viz., the fact that we apply one and the same predicate 'language' to different sorts of activities. Wittgenstein's doctrine of family resemblances, extended as it is from games to language, is then the most powerful argument that one could think of in favor of not the weaker thesis that there *is* nothing common to all that we call language but the stronger thesis that it is *not necessary* that there be something common to all that we call language. Or perhaps more accurately, the stronger thesis that it is not necessary that there be something common to all that we call language might be said to follow from the weaker thesis that there *is* nothing common to all that we call language, taken together with the doctrine of family resemblances as an account of the very fact that we do apply one and the same predicate 'language' to many different kinds of structures/activities.

What Wittgenstein, then, achieves might be expressed as follows: All that we call language is a vast multitude of *language games* with family resemblances as the only thing to characterize them, with these family resemblances themselves determined by the ordinary natural language(s) as the *paradigm*:

> I want to say: It is *primarily* the apparatus of our ordinary language, of our word-language, that we call language; and then other things by analogy or comparability with this.[87]

The vast multitude of language games, that language is now *reduced* to, is not however something fixed once for all. According to Wittgenstein (PI: §§ 18, 19, 23), it is an ever-expanding multiplicity, an ever-growing family, with each language game having a distinctive individuality of its own. Once we grant individuality to each language game, two things follow immediately. First, the unit of meaning/communication is shifted from the individual word/sentence to a whole language game. Secondly, the subject-specific concepts find a readily available anchorage in language games, thereby getting relativized to different contexts. As a consequence there are no such things as absolutistic concepts whatever – one and the same expression will mean different things depending on which language game it is used in. This is indeed what Wittgenstein shows in (*PI*: §§ 39–64).

Is the vast multitude of language games, which (*PI*: §§ 23, 24) stresses, ever-growing limitlessly? On the one hand, we may imagine simplest language games as simplest systems of communication, e.g., of the primitive tribes. And on the other hand, we may imagine highly complex systems of communication, such as those which may resemble our language. But are there no limits to this vast multitude on either side?

Wittgenstein's theory of language games is not by itself a self-sufficient doctrine. For sooner or later one must face the question: How do we *delimit* the ordinary languages which, according to later Wittgenstein, serve as our *paradigm* for determining the family of language games? This question is important, especially in so far as it is true that, whatever be our paradigm, it is not 'God given'; we can change it. After all, did not Wittgenstein himself come a long way from the *Tractarian* paradigm for all languages to the one propounded in *PI*?

In other words, the question above relates to the character of the paradigm itself. For isn't it necessary to know what a given paradigm – once chosen – dictates? Unless we know what the character of the chosen paradigm for all languages is we cannot think of the relevant patterns of family resemblances in terms of which, in ultimate analysis, the family of language games is to be determined. I think that it is essentially to a question such as this that Wittgenstein seems to be addressing himself when he says (*PI*: § 19):

> It is easy to imagine a language consisting only of orders and reports in battle . . . And innumerable others . . . And to imagine a language means to imagine a form of life.

This may be taken to suggest that (a) the *individuality* of a language game is a function of the culturally significant setting – the form of life (*Lebensform*) – in which it is deeply embedded and that (b) the relevant patterns of family resemblances between language games must be sought at this deeper level. But is it possible for Wittgenstein to make this suggestion without bringing essentialism back in? The answer is in the negative. If any sense is to be made of what Wittgenstein says, it is necessary that we seek a proper answer to the question: What is a form of life? Any attempt to seek an answer outside the framework of methodological essentialism will land us in an infinite regress. And this only shows once again the correctness of my thesis above that essentialism suppressed at one level of analysis (of language) raises its head at another level of analysis (of forms of life).

But how innovative and illuminating is the idea of a language game itself? Given Wittgenstein's approach to language, it may be correct to speak of language games as the members of a vast family. That is, it may be correct that languages/linguistic activities have game-like properties. If they have, then language is game-like, but certainly much more besides. But what are its game-like properties? If 'being rule-governed' is singled out as a game-like property, any talk of language as being game-like is *unilluminating*; for we already know that language is rule-governed. If 'game-like' means 'being governed by inexact rules', then certainly not all language (e.g. in mathematics and science) is governed by 'inexact rules'. If, however, 'game-like' means 'being capable of invented', then certainly ordinary language is not something we have invented on the model of games which are invented and intended as entertaining games of strategy.

I think that it can be shown that Wittgenstein's approach to language leads to what I propose to call the *paradox* of family resemblances. It is the kind of paradox which radicalizes the consequences of the strategy of so changing the subject of predication/appraisal as to suit one's pre-conceived plan. The paradox may be simply stated as follows. Depending on our human interest, and the paradigm or the *family* we choose as our preferred family, anything complex can be so viewed as to make it *resemble* the given family. For example, if we start with games as a *family* in Wittgenstein's sense, we can look at as complex a 'thing' as language in terms of its resemblances to games and, in ultimate analysis, in terms of its membership (howsoever loose or strong) of the family of games. The result will be exactly as paradoxical and devastatingly innovative as Wittgenstein's conception of 'language games'. Wittgenstein's doctrine of family resemblances leads to this paradox of bringing *anything* whatever under what might be termed the 'preferred family' or paradigm, taking advantage of the important fact that there is generally greater complexity to the subjects/domains of investigation than one can imagine or explore. Given Wittgenstein's theory of family resemblances, which is propounded with the help of the example of games, and given his consequent assumption of games as the preferred family – as the family about which, unlike language, we have greater clarity to begin with – one can look at language as resembling games sufficiently enough to warrant the former's inclusion as a member of the latter. This process can repeat itself with respect to language with every change in the preferred family. If the preferred family is shifted from games to tools, language can start looking as a tool.[88] And there is no end to this process.

Some aspects of the paradox can be restated as follows. Given the family resemblance theory, and the family of games as the *preferred family*, language and games are either analogous activities or they are not. If they are, then they share enough similarities to be able to form a common family in Wittgenstein's sense. Since 'family' in the amended sense suggested above means 'paradigm-determined family', in the context imagined it makes sense to ask: What is the paradigm that determines the 'fact' that language and games form a family? Obviously, it has to be something which is, among other things, both a language and a game at one and the same time; hence in one important sense it has to be something essentially inconceivable – something like Wittgenstein's 'language games'. However, for it to be conceivable as something which is both a language and a game, it is *necessary* that we are already equipped with *theories* of language and games other than the family resemblance theory. It follows that games and languages *cannot* form a common family in the sense of Wittgenstein without landing us in an enterprise which is full of self-contradiction. Since the paradigm around which alone games and languages could form a family (in the amended sense of family resemblance theory) is inconceivable or at least not forthcoming, the two are not analogous in the present sense of the term. Now if they are not analogous, then there is no point in drawing parallels between them, or between our very learning of them, etc. And hence expressions such as 'language game' are not just misleading and misconceived but full of contradiction.

There is, no doubt, a range of *trivial* similarities between language and games. These relate to their character as rule-governed activities and we have already considered them above, if only to point out how unilluminating it is to draw one's attention to these. I think that the only non-trivial sense in which one could perhaps say that languages are game-like is brought out by the suggestion that language and games form a family in the amended sense of family resemblance theory. But we have already seen that any such suggestion is full of paradox, giving rise to unanswerable questions relating to the paradigm which could be said to determine languages and games as forming such a family.

In (*PI*: § 82) Wittgenstein asks a basic question concerning the nature of a rule (in ordinary natural language context or in day-to-day life): "What do I call 'the rule by which he proceeds'?" And, after considering various possibilities, he goes on to ask (*PI*: § 83): "Doesn't the analogy between language and games throw light here?" Obviously, Wittgenstein does this without realizing how perverse it is to assume such an all-pervasive analogy between games and language as to radically affect one's very approach to concepts and rules in the context of language and science. This perversity manifests itself quite clearly in Wittgenstein's own strategy of changing the subject of predication from language to what he calls "language games" (see e.g. *PI*: § 7). It is in this way that Wittgenstein's family resemblance theory, as a generalized account of most of our concepts/rules, can be seen to lead to the consequence of *chaotic commensurability* between subjects as disparate as language and games, for which "family resemblance" is just another name. In the framework of this theory, *anything can be made to look like anything else*, with Feyerabend-type doctrine of "Anything goes" following as a corollary.

Let us consider more closely how far games may be compared with language. No doubt, all games are strictly rule-governed activities. But the kind of rigidity which characterizes them and their rules is not to be found either in our language or in science. Once they are designed, games take shape as *rigidly* rule-governed activities, as rigid as calculi. In this sense, games as an analogy hardly fit into *PI*, since it only reminds us of the *Tractarian* view of language. Wittgenstein (*PI*: § 81) himself reminds us "that in philosophy we often *compare* the use of words with games and calculi which have fixed rules, but cannot say that someone who is using language *must* be playing such a game". He recognizes, in *PI*, that his earlier *TLP*-view of language, based on assumptions subliming its very logic, was a mistake. There it is held that to utter a sentence, to express a thought in it, to understand what it means, is to operate a calculus according to definite rules.

Games serve as a good example of those rule-governed activities which have a maximal rigidity but which lack creativity. I have here in mind the kind of creativity we are familiar with in our language on the one hand and in science on the other. As regards our language, it maintains itself in states of rule-governed creativity once the non-verbal human infant has been transformed by language-learning into a fluent speaker of his native language. Creativity of language may extend itself in all directions including conceptual innovation which is not based on any violation of the rules of the game. It is in science that a transition from rule-governed creativity to rule-enriching creativity is necessitated as and when the extant rules are found to have been violated. Thus, what is conspicuous by its very absence in the sciences is any attempt either to "play safe" (as the inductivists have been again and again insisting and wanting us to believe) or to "play a risky game" as a game of strategy against one's own rivals (as Wittgenstein and his followers would like us to believe). On the contrary, creativity in the sciences involves a tortuous process of risking everything in our search for *knowledge*, letting it grow as an orientating variable of creative human interaction (in the sense already discussed in Chapter 2 above). It involves the search-cum-discovery procedures of *working backwards* from problems to theories and *working forwards* from theories to problems.

Now, even if Wittgenstein's family resemblance theory, along with his language-game analysis of language, is reinterpreted, or extended, in order to bring it under the more respectable and fashionable approach of game theory, it will not serve the anti-essentialistic Wittgensteinian enterprise as has been conceived in *PI*. Thus, any Hintikka-type enterprise of 'game-theoretic semantics' cannot be a proper extension of Wittgenstein's own enterprise, without defeating the whole original purpose. At best, it would only serve to distort the main thrust of what Wittgenstein himself had aimed at. For it can be argued at length that the game theory is itself no less essentialistic in character than any other theory. Thus what it asks or presupposes are the following questions: What is a game? How best can one play a game of strategy, with his rivals set against him, with a view to maximizing his gain or with a view to either minimizing one's loss or winning the game? What are the essential features in different game-situations? Is it winning/losing as possible outcomes? Is it the winning strategies that help maximizing one's gain or minimizing one's loss? Is it the conflict of interest between different parties playing the game? The kinds of essentialistic answers that game theory assumes or gives to these and similar questions can be guaged from how, in the mathematically extended sense of the term 'game', all games are classified on the continuous scale of complexity as: (a) zero-sum games; (b) non-zero-sum games; (c) n-person games, and so on. The one fundamental assumption that the theory makes is the assumption of rationality: The assumption that the principles of rationality that are at work in the playing of games with one's rivals can be formalized. In this sense, game theory can be understood as concerned with construction of what I propose to call formalizing models of games conceived as rational-decision making activities and strategic interactions in situations of conflict of interest between rival individuals or groups.

In the strictly game-theoretic sense of the term 'game', a game, such as two-person zero-sum game, turns out to be a formalizing model (in the sense of § 5.3 below) of the resolution of a conflict-situation of a given type. One cannot therefore talk of language games if it is intended in a sense even remotely covered by game theory. For as a formalizing model of conflict-situations of certain category and resolution thereof, a game turns out to be a *calculus of strategies* of resolution which none of the players can ignore. Wittgenstein's introduction of the 'language-game' notion and his followers' (e.g. Hintikka's) further extension of it thus remains an enigma. For neither language itself, nor language-use can be reduced to a calculus. No doubt, an attempt to do so was made in *TLP*, an attempt which is based on methodological essentialism but which raises the wrong question: What is the *simplest* structure with which a language can describe the world of facts? An over-simple molecular model – an ideal-type formalizing model – in terms of elementary sentences and truth-functional compounds thereof was Wittgenstein's answer to this question.

Under the *TLP*-molecular model, a language with a molecular structure must be constructible essentially as a calculus (Kalkül): given the basic elements of the vocabulary, logical syntax and rules of semantical interpretation, everything else in language should be constructible as a calculus. Given this model, however, ordinary natural language can be seen to deviate from it by way of the complexities of its structure, flexibility of its grammatical and logical syntax and the like. Since it was difficult to reject it completely from the class of languages, it had to be subjected to a reductive analysis in terms of the *TLP*-molecular model. The aim of analysis was to identify the factors (such as those of grammar) complicating the structure of ordinary natural language so as to show it as reducible (in principle) to a calculus. In *PI*, however, Wittgenstein recognizes the legitimate and irreducible character of the complexities in our natural languages. But he fails in an important respect. In order to give up the *Tractarian* oversimple view of language as a calculus, and in order to recognize the complexities in our ordinary natural language as irreducible, it is not at all necessary to give up the basic framework of methodological essentialism, or of appropriate formalizing models. It follows yet in another way that the later Wittgenstein's introduction of the 'language-game' notion must remain unexplained.

5.3. MODELS THAT FAMILIARIZE AND MODELS THAT FORMALIZE: METHODOLOGICAL ESSENTIALISM IN RETROSPECT

The existing literature on the subject of models is, I believe, full of confusions.[89] In it exclusive attention is paid to models in the sciences, the fact that type-distinct models are resorted to both in the sciences and in philosophy not-withstanding. In my (1983), I myself considered the role of models in the sciences, characterizing them as reduction-strategies. I now think it necessary to draw a basic distinction between what might be called (1) the familiarizing models and (2) the formalizing models, if only to recognize models, both in the sciences and in philosophy, other than those of the category (1) that I considered earlier essentially as reduction-strategies. My view of models of the category (1) remains essentially the same. Thus the familiarizing models are the ones which serve essentially to *familiarize* us – in the subjective epistemic sense of the term – with the newly postulated theoretical entities/properties or structures in terms of the network of concepts/laws which we happen to be already familiar with. Consider, e.g., the classical (rediscovered Mendelian) genetics. It is characteristic of it to employ the (unitary) model of the 'beads on a string' to understand the *chromosome* and the *gene* in relation to each other. What the model brings out is, among other things, (a) the particulate nature of genes as discrete entities and (b) their linear arrangement on structures called chromosomes.[90] Although the model may be considered a step towards the asking of the type of questions concerning the molecular structure of the gene which modern molecular biology seeks to answer, it is essentially a familiarizing model seeking to reduce the unfamiliar to the familiar. This should be clear from the affinity between it and Bohr's (miniature) solar-system model of the atom, and other mechanical models in classical physics generally. 'The beads on a string' model is a very good physical/pictorial model of the gene viewed in relation to the chromosome. Recent developments in molecular biology confirm, I believe, *the methodologically unprogressive character of this model*. It is exposed as to its incompetence to suggest any new methodological rules of problem-formulation and theory-construction as regards our understanding of what genes are and how they behave at the bio-chemical level. Thus the real breakthrough in *methodological variance* is seen to have come in the form of Watson-Crick (1953) double helical, non-unitary (*interactive*) model of the gene, elucidating in newer, unfamiliar terms the molecular structure of the DNA molecule and explaining how the genetic information is carried by it.[91] This takes us to the subject of formalizing models in the sciences and in philosophy.

The formalizing models, on the other hand, serve to formalize, at the subject-specific levels of theory-construction, the invariance relations/laws involving the causally relevant variables, capturing the intrinsic/essential properties of structures/functions in its fold in the subject-specific domains in the individual sciences. Consider, e.g., the Watson–Crick (1953) model of the DNA molecule in the gene-nucleus in molecular genetics. In the subject-specific problem-situation here, the structure of DNA is captured as one of double helix: "a ladder-like molecule twined around itself in a spiral". The current computerized modelling of the electron density of protein molecules, that is being used to predict the X-ray diffraction patterns thereof, is, I think, another good example of what I call formalizing models. Such models, which capture/formalize the invariance relations at the subject-specific levels, know no distinction between different disciplines including philosophy. That is, any discipline may employ such models, so long as they aim at formalization and not familiarization. (The familiarizing models, on the other hand, seem peculiar to certain types of strategies resorted to in the sciences). Thus, consider, e.g., the Deductive-Nomological model of scientific explanation, also called the covering-law model, as proposed by Hempel (1965). What the model seeks to formalize, at the level of logico-philosophical analysis, are the properties of non-probabilistic causal explanation in the sciences.

In so far as the formalizing models, in science and philosophy alike, seek subject specific theoretical identification of relevant systems, processes or structures, they are constructible only within the broadly conceived framework of methodological essentialism. This particular feature distinguishes formalizing models sharply from those which we called earlier familiarizing models. The latter are thus non-essentialistic models.

Analogy is, I think, the essence of all familiarizing models. If this is so, such models at best fall within the rule-governed creativity in science: that is, they appear as strategies seeking to extend the antecedently available frameworks of subject-specific problem-formulation and theory-construction to newly encountered domains.[92] But the growth of objective knowledge is not just a function of rule-governed creativity or of strategies appropriate to it.[93] What distinguishes the growth of objective knowledge from system-growth elsewhere is its being a function of rule-enriching creativity.[94] It is here that formalizing models can be recognized as playing an important role in methodological variance of the kind that leads to *CSD*-variance in the sciences in the sense of my (1983). But, unfortunately, such models are still considered an exception rather than a rule.[95] This is particularly thought to be true of 'episodes' such as Dirac's relativistic theory of the electron, "which was hypothesized directly as a mathematical equation without previous physical analogies".[96]

I have characterized what I call formalizing models as those which serve to formalize the invariance relations/laws involving the causally relevant variables, thereby capturing the unique properties of the very structures/systems postulated in the subject-specific domains of scientific inquiry. This is excellently borne out by the example of Watson–Crick double helix model of the DNA molecule in the gene-nucleus (for detailed discussion see Chapter 6, § 6.2.1 below). The entire quest for this model was premised on structural-dynamical assumptions of a kind that ruled out the possibility of the molecular biologists resorting to strategies of familiarizing models in order to unpack the structural and functional (replicative) invariances in the gene already theoretically *identified* (by Mendel) as the unvarying bearer of hereditary traits. Resort to such strategies would have, of course, made it necessary for the molecular biologist to fall back upon such background resources as the familiar framework of statistical mechanics with its laws applicable to microscopic entities like gas molecules, etc. Perhaps, 'the beads on a string' model was a step in this direction. This example shows yet in another way the importance to every science of subject-specific assumptions of the type to be distinguished in Chapter 6 and of methodological essentialism as the framework for subject-specific questions that must be based on such assumptions.

To sum up our discussion in this Chapter: Methodological conventionalism, unlike methodological essentialism, leads to oversimplifications of the kind which give rise to the strategy of so changing the subject of predication/appraisal as to suit one's pre-conceived plan. What this strategy does in its turn is this: (1) It leads to consequences as radical and sceptical in nature as *incommensurability inside/within* the domain of objects that one is investigating, making these objects incommensurable with one another; and (2) it leads to consequences as devastating as *chaotic commensurability* between a given domain and all other domains whatever, rendering the distinct domains as indistinct ones. That is, the strategy leads to the consequence of chaotic incommensurability at one particular level of analysis and to the consequence of chaotic commensurability at a higher level of analysis. The actual examples are not far to seek. They are part and parcel of the received views in the recent philosophies of science and language. Take Wittgenstein first. The domain of language is the domain which ranges over essentially incommensurable language games. But taken as a whole, i.e. at a higher level of analysis, the domain of language is not distinct from but analogous to those of games, tools, etc. Coming to T. S. Kuhn and Feyerabend, we notice the following. First, the domain of an *empirical science* is the domain which ranges over incommensurable paradigms or paradigm-determined puzzle-solving enterprises, making each theory in it incommensurable with every other. Secondly, while science as a whole is comparable to anything ranging from games (Kuhn) to religion, magic, witchcraft, etc., scientific revolutions are comparable to political/ideological revolutions of all sorts whatever.

CHAPTER 6

OF VARIANCE AND INVARIANCE IN SCIENCE: EMPIRICAL SCIENCE AS AN ENTERPRISE COMPRISING *NFCPS* SYSTEMS

No fairer destiny could be alloted to any physical theory, than that it should of itself point out the way to the introduction of a more comprehensive theory, in which it lives on as a limiting case.

Albert Einstein (1960)

6.1. THE PROBLEM OF THE CONDITIONS OF OBJECTIVE KNOWLEDGE

The success of a sound philosophical theory of scientific knowledge depends, I believe, as much on its principles of formalization as on the availability of good examples that are of crucial relevance to such an enterprise. Consider, e.g., what Immanuel Kant (1724–1804) accomplished in the philosophical climate of his time by letting the principles of his philosophy spiral around the scientific theory dominant at that time. Who can say that Kant's philosophy, with Newton's physics as one of its ingredients, did not, however unintendedly, lead to better examples in our own time, say in the shape of Albert Einstein's special theory of relativity (STR) and general theory of relativity (GTR) on the one hand, and quantum mechanics on the other. Wherein lies, then, one might ask, the basic rationale for the philosophy of science within the broad framework of methodological essentialism? The answer is, I think, as follows. In the very process of appealing to the extant scientific theory in all its relevant aspects as a good example for the larger philosophical purposes of elucidating and criticizing the most general ideas ingredient in and presupposed by our extant systems of knowledge, the philosopher might, intendedly or unintendedly, trigger processes of knowing that lead to still better examples, not only of scientific knowledge but of methodological variance. There was a time when one could think of Einstein's STR as a good example of such a scientific theory forcing us, as it did, to renounce some of our deep-rooted methodological prejudices that had become ingredient in the Newtonian tradition of physical problem-determination and theory-construction. Yet sooner or later one learns, with Einstein himself, that the STR is not after all so revolutionary as the GTR is. Thus, Einstein observes:

> You consider the transition to special relativity as the most essential thought of relativity, not the transition to general relativity. I consider the reverse to be correct. I see the most essential thing in the overcoming of the inertial system, a thing that acts upon all processes, but undergoes no reaction. This concept is in principle no better than that of the centre of the universe in Aristotelian physics.[1]

I believe that the question of the conditions of objective knowledge is, therefore, philosophically more relevant in today's scientific climate than it was in the time of Immanuel Kant. For to ask such a question *afresh* is to ask: How is empirical science – conceived not as a static corpus or a structural hierarchy of epistemic structures but as a growing system of the developmental structures[2] of knowledge – possible? It seems quite reasonable to assume that every science *begins* and *grows* as a theoretical problem-solving enterprise. The question arises: What makes such an enterprise possible? For an adequate answer, I believe, we must turn to the kind of *principles/rules* without which science in its present sense would be not only indistinguishable from other types of enterprise but just inconceivable. As a fundamental constraint on any adequate characterization of science, or of the principles underlying its structural growth, it is necessary that it does not stretch the similarities between science and other enterprises to the point of the trivialization of its unique features whatever. For it can be shown that the character and scope of these principles/rules is such that it is not possible to account for them in a narrowly conceived empiricist/reductionist fashion or to trace them simply to 'experience'/'observation', or to have them eliminated in the current Kuhn–Feyerabend type of relativist fashion[3] by reducing them to Wittgensteinian family resemblances. On the contrary, they are the type of principles which at least some philosophers would see as raising afresh fundamental questions of invariance in science. In any case, the real test of their significance and relevance in the present context will depend on their capacity to account for both *variance* and *invariance* in science and to throw light on the fundamental presuppositions of all science conceived not as a hierarchically organized theoretical problem-solving enterprise but as a more complex kind of enterprise, viz. an enterprise comprising *NFCPS* systems in the sense of my (1983).

In what follows, I intend to answer the question raised above – viz. the question of the very conditions of empirical science – by identifying (a) the kinds of *subject-specific* theoretical assumptions/rules which, I believe, the individual sciences must invariably follow for purposes of problem-formulation and theory-construction; and (b) the kind of *subject-neutral* abstract principles which operate at a more fundamental level making it possible for the individual scientist not only to *choose* but also to *vary* his subject-specific assumptions in a rational, though essentially unpredictable, manner.[4] Under the category (a) of the subject-specific assumptions, I shall distinguish, with the help of examples from various sciences, two chief types, namely those that might be called (1) *the structural-dynamical assumptions of complexity* and (2) *the ideal-type assumptions of simplicity*, respectively. Under the category (b) of the subject-neutral (abstract) principles, on the other hand, I shall argue for a thesis of *theoretical universals* which are conceived as the invariant properties of all science – i.e., those properties whose existence (of course in the Popperian $world_3$ of theories and problems) must be granted on strong epistemological and methodological grounds.

6.2. THE STRUCTURAL-DYNAMICAL ASSUMPTIONS AND THE IDEAL-TYPE ASSUMPTIONS IN THE INDIVIDUAL SCIENCES

I think that it is not only a useful and reasonable strategy but quite *necessary* for every science to begin with sufficiently rich, though essentially variable, subject-specific assumptions about those aspects of the world that are of interest to it – a world which becomes available to the thinking human *mind* as much as is permitted by the assumptions/theories that *it* itself must construct. For only in this way is it possible for a science to lay down its own foundation not once for all but till further notice. At the very outset, then, we may distinguish between two chief types of subject-specific assumptions, both of them being crucial to the procedures of problem-formulation and theory-construction in the individual sciences, as follows:

(a) The structural-dynamical assumptions; and
(b) the ideal-type assumptions

Depending on the subject-specific contents of any such set of assumptions, it is these which will determine and restrict in significant ways the *kind* of theory that a given science may build up to account for the (observable) complexities of the natural phenomena falling in its domain, thereby ruling out certain other possible types of theory as impermissible. It is, I think, a distinguishing mark of what I call the structural-dynamical assumptions to identify the vast bewildering complexities in the (observable) course of the natural phenomena so as to view these as a function of the causally relevant mechanisms/structures with relatively simple and theoretically discernible properties. With such assumptions in the background, the search for the empirically testable scientific theory assumes significance as a search for the relevant types of parameters or variables with the relevant types of properties. The structural-dynamical assumptions may accordingly be construed as *assumptions that view the (observable) complexities in the natural phenomena as a function of the relatively simple properties and relations of the relevant types of (causally relevant) variables in a given science*. Theoretical identifications thus made are tested by the postulated causally relevant variables showing relatively simple properties. But these assumptions also operate from the core of scientific creativity by posing problems of complexity at an altogether different level, which introduces still deeper structural-dynamical assumptions of its own. For this very reason it is better to view the structural-dynamical assumptions generally as assumptions of complexity.[5] It is, then, only against the background of such types of subject-specific assumptions that a science can proceed further to raise essentially simple problems and essentially simple theories concerning an essentially complex world. That is, for the individual science to be able to raise specific theoretical problems and to construct specific theories in its domain, its subject-specific structural-dynamical assumptions – whose detailed nature is a matter of further investigation – *must* be accompanied by another type of assumptions, viz. the ideal-type assumptions. Conceived as assumptions of 'the pure cases' which are so simple as to closely approximate to what might be regarded as *self-explanatory cases*, such assumptions are distinguished, I believe, by their unique resolving power (T_{RP}) on the one hand and their characteristic lack of any explanatory power (T_{EP}) on the other.[6] Among other things, then, it is essentially these two types of subject-specific assumptions which codetermine the orientating theoretical framework/basis for a science enabling it to operate by means of the subject-specific methodological$_1$ rules of problem-determination and theory-construction. Such rules will be distinguished, in § 6.5.4. below, from another type of rules, viz. those that might be called the subject-neutral methodological$_2$ rules of theory-appraisal or, more generally, of epistemic appraisal, that are common to all science. Rival theories, T_1, T_2, . . ., in the individual sciences (and, of course, theoretical problems that correspond to them), can be constructed according to methodological$_1$ rules and the two types of assumptions distinguished above. They are, then, in their capacity as epistemic structures, invariably subject to characterization and rational appraisal (for discussion turn to Chapter 9) in terms of their T_{EP} (explanatory power) and T_{RP} (resolving power) as follows:

$$T_{EP}(T_2) > T_{EP}(T_1)$$
$$T_{RP}(T_2) > T_{RP}(T_1)$$

This sets these structures apart as a category distinct from their *necessary framework* of (the subject-specific) theoretical assumptions and methodological rules.

Thus, one essential function of the subject-specific structural-dynamical assumptions of complexity is (1) to identify the complexities of the observable course of natural phenomena in a given domain where everything would, otherwise, seem so simple and so unproblematic to the unsuspecting eye as to dispense with any inquiry whatever; (2) to indicate the *types* of causally relevant structures with relatively simple properties in terms of which the complex observable phenomena could be explained adequately and, if possible, also predicted; and (3) thereby to raise the possibility of newer and newer levels of complexity being theoretically identified, everytime freeing science from the kind of recurrent *oversimplifications* of reality in which it gets entangled from time to time. On the other hand, the subject-specific ideal-type assumptions – presupposing as they do the necessary setting of the subject-specific structural-dynamical assumptions of complexity – involve the laying down of the laws concerning what might be termed the *simplest cases* of the phenomena to be explained. Such assumptions effect a measure of simplification of the domain of one's interest – of the very subjects of objective knowledge – chiefly as a strategy for problem-formulation and theory-construction. The kind of simplification this brings about in the given science – which is necessary for the specialized problem-formulation and theory-construction – is best viewed as a function of the unique resolving power of the given set of ideal-type assumptions.[7] As a rule, it is brought about by so abstracting from the immense complexities of the natural phenomena as to derive the domain of specialized scientific investigation as one in which all relevant phenomena are *explainable* as involving deviations from what the subject-specific ideal-type assumption requires.

6.2.1. *The Subject-Specific Assumptions in the Individual Sciences: Some Case-Studies*

As an example of a subject-specific structural-dynamical assumption identifying the problematic complexity in the observable natural phenomena as a candidate for theoretical explanation consider the following from synchronic linguistics:

> Every adult speaking a given language is, at any moment, capable of spontaneously uttering/perceiving and understanding an indefinite number of sentences which, for the most part, he has never previously uttered or heard.[8]

In other words:

> Every *fluent speaker* of a given language has got what we may call his *linguistic competence* – i.e., his knowledge of his language – which he has *acquired* during his brief (childhood) exposure to the limited number of utterances of others and which he puts to use in his *linguistic performance* in familiar as well as unfamiliar/novel situations.

The whole complexity being identified here might perhaps be summed up in a single universal generalization to the effect:

> All non-verbal human infants turn into fluent speakers in their native languages when exposed to (finite number of) utterances of fluent speakers (such as their parents) of those languages.

The magnitude of complexity thus tentatively identified in the domain of language and human behavior can be gauged from the corresponding subject-specific questions involving considerable degrees of abstraction in the following manner:

(a) What is *known* by a speaker who is fluent in a natural language? Or, what is the exact nature of the *linguistic competence* which underlies the ability of a fluent speaker to communicate with others in that language?

(b) What is the *nature* of the principles/mechanisms which a speaker must employ in order to put his linguistic knowledge to use in concrete though indefinitely varying problem-situations? Or, simply, what is the nature of linguistic performance?

(c) How does one acquire one's linguistic competence? Or, what (innate) mechanisms/causally relevant parameters and developmental processes are responsible for transforming a non-verbal human infant into a fluent speaker in the sense of (a) and (b)?

A very popular version of the kind of framework to which these questions may be traced is available in Chomskyan mentalism.[9] To answer each of these and similar questions within the framework of their respective theoretical assumptions would amount to so identifying the relevant parameters along with their causally relevant properties and relations as to explain the given observable complexities as a function of these parameters. Thus, an answer to the question (a) would require the construction of the models of the grammars of the individual languages. Similarly, an answer to the question (b) would require the construction of the models of linguistic performance, identifying the principles by means of which a speaker's linguistic competence is put to use as and when warranted by the indefinitely varying problem-situations. And, finally, an answer to the question (c) would demand the construction of a whole theory of language-acquisition, identifying the underlying developmental processes and, as per Chomskyan programme, the distinctive *innate* and the *acquired* components in the linguistic competence of a speaker. That the construction of such a theory, as also of the two types of models just mentioned, would presuppose the construction of a general linguistic theory identifying what Chomsky calls 'language universals' will be of considerable methodological interest later on in § 6.3.[10] The Chomskyan subject-specific assumption of a deep and a surface structure of language – with the postulated language universals having more to do with the former than the latter – can in this context be taken as yet another very interesting example of a structural assumption from synchronic linguistics.[11]

Consider now a whole series of subject-specific assumptions (1–12) in the field of biological theory, which are made not only at the observational level but at the deeper theoretical level giving rise, by virtue of their T_{RP}, to the subject-specific problem-determination, problem-differentiation and problem-proliferation. Each of these assumptions could be related to a corresponding set of basic assumptions of the type considered above:

(1) The assumption that there is a structural invariance of a fundamental kind underlying the great diversity of the macroscopic structural patterns/biological organizations ranging from living forms as simple as the bacterium to those as complex as man, giving rise, by virtue of its T_{RP}, to the problem: what is the nature of such an invariance? Or, what *kind* of a structure can serve as the basic/invariant structure of all forms of life?

↓

(2) The assumption – arising out of the discovery of the cell – that the cell is the basic structural unit of all forms of life, giving rise, by virtue of its T_{RP}, to the problem: what kind of a structure is the cellular structure conceived as such a basic unit?

↓

(3) The assumption that the cell must have a *fine-structure* of its own with a functional organization, giving rise, by virtue of its T_{RP}, to the problem: Into what kinds of *structural* and *functional* organizations is the cell-structure analyzable?

↓

(4) The assumption that the cellular fine-structure is analyzable into the two chief types of functional organizations – viz., (a) the *cytoplasm* conceived as the seat of metabolism and (b) the *nucleus* as the control centre, and as the source of information that governs the morphology and the metabolic mechanisms of the cell – giving rise, by virtue of its T_{RP}, to the following problem: what is the nature of the intra-cellular and inter-cellular interactions underlying the developmental processes of a multi-cellular organism? What is the precise structure of the control mechanisms in the nucleus? And so on.

↓

(5) The assumption – owing mainly to the advances in biochemistry during the second quarter of the twentieth century – that, contrary to the assumptions made earlier, there are, at the microscopic level, more profound kinds of structural and functional invariances, in terms of the very chemical machinery,

which characterize the intra-cellular and inter-cellular interactions and which explain the macroscopic complexities and diversities of biological organization, giving rise, by virtue of its T_{RP}, to the problem: what is the detailed nature of such chemical invariants?

↓

(6) Employing the general systems theoretical background of cybernetics, the assumption of *the complex cybernetic system of the simplest cell* to the effect that within each cell there is a cybernetic network – surpassing in complexity, richness and potency "anything that might be predicted by the study of the overall behaviour of whole organisms"[12] – which guarantees the functional coherence of the intra-cellular chemical machinery, giving rise, by virtue of its T_{RP}, to the problem: what is the detailed nature of such a cybernetic network?[13]

↓

(7) The two-fold theoretical identification of the cellular chemistry – owing to the discoveries made around the middle of this century – to the effect (a) that "all living beings, without exception, are made up of the same two principal classes of macromolecular components: proteins and nucleic acids . . ."[14] and (b) that "the same reactions, or rather sequences of reactions, are used in all organisms for the essential chemical operations: the mobilization and storing of chemical potential, the biosynthesis of cellular components."[15]

↓

(8) The assumption, deriving from (7) above, that the structural elements/components of the cellular chemistry are the same and are synthesized by the same processes in all living beings, giving rise, by virtue of its T_{RP}, to a radical *modification* (*differentiation*) of the *antecedently determined* problem of reproductive invariance:[16] (a) "what is the source of their prodigious morphological and physiological diversity?" And (b) "how does each species, using the same materials and the same chemical transformations as all the others, maintain, unchanged from generation to generation, the structural norm that characterizes it and differentiates it from every other?"[17]

↓

(9) The assumption that there are deeper parameters that range

over what might be designated as the fundamental *embedding processes*[18] of the common developmental pattern of the multi-cellular organisms, giving rise, by virtue of its T_{RP}, to the problem: what kinds of embedding processes characterize this developmental pattern?

↓

(10) The assumption that the common developmental pattern of the multi-cellular organisms is, among other things, characterized by the embedding processes of *genetic replication, cellular differentiation* and *histo-organo-genesis*, in that order[19], giving rise, by virtue of its T_{RP}, to problems such as the following: how do the adjacent genetically identical cells give rise to cell lines that are morphologically, physiologically, and bio-chemically different?

As examples of subject-specific structural-dynamical assumptions of complexity not at the observational but at the deeper theoretical level in the biology of the cell, consider the following:

(11) The assumption that there are variables ranging over the control mechanisms in the cellular structure in which genes play a crucial role, giving rise to the problem: what are these variables and what role do the genes play in the mechanisms controlling development?

↓

(12) The assumption that genes have an internal structure of their own which can be rearranged by genetic recombination, giving rise to the problem: what is the structure of the gene – the unvarying bearer of the hereditary traits – in bio-chemical terms?

The assumption (12) marks the transition from the conception of the gene as a hypothetical construct posited in order to account for the laws of Mendelian genetics to one which theoretically identifies the gene with what the chromosomes in the nuclei of cells carry as the unvarying hereditary entity "which can be replicated and transmitted intact (barring mutation) to all subsequent generations".[20] Without an assumption of complexity as abstract and fundamental as this, molecular genetics would not have been possible. The attempts to answer the question of the bio-chemical structure of the gene led first to the discovery of DNA as the essential genetic material and then to the Watson–Crick (1953) model identifying the structure of the DNA molecule in the gene-nucleus.[21]

A cell is a complex system consisting of a nucleus surrounded by cytoplasm, with the former as the source of cellular control. Precise theoretical identification of the principal source of control within the chromosomal gene came only after the rediscovery of Mendelian genetics in 1900. Thus, the identification of the gene as the unvarying bearer of hereditary traits may be traced to Gregor Mendel. But it is Oswald Avery's experiments which, in their turn, suggested the identification of its chemical structure with the DNA molecule; Al Hershey and Martha Chase whose work around 1952 confirmed it experimentally;[22] and Watson and Crick who elucidated in 1953 the physico-chemical basis of its replicative invariance. In the language of the general molecular theory of the genetic code, DNA is a double helical strand or molecule. Outward *rigidity* of DNA as a physico-chemically organized molecular structure notwithstanding, its internal complexity and stability as an information-carrying system is what makes it a *multipotent* interactive element capable of interacting with other elements in various ways. This has to do with its capacity (a) to make copies of itself in such a way as to give rise to replicative invariances and (b) to transfer its information to other parts of the cell – in which process it makes several kinds of RNA that pass to the cytoplasm to participate in the formation of proteins. Thus, the questions which molecular genetics seeks to answer relate to (1) theoretical identification of the chemical and physical structure of the DNA molecule; (2) its properties and functions as the information-carrying system constraining the intra-cellular and inter-cellular interactions; and (3) the causal, or causally relevant, mechanism of information-transfer from DNA to other parts of the cell.

Thus, molecular genetics has been asking a very fundamental question as follows: What chromosomal molecules/components carry the hereditary material – the genetic code – which is responsible for both biological continuity and diversity? The problem is developmentally traceable to the work of T. H. Morgan, Max Delbrück and others starting in the late 1930s. It was Delbrück whose experimental study of viruses, such as bacteriophage, for answering questions relating to the hereditary material responsible for self-replication proved quite decisive. It was around 1952 that his work culminated in Al Hershey's and Chase's experimental confirmation of Avery's identification of DNA, instead of protein molecules, as the hereditary material.[23] From here onwards what we notice is this. The problem, in molecular genetics, *shifts* itself to a different framework sufficiently rich to allow problem-determination with respect to the very structure and function of DNA, within the theoretical constraints of the assumptions about reproduction, replication and transmission of genetic information:

> What properties, structural and functional, must the DNA molecule have in order to carry out jobs of replication and transmission of information to other parts in the cell? Or, how do genes replicate and carry information?

Answers to these questions became available only in 1953 with the publication of the results of James D. Watson's and Francis Crick's work on the structure of the DNA molecule (in *Nature*, April 1953). Interestingly enough, this work was carried out within the framework of theoretical assumptions about DNA's structural, genetic/informational and bio-chemical aspects. Thus, the kind of molecular model of DNA, which was being sought, was required to (a) be in agreement with the X-ray diffraction data already available then and (b) explain replication and transfer of genetic information from the cell nucleus to cytoplasm (called auto-catalysis and heterocatalysis respectively). The X-ray diffraction data available by this time were structurally significant in so far as these pointed to a remarkable structural regularity and chemical stability of the DNA molecule.[24]

Among other things, it is noteworthy from a methodological point of view that these data provided a solid ground to understand the problem of solving the structure of DNA as one of theoretically identifying the deeper *symmetry* in terms of which one could finally decipher the X-ray diffraction patterns of crystalline DNA pointing to its structural regularity and chemical stability. As a consequence of this, they ruled out certain theoretically potential clues as false starts that were otherwise tempting both to the crystallographer and the more theoretically oriented scientists active in this problem-area. But in the very process of doing so they restricted the potential alternative molecular models of the DNA molecule in a significant manner, without however uniquely picking out anyone of them.[25]

The discovery by Watson and Crick of the spiral-like double helix as the correct solution of the problem was not to be, however, as simple a procedure as the theoretical identification of the primary genetic material with the DNA molecule (instead of the protein molecule) might have suggested initially. Not only was further X-ray work needed, the theoretically correct and empirically adequate model had to be such as could predict the kind of diffraction patterns that would be favorably comparable with the experimental X-ray data. The fact that the same X-ray data were often subjected to varying interpretations by different scientific groups does not imply that they were working under some sort of incommensurable paradigms. These were, on the contrary, a natural consequence of the varying methodological approaches to attacking èssentially one and the same problem.[26] Following Watson (1968), one may be tempted to say that there were mainly two approaches at work here, the pure X-ray crystallographic approach on the one hand and the approach of (Watson and Crick) theoretical model-building on the other. But I consider this misleading. The two approaches appear to me to have served as complementary to each other ultimately. The real differences, if any, lay, I think, in working with different structural hypotheses or assumptions. For example, the important work which Rosalind Franklin did in the year 1952 on the X-ray diffraction of crystalline DNA could not have been possible, I think, without the support of necessary background structural assumptions.[27] It is impossible to let one's experimental interactions go on and on in absence of the relevant theoretical assumptions and questions in the background. I would now conclude this discussion with a statement of what appear to me as most important theoretical assumptions which were, I think, not only intimately intertwined with the problem of the solution of the structure of DNA but *necessary* for the solution of this problem. And it is these with which Watson's and Crick's search for a solution was impregnated.[28] These might be stated as follows:

(1) All genes are composed of *crystalline* DNA (deoxyribonucleic acid), the earlier belief that genes were special types of protein molecules not-withstanding.
(2) Any molecular model of the chemical structure of DNA *must* incorporate features *causally* relevant to the dynamics of (a) gene (DNA) replication and (b) DNA's controll of cell-biochemistry.[29]
(3) Since DNA molecules form crystalline aggregates capable of being studied by techniques of X-ray cyrstallography, DNA contains a very large number of nucleotides linearly linked together in a *regular* way.
(4) The *simplest form* for any regular polymeric molecule is a com-

pound helix composed of several polynucleotide chains twisted about each other, forming a spiral-like structure.

The double helix,[30] the most likely of all candidates once it was conceived, turned out to be the most acceptable solution, relative to the background knowledge and the latest X-ray diffraction data. To these assumptions (1)–(4) one might also add all those assumptions that must underlie the cyrstallographer's techniques of X-ray diffraction. More interesting, however, are those assumptions which a scientist abandons or varies so as to get at the correct solution. Consider, e.g., the following assumption (5) with which Watson worked at a particular stage of investigation to arrive at the correct solution of the structure of the DNA molecule:

(5) Each DNA molecule consists of two chains with identical base sequences held together by hydrogen bonds between pairs of identical bases.[31]

The assumption (5), which is essentially a like-with-like structural assumption concerning the four base pairs (here envisaged as adenine with adenine, cytosine with cytosine, guanine with guanine, and thymine with thymine) as these build up a DNA molecule, had to be varied radically so as to arrive at a solution which did not complicate the structure of DNA, as this assumption did, by depriving it of a regular backbone.[32] Interestingly enough, the resulting new assumption adequate to its task had to be an unlike-with-unlike structural assumption or requirement as follows: "Two irregular sequences of bases could be regularly packed in the center of a helix if a purine always hydrogen-bonded to a pyrimidine" – i.e. if adenine would always pair with thymine, and guanine only with cytosine.[33] With this, the question of which type of double helix would be causally adequate and simpler to unravel the scheme of replication assumed more significance. To Watson the answer that suggested itself was this: not the double helix with a like-with-like pairing but one with an unlike-with-unlike pairing would serve the purpose.[34]

To turn now to subject-specific ideal-type assumptions such as, for example, Aristotle's assumption of circular motion as the most perfect celestial motion, Newton's 1st law of motion, the assumptions of the rational problem-solver/decision-maker in the social sciences and the (Chomskyan) assumptions about the ideal speaker-hearer's linguistic competence.[35] As I have argued in my (1983), a subject-specific ideal-type assumption is characterized by its unique resolving power, while being without any explanatory power. Thus, what is noteworthy is how Newtonian mechanics aims at scientific explanation of the theoretically resolved phenomena of motion by investigating the complex behaviour of the material bodies in motion as involving deviations from what one would expect if the domain of investigation were as *simple* as described by the ideal-type law of inertia. This law states that material particles sufficiently far removed from other material particles continue to move uniformly in a straight line or continue in a state of rest. One of the greatest theoretical advances attributable to Newtonian mechanics is thus the kind of simplification which it brings about at the fundamental level of problem-formulation. And one of the ways in which this is done is by varying the earlier subject-specific assumptions and replacing these by new ones. In Aristotelian physics, the path of a heavenly body calls for explanation if and when it departs from perfect circularity, its circular motion being its natural motion. In Newtonian physics, it is departures from perfect rectilinearity that must be explained:

> What kinds of forces must be invoked to explain all complex motion as involving deviations from what the ideal-type law of inertia requires or describes?

Thus, it is the underlying, though mostly unstated, structural-dynamical assumptions together with the ideal-type law of inertia which make Newtonian mechanics and his theory of gravitation possible. This they do by giving rise to the problem of fundamental methodological significance to physics: what kind of a physical theory must physics aim at? What is true of Newtonian mechanics is, I think, also true of Einstein's special theory of relativity (STR) in so far as the latter's validity is assumed only for the reference-bodies Ḱ which are in a state of *uniform, rectilinear* and *non-rotary* motion with respect to the (Galilean) reference-body K – a body whose motion is such that Newton's law of inertia holds with respect to it and hence with reference to which the laws of nature are to be as simple as possible.[36]

What is true of the natural sciences like physics and biology is also true of the social sciences generally and of linguistic theory in particular; although the latter are still struggling against the predominant taxonomic view of their basic scientific tasks. Let us consider the scene in linguistic theory. It follows as a consequence of my thesis concerning the types of subject-specific assumptions – on which the methodological rules of the individual sciences must rest, making these sciences possible – that any domain of human behavior and learning cannot be studied scientifically or even philosophically if the very framework of one's basic assumptions oversimplifies the domain of interest from the very outset, i.e., if it is not the kind of framework consisting of type-distinct assumptions considered above. This is especially true of language, among other subjects. Thus, here, as in any other domain, it is necessary to assume a framework which recognizes explicitly that human behavior in the domain of language-use and language-learning is immensely complex; that it is not as simple as it might seem at first sight to one inclined to account for it mechanistically; and that, on the contrary, there are structural-dynamical variables ranging over the underlying complex cognitive structures and developmental processes. With the subject-specific assumptions such as these granted, it would seem to follow that any framework which employs highly oversimplifying assumptions while ignoring complexity altogether in its declared programme of language-investigation must be ruled out as unacceptable in this domain, as also in any other domain.[37] The distinguishing mark of a legitimate methodological framework is thus its ability to employ from time to time the two types of subject-specific assumptions considered above so that every individual science keeps asking: what kind of a theory must it aim at? Only a specification, from time to time, of the subject-specific methodological rules of theory-construction and problem-determination can constitute an adequate answer to such a fundamental question.

I think that it goes considerably in favor of the Chomskyan methodology in the field of linguistic theory that, in and through its subject-specific assumptions[38] of complexity and those concerning the ideal hearer-speaker's linguistic competence, it raises the fundamental methodological problem: what kind of a linguistic theory must linguistics aim at? Or what kinds of grammatical models of the individual languages are permissible as legitimate models? In this way, it exposes the scientifically hollow character of the behavioristic and the empiricist frameworks as nothing but a trap of highly oversimplifying assumptions – i.e. assumptions to the effect that human behavior/performance in different domains of learning is nothing but a special case of animal behavior in general. (This is what is advocated by Quine and Skinner and generally by all disposition-theorists). Thus, the Chomskyan enterprise is informed at every point by philosophy – traceable to the classical rationalist tradition of Descartes and his followers – which assumes the following methodological requirement. In the field of language-study, one should avoid all oversimplifying assumptions of the kind embodied in Skinnerian behaviorism, Neo-Bloomfieldian taxonomic linguistics, Quinean empiricism and the like. Such a methodological prohibition follows, I believe, from the subject-specific assumption of complexity above that there are structurally-dynamically relevant variables which range over the complex, but distinguishable, cognitive structures and developmental processes underlying the vast and the bewildering complexity of human behavior in the domains of language-learning and language-use, among other domains. In the present context, such an assumption is not as novel to make as it might seem at first sight. I think that it is again traceable to the positive contributions of the classical rationalist tradition of Descartes – the philosopher of mind who founded the framework of the study of human behavior in different domains in terms of an appropriate mental organ in the postulated complex mental 'substance' called mind.[39]

6.3. THE SUBJECT-SPECIFIC METHODOLOGICAL PROBLEMS OF THEORY-CONSTRUCTION AND PROBLEM-FORMULATION

In any case – and this is of great methodological significance – the Chomskyan enterprise of general linguistic theory is comparable in philosophical significance to the Cartesian programme of laying down the philosophical foundations for physics on the one hand and for psychology on the other.[40] For characteristic of the latter is to define matter as *extension* and mind as a *thinking thing* which is not extended. In so far as the former definition contains the essentials of a methodological framework for building up physical theory concerning material bodies that can act upon one another only by a mechanical push it answers, however crudely, the important methodological question: what kind of a physical theory must physics aim at? The latter definition similarly provides a parallel but radically different methodological framework for building up a cognitivist psychological theory of human behaviour and hence it constitutes an answer to the question: what kind of a theory must psychology and the related disciplines aim at? Since it is quite a reasonable assumption to make that language involves multi-variable interactions between systems of cognitive structures, its scientific study within the broader framework of such a psychological theory would make excellent sense.

If the two types of subject-specific assumptions distinguished above did not pose further methodological problems for the individual sciences, they could in no way be said to render these sciences possible. These assumptions, however, invariably raise problems such as: what kind of a theory must physics/linguistics aim at? It is primarily in answer to such questions of fundamental methodological importance to the individual sciences that one thinks of the subject-specific methodological rules of theory-construction and problem-formulation that are appropriate to each science. It is in this context, then, that the problem of language-universals in the sense of Chomsky assumes great methodological significance to linguistics. The problem, as formulated by Chomsky, concerns the class of permissible grammars – or models thereof – of the individual languages, i.e., their delimitation in terms of certain subject-specific principles of invariance. It is the task of the general linguistic theory to discover/identify such principles as the principles of what Chomsky calls *universal grammar*. The basic problem here is simply this: what kind of grammars must linguistics permit as the grammars of the individual languages? And at least one adequate way of solving it is to construct a general linguistic theory that identifies the language-universals as the very invariant properties which every grammar that a linguist constructs must reflect; a grammar thus constructed being a 'scientific theory' "that purports to set forth explicitly the principles of the grammar represented in the mind of the person who knows the language in question".[41]

I believe that every science must sooner or later face the same kind of methodological problem. This is bound to be so even if a science does not raise such questions explicitly. And such questions may remain essentially unasked while a science is engaged in the task of theory-building and problem-determination according to the methodological rules which it follows quite implicitly. But that this is possible does not in any way detract from the philosophically significant thesis above that it is these rules which make a science possible.

6.4. TWO CONCEPTS OF INVARIANCE: THE PROBLEM OF THEORETICAL UNIVERSALS

Two senses of 'invariance' must, however, be distinguished in the present context. A grammatical property may be invariant and, therefore, of crucial significance to the very concept of a *methodologically* permissible grammar if it is a biologically necessary property in the precise sense – which we owe to E. Lenneberg and N. Chomsky[42] – that it is, in ultimate analysis, traceable to the genetically determined and biologically innate *language-responsible cognitive structure/state* (*LRCS*). In other words, *any* explicitly constructed grammar of a given language must, as a matter of biological necessity, meet the general conditions fixed by *LRCS*. But we may also speak of a grammatical property as being invariant in a different sense altogether. In this second sense, invariance is essentially an epistemological-ontological concept of structure-property correlations in a grammar as a special case of such correlations in the objective knowledge systems in the sciences such as a scientific theory. For example, the property, among other possible properties, which would distinguish a grammar of English as a methodologically permissible grammar has to do with the fact "that its rules and principles actually generate sentences".[43] But for its epistemologically relevant structure-property correlations we will have to look for other kinds of properties such as the simplicity and the falsifiability of its basic assumptions.

In terms of the distinction made above, an important clarification follows as regards the question of theoretical universals I have raised in my (1983). The question, as I understand it now, relates to those properties[44] which could yield the concept of an epistemologically permissible scientific theory/problem for the individual sciences. For example, the *falsifiability* of a scientific theory seems to me to capture one such property, or in any case to *presuppose* one (for more discussion on this subject, see Chapters 7 and 8 below). Its *simplicity* might be said to capture another such property. More precisely, then, theoretical universals may be understood as follows. They relate to the methodologically invariant properties in the objective knowledge systems in the sciences such as theory-problem interactive systems. They can be studied by specifying first the epistemologically-ontologically relevant structure-property correlations in these systems.

In this way, we arrive at an important distinction between (a) the concept of a theory-problem interactive system as being methodologically permissible in *a* science and (b) the concept of a theory-problem interactive system as being epistemologically permissible for *every* science. It is the subject-specific methodological assumptions and rules of theory-construction and problem-determination which yield the former concept. It is no surprising consequence thereof that every individual science is distinguished by its own concept of a methodologically permissible theory-problem interactive system. The concept of a transformational generative grammar of an individual language, which meets all the conditions fixed by *LRCS* in the sense of Chomsky, is a very good example of how the methodologically permissible character of a scientific theory in linguistics is understood presently. But sooner or later, we must face the problem of demarcation: what is it which makes the individual sciences as rule-governed enterprises possible? Each science is, no doubt, characterized by its concept of a methodologically permissible theory-problem interactive system. It is, however, not this concept but the concept of an epistemologically permissible theory-problem interactive system which every science presupposes at a much deeper level. It is this which is presupposed by the individual scientist's search-cum-discovery procedures that are employed in search of such systems. Theoretical universals being the methodologically invariant properties of such systems in and across the individual sciences, they must be admitted as fundamental constraints on the scientist's search-cum-discovery procedures themselves.

This condition on science in terms of theoretical universals amounts, then, to constraining/restricting the individual sciences, their subject-specific theoretical assumptions, their rules of theory-construction and problem-determination and, therefore, their theory-problem interactive systems in significant ways. At the same time, however, this makes it possible to explain how the individual scientist may choose and vary them all from time to time in a rational, and not in an arbitrary, manner. Thus, it is quite reasonable to require that the subject-specific assumptions and rules, which a science may employ in its search-cum-discovery procedures, meet the set of methodologically invariant conditions in terms of theoretical universals. For example, some, if not all, of these assumptions must be falsifiable, if not false. This particular condition ensures the falsifiability of the epistemic structures of more specific theories which presuppose the assumptions in question. Without presupposing the falsifiable assumption that genes must have a structure of their own which can be rearranged by genetic recombination, neither the question "what is the structure of a gene in bio-chemical terms?" nor the discovery of DNA and its double-helix structure as the bio-chemical basis for heredity would have been possible. Similarly, without presupposing the falsifiable assumption that all non-verbal human infants are born with *LRCS*, Chomskyan mentalistic linguistics would not be possible in its present form. The questions such as "What kind of structure is *LRCS*, at the biological and psychological levels?", "With what other factors/structures must *LRCS* interact in significant ways before a child is able to acquire the knowledge of his language and perform linguistically?", or "What are the universal properties of *LRCS* and how does it contribute to language-acquisition?" would be rendered unaskable questions in absence of their necessary presuppositional settings in terms of appropriate subject-specific assumptions already constrained by theoretical universals in significant ways.

The question that I would therefore like to consider is one which – anticipated in the very title of the present essay – is raised by the two types of subject-specific assumptions distinguished above and hence by the subject-specific methodological rules of theory-construction and problem-formulation that rest on these assumptions:

> What universally valid constraints must be admitted so as to delimit in methodologically significant ways the range of the legitimate subject-specific assumptions/rules which the individual sciences follow from time to time? With the subject-specific assumptions and methodological rules in the individual sciences being essentially variable, can one choose and vary them in an arbitrary or random manner? Or, are there any principles of invariance which delimit the range of these assumptions and rules in significant ways, making it thereby possible for the individual scientist to choose and vary them in a rational manner?

I believe that the problems of variance and invariance thus alternatively formulated must be faced boldly if science, and indeed our very understanding and characterization of it, are to be saved from the Popperian or Kuhnian methodological conventionalism on the one hand and the Feyerabendian epistemological anarchism – a variation on the former – on the other.[45]

If it is, then, the two types of subject-specific assumptions distinguished above and the subject-specific methodological rules based on such assumptions which make the individual sciences possible, what makes these subject-specific assumptions and rules themselves possible? I think that this question presupposes that there are or must be principles (of invariance) in terms of which alone is it possible to delimit the permissible range of the subject-specific assumptions in question. The question itself concerns the identification of such principles. Hence for an answer to it we must look for the type of *subject-neutral* (and not subject-specific) principles of which we could say that in their capacity as the source of a universal constraint on science they (1) pre-determine the scientist's choice of and variation in the very assumptions/rules of his enterprise from time to time and (2) provide the *background invariance* necessary for critical appraisal of scientific/methodological variance at various levels. Such principles are so fundamental that they are *presupposed* by *all* science irrespective of their individual differences. That is, they are presupposed at every stage of subject-specific theory-construction and problem-determination and also of scientific appraisal. And without such principles a scientist would not know what types of subject-specific assumptions to make or which subject-specific methodological rules to vary and which to lay down. It is only proper to designate the kinds of properties which are implied by such principles as *theoretical universals*; and it is only in terms of such properties that science can be adequately demarcated from non-science, or the epistemic structures of the former from the non-epistemic structures of the latter, without getting entrapped in methodological conventionalism (Popper and Kuhn) or epistemological anarchism (Feyerabend).

In any case, it seems a better alternative to both methodological conventionalism (Popperian and Kuhnian version) and epistemological anarchism (Feyerabendian version) to hold that the individual sciences can frame their subject-specific theoretical assumptions and vary as well as lay down their subject-specific methodological rules of theory-construction and problem-formulation only within the limits set by theoretical universals. The very assumption that there are theoretical universals as the methodologically invariant properties of theory-problem interactive systems in the sciences, or as structure-property correlations in the objective knowledge systems, makes it possible to view science as an enterprise comprising *NFCPS* systems. The identification of these theoretical universals is a task which might be quite safely left to epistemological ontology, itself an enterprise which inquires into the structure and growth of objective knowledge systems in the sciences.

Discussion on the problem of theoretical universals as a problem of invariance will be resumed in Chapters 7 and 8 below. In what follows, I propose to discuss the nature of methodological variance the most significant phases of which appear as scientific revolutions (= *CSD*-variance).

6.5. THE NATURE OF METHODOLOGICAL VARIANCE: SCIENTIFIC REVOLUTIONS AS A FUNCTION OF WORKING BACKWARDS FROM RULE-ENTANGLEMENT

There is no agreement between the historians and philosophers of science on how to test and find out objectively whether a scientific revolution has really occurred at a particular point and time in the developmental history of the sciences. Yet, if we want to understand the nature of methodological variance in science, we must sooner or later ask the following questions which are anticipated by our discussions above and in Chapter 5:

> What is the nature of scientific revolutions? Do they follow any rational pattern? Are there deeper principles of scientific change, or of methodological variance, in terms of which alone is it possible to understand their nature and their complexity? Or must one rely on Wittgensteinian 'family resemblances' between science and other forms of activity for this purpose?

The recently formulated rival theories on the subject of scientific revolutions invariably fall back on either weaker or stronger assumptions seeking to assimilate science and scientific revolutions to subjects that are *essentially* different. Popper's methodological conventionalism, Kuhn's methodological anarchism and Feyerabend's epistemological anarchism are only consequences of such assumptions. In what follows, I shall consider certain aspects of Kuhn's and Popper's approach if only to point out their fundamental inadequacies and to give my own account in terms of principles which are, I believe, inherent to science. I shall deliberately refrain from going into matters of detailed argument, since I have already dealt with them in my (1983).

6.5.1. *Of the Asymmetry between Scientific and Political Revolutions*

If the recent and current controversies concerning scientific revolutions are any indication, it would seem quite tempting to think of these revolutions as repeating the pattern(s), if any, of what are called political revolutions or, more precisely, revolutions in the sense of radical changes in a people's allegiance to one sort of ideology as against some other. In this picture – which takes political revolutions for its paradigm – scientific enterprise will always apppear to assume either of the two phases: the pre-revolutionary 'normal' phase and the post-revolutionary 'extra-ordinary' phase.[46] Moreover, whatever passes for science under the one will be *incommensurable* with whatever passes for science under the other, incommensurability here being simply a function of the change from the pre-revolutionary to the post-revolutionary ideology or ideological commitment.[47] The unquestioning commitment to a given 'paradigm', and the resultant ideology, is what makes 'scientific' practice in either of its two phases possible, 'scientific' practice being simply a practice of 'solving puzzles' within the *limits* set by the given paradigm.[48] It is always the 'dominant scientific theory' of the time – dominant in terms of its general acceptance on the basis of its successful applications – that is most naturally chosen as the paradigm and hence as the framework for 'scientific practice'.[49] The paradigm is thus found to change and knowledge to grow with every radical change that replaces the dominant theory of the time by its rival. And, to complete the underlying analogy with political/ideological revolutions, the *subject* of *scientific appraisal*, if any, is not the kind of structure called 'scientific theory' but the *scientist's skill* in the 'puzzle-solving' enterprise called 'normal science'.[50] The structure of a scientific revolution, on this picture, appears as a pragmatically complex structure of a given scientific community shifting its allegiance to a paradigm which is incommensurable with that of the pre-revolutionary scientific practice.[51] Consequently, the growth of science too would appear to follow a pragmatically complex pattern, involving, among other things, a (cyclical) *movement* of the scientific community from its collective commitment to a given paradigm, or to what Kuhn (1974) now calls a disciplinary matrix, to similar commitment to its incommensurable 'successors'. It is a pattern of ideological variance in which incommensurable paradigms are embraced one after another in a manner analogous to episodes such as religious or political conversions.

Such a conception of scientific revolutions, of the growth of science and of scientific appraisals, informed as it is deeply by the analogy of political/ ideological revolutions – and which is in sum what T. S. Kuhn advocates in his (1971a) and elsewhere – is open to a number of decisive criticisms from the very objectivistic point of view which it seeks so persuasively to undermine.

First of all, how far is it even plausible to assume political/ideological revolutions as following a definite pattern and to dogmatically let one's characterization of scientific revolutions be guided by whatever picture one may draw of the former? I think that it is not plausible at all even to assume that our understanding of the former is more perfect/clear than our understanding of the latter; or that there is any significant family resemblance/parallelism between them. Consider, e.g., the fact that political/ideological revolutions are often mediated and followed by bloodshed, mass killings, elimination of the rival groups *wholesale*. In a sense, it is true to say of such revolutions that they are a function of ideological intolerance among the rival political/ideological groups where intolerance is directed not so much at the rival ideology as the group that upholds it. Thus, what matters in such revolutions are our usual answers to the following questions:

> Who are the persons, or groups,
> with such and such ideology, who
> hold such and such views? And what
> is their numerical strength?

Since the questions of group-identity and numerical strength are so crucial here, no single individual can afford to be a revolutionary in the political sense of the term and no single individual can bring about a political/ideological revolution. Such revolutions are, therefore, inconceivable without a mass movement, whether violent or non-violent.

Political revolutions are, in an important sense, a function of ideological intolerance. They have invariably a violent aspect, even when they may be non-violently intended or projected. Thus, we hardly show any surprise on learning of a violent political upheaval overtaking a whole people, as is amply borne out by recent events in world political history. Ideological intolerance, being itself a game of numbers and group-identity, calls for study in respect of its potential for violence. As a point of significant contrast with science, this would warrant the view that politics is certainly a many-persons game in the sense of game theory.

When this many-persons game is played by resort to violence or by demonstration of the numerical strength, you get political or ideological revolutions as a result. Questions of time-scale are, of course, very important here as in scientific revolutions. But I shall refrain from going into these here. There must, therefore, exist a fundamental asymmetry between political and scientific revolutions. An argument to this effect showing that science is neither a game among games nor amenable to a game-theoretic analysis may be developed as follows.

Our everyday life is, no doubt, characterized by recurrent situations of *strategic interactions* in which the chief variables of interaction range over different people with divergent interests and different means of attaining different goals. Essence of such situations of strategic interaction is conflict of interest between different people/players as the values of the chief variables of interaction. Examples of such situations range from relatively simpler games (in the ordinary sense of that term) to relatively more complex cases of the vendor wanting to sell dearer and the buyer wanting to buy cheaper or of each political party in a multi-party system wanting to so turn the tide of election campaign in its favor as to win the elections. Common strategic aspects of conflict situations such as these constitute the subject-matter of the mathematical theory of games. The theory seeks to analyze these aspects in terms of simplified theoretical 'models' called 'games' (in an extended mathematical sense of the term), viz., zero-sum/non-zero-sum games on the one hand and two-person/n-person games on the other.[52] I think that while every strategic interaction in the game-theoretic sense may be viewed as a multi-variable problem-solving situation, not every multi-variable problem-solving situation can be assimilated to an n-person game in the game-theoretic sense, where the number of variables of interaction is $n > 2$. Now, objectively viewed, science can be shown to be a growing multi-variable problem-solving system; yet it cannot be called a game in the specific game-theoretic sense just considered, simply because here the game, if any, is not one of strategic interaction between different persons/groups, with each person's/group's *interests* set against those of the other. On the contrary, it is only proper to describe science as a *no-person game*.

However, of an ideological revolution, unlike a scientific one, it is true to say that there is always an *inside view* and an *outside view* of it, with the former frequently clashing with the latter. Thus we are not surprised if the members of a particular political or ideological group consider as a (long awaited) revolution those very changes which are 'seen' by others either as a degeneration or as being dangerous for the growth of the very social structures which they consider socially relevant. Whichever way described, such changes owe their character and significance essentially to the conflictual perceptions of the warring political and ideological groups in a society itself being pulled in different, or opposite, directions. All ideological or political revolutions may be, therefore, beyond rational appraisal in the following sense. Being essentially a function of group-ideology, always relative to the conflictual situations of several such ideologies set against each other in the same society, it will be futile either to look for or formalize a single set of inherent features as a criterion of such revolutions. However, if any one characteristic could be singled out as being common to them, it is the kind of incommensurability they bring about between the 'pre-revolutionary' and the 'post-revolutionary' states of political and ideological change. That this is, in fact, generally so is supported by my observation just made that such revolutions are a function of group-ideology, and of intolerance relations between the conflicting ideological groups in the same society. No two or more groups will generally agree on one and the same thing as constituting such a revolution. What is 'post-revolutionary' for one group may be 'pre-revolutionary' for another. Moreover, one and the same group may refrain from calling 'revolutionary' what, at an earlier time, appeared to it as 'revolutionary'. One might, therefore, legitimately wonder whether one needs to be reminded in this context of those *trivialities* to which our attention has been drawn by Kuhn, Feyerabend and Hanson. Thus, it may be trivially true that under the burden of different ideologies one and the same thing appears as a duck to one group but as a rabbit to another. Similarly, in more extreme cases of religious belief, one and the same animal (species) may be labeled and perceived by one group as a symbol of great cultural significance to its members while others do quite the opposite, or just refuse even to recognize its place among natural kinds.[53]

In sharp contrast to this, any suggestion that scientific revolutions might be possible by resort to violence, or a game of numbers, is to render them essentially inconceivable. No violently or non-violently planned mass movement could conceivably have facilitated the revolutionary changes from the system of Newton's mechanics to Einstein's special and general theories of relativity. The same is true of the earlier

Copernican revolution. The game of numbers is simply ruled out as irrelevant in such and similar contexts. For science, both in its pre-revolutionary and post-revolutionary phases, to be out there, it is quite irrelevant to ask: *Whose* theory is this? Or, *how many* subscribe to this theory? On the other hand, such questions assume a crucial importance in the context of political and ideological revolutions. It is, therefore, mistaken to extend one's picture of such revolutions to the field of scientific revolutions, or to let one's characterization of the latter be informed by the former as the paradigm of all types of revolutions. Kuhn's characterization of scientific revolutions is, as one might conclude, unacceptable in so far as it is based on his unargued, and hence dogmatic, assumption that they are either analogous to or a species of political (and ideological) revolutions.

What sorts of questions are then relevant in the context of scientific revolutions? How do such revolutions differ from political or ideological revolutions? To ask these and similar questions is, of course, to underline the important point that in the context of science the problem is not one of tracing scientific revolutions to other types of revolution. Nor is it one of tracing their family resemblances to them, if any, in the sense of Wittgenstein. The problem is one of demarcation instead of assimilation. Thus, for purposes of characterizing such revolutions, we must look for those features or principles which are inherent in science itself and not foreign to it. As a first step, it is necessary to recognize that there is an intimate connection between them and *methodological variance* in science without incommensurability. They have no connection whatever with ideological variance which may involve mass conversion experiences of the type Kuhn has been talking of. Thus, as one might conclude from this, there is an intimate connection between the two problems of demarcating (a) science from non-science and (b) scientific revolutions from non-scientific revolutions.

It is at this point, then, that Kuhnian approach fails where the Popperian approach succeeds to a great extent. For it fails to address itself to the right sorts of questions concerning science and scientific revolutions. The result is not at all surprising. The Kuhnian characterization of science and scientific change in terms of *states* of 'normal science' interrupted by those of scientific revolution commits oversimplifications such as the following. Scientific revolutions are, we are told, a species of political or ideological revolutions, the latter being taken as the paradigm of all that can be called 'revolutionary'. One could, of course, be persuaded to think so only if science did not possess distinctive features and complexities of its own, sufficient enough to single it out as a distinctive theoretical problem-solving enterprise. The oversimplification thus committed concerning scientific revolutions is, however, rooted in another concerning science as a whole or what Kuhn calls 'normal science'. Interestingly enough, it entails a radical change in the *subject of predication*, i.e., in the *very subject of scientific appraisal*. That Kuhn himself is aware of this is obvious from his thesis in this context that it is never the structure/corpus of scientific theory but the scientist's skill at puzzle-solving which is subject to 'scientific' appraisal/evaluation.[54] Of such and similar oversimplifications in the philosophy of science, then, it can be safely said that it is the best strategy of so changing the subject of predication as to obtain the kind of consequences that are desired from the very beginning. In the present context, it is what Kuhn calls 'incommensurability' between the 'successive' paradigms of puzzle-solving scientific practice which constitutes the 'desired consequence'. Incommensurability in this sense is easily obtainable as a consequence if we are successful in pursuading ourselves that in 'normal science' what ought to be subject to scientific test/evaluation is not the structure of scientific theory itself but the individual scientist's ability at puzzle-solving. The kind of oversimplification that alone could do the trick of pursuation is, then, best viewed as the Kuhnian strategy of changing the subject of predication – i.e., the strategy of changing the subject of scientific appraisal – from the (epistemic) structure of scientific theory itself to the scientist's ability at puzzle-solving. Against the Kuhnian strategy it should suffice to remark here that even the most primitive form of science could not be, without committing an error, subjected to such an oversimplification.[55]

All this, however, must be seen in the light of Kuhn's firm belief that for purposes of understanding science one 'blunder' which one must avoid – something where Popper fails according to him – is to look "to the extraordinary or revolutionary episodes in scientific development".[56] That is, in order to understand and characterize science as a whole what is most important is this. One must attend to 'normal science' and not to rare episodes called scientific revolutions. A careful

> look at the scientific enterprise suggests [says Kuhn] that it is normal science, in which Sir Karl's sort of testing does not occur, rather than extraordinary science which most nearly distinguishes science from other enterprises. If a demarcation criterion exists (we must not, I think, seek a sharp or decisive one), it may lie just in that part of science which Sir Karl ignores.[57]

Normal science being defined by Kuhn as a puzzle-solving enterprise that does not involve any of the elements (corroboration, verisimilitude, etc.) intimately or remotely connected with the Popperian falsificationist methodology, it follows that this demarcation criterion lies in the complex activity of puzzle-solving itself.[58] That is, not the Popperian testability or testing which is directed at theories, but puzzle-solving activity with the scientist's *skill* always under test provides the requisite criterion that distinguishes science from other enterprises. But in the process, and this is what Kuhn fails to realize, science is rendered indistinguishable from all those activities or human interactions that are amenable to game-theoretic analysis. However, quite consistently with this, Kuhn arrives at the conclusion that Popperian tests play no role in so far as scientific revolutions are concerned, if only because they play none in normal science:

> with or without tests, a puzzle-solving tradition can prepare the way for its own displacement. To rely on testing as the mark of a science is to miss what scientists mostly do and, with it, the most characteristic feature of their enterprise.[59]

It follows, yet in another way, that it is normal science in the sense of Kuhn that must be taken to distinguish science as a whole from non-science. But what sense is one to make of Kuhn's remark that a puzzle-solving tradition – a particular phase of normal science – can prepare the way for its own displacement? I can think of only the following. Strong ideological commitment to the dominant theory is what distinguishes the Kuhnian normal phases from the revolutionary phases of science. But such commitment always renders a given system, whether of theories or of beliefs in them, so *closed* and *windowless* as to make it collapse under its own tendency to entropy. It is for this and similar reasons that I was led, in my (1983), to put forth the idea of a *physical theory without pragmatical imperatives*.

In order to understand Kuhn's position more precisely, it is helpful to distinguish between three things: (a) science (as a whole); (b) normal science; and (c) extra-ordinary (revolutionary) science. Normal science is science really in a steady-state in which scientists are committed to a given paradigm-based puzzle-solving tradition in the sense of Kuhn. What is under test in science, in its states of normal science, is neither the scientific theory nor its associated paradigm itself. It is, on the contrary, the individual scientist's skill at puzzle-solving strictly according to those commonly agreed upon rules of the *game* which Kuhn

(1974) calls 'disciplinary matrix'. But what is it then which we may call revolutionary science? Well, it is just a very *rare episode* which is always of a temporary duration. As a state of scientific change, it interrupts the game of science, in its pre-revolutionary states of normal science, *suddenly*, if only to let it settle down to becoming the same kind of game again, except for an incommensurability between the pre-revolutionary and post-revolutionary states of normal science.[60] But what is it then which we call science? Well, all science is really *normal science* interrupted by the rare episodes of extra-ordinary science, with pre-revolutionary and post-revolutionary states of normal science always incommensurable with one another. Not only is there science then, but there are also normal and extra-ordinary science. And, over and above them, there is an infinitude of different states not only of extra-ordinary science but also of normal science, with the two kinds of states always incommensurable with one another.

As a result, one can now see more clearly that the problem of demarcation in the sense of Popper is not solved but *dissolved*. But one can also see that, when it comes to actually solving it, Popper's position becomes indistinguishable from that of Kuhn. Now, when we say that a given theory is scientific, or a theory T_2 scientifically better than another theory T_1, what does 'scientific' or 'scientifically' mean or signify here? According to Popper (1968), whatever we may mean by it, it is certainly a matter of convention/decision how we define science. This suggests a certain flexibility in doing so, analogous to flexibility in games, depending on the scientific community. According to Kuhn, on the other hand, whatever it may mean, it is bound to vary from one community/ideology-based paradigm to another, i.e., from one revolution to another. Hence there is nothing universally valid about it.[61, 62] I believe that both these answers lead to methodological anarchism, as should be clear from the manner in which Paul Feyerabend cultivates it in his (1978a, 1978b). Under the Popperian and Kuhnian conceptions, simultaneous cultivation of contradictory ideas of or decisions concerning what is and what is not 'scientific' are permissible. For suppose revolution is taking place simultaneously in more than one individual science. It is quite permissible that we understand the predicate 'scientific' in quite contradictory ways. And this is no less anarchistic than what is permitted by Popper. On the one hand Popper conceives of science without inductivist methods; but on the other hand the cultivation of science in radically different ways is permitted by his concept of science as a game whose rules we (different groups) can lay down as we please.

Coming back to scientific revolutions, which on Kuhn's view do not matter so much as does 'normal science' in its infinitude of incommensurable phases, Kuhn's approach can be seen to be beset with some serious difficulties. *First*, it may be a matter of historical fact that – and that is the point of Kuhn's historicist approach – the pre-revolutionary phases of 'normal science' are always found to be incommensurable with its post-revolutionary phases, with each phase bringing in its fold new demarcation criteria every time. Even if one accepts a historical generalization as false as this, one does not thereby rule out the future possibilities of such phases appearing without incommensurability at all the relevant levels. Stronger arguments would be needed to do so, not just the historical ones appealing to historical case-studies. But before one is able to frame philosophical arguments to derive Kuhnian doctrines and characterizations as a consequence, one must work out purely in philosophical terms conditions necessary and sufficient for any pre-revolutionary and post-revolutionary paradigms (or even theories themselves) to be *commensurable* at all the relevant levels: semantical, methodological, epistemological and perhaps ideological. Once this is done, one is most likely to find that neither scientific revolutions nor the commensurability/incommensurability of theories are a matter as simple as these have been made out to be. Or rather, if the matter of characterizing science as a whole is as complex as should be obvious from its historical, sociological, ideological and methodological aspects, so is the commensurability/incommensurability between rival theories or paradigms. Thus it would become clear that one blunder which philosophers of science must not commit is to derive consequences such as Kuhn Feyerabend incommensurability by means of a simplistic appeal to the semantical aspects of science (the predicate 'mass' *means* one thing in Newtonian mechanics and quite another in Einstein's STR) while ignoring other deeper aspects. It would also become clear how tempting it can be, in absence of an adequate philosophical theory of the sort indicated above, for one committed to the historicist approach to rush to philosophical conclusions from hasty historical generalizations. I think that much of what is seen as semantic incommensurability, or incommensurability as a result of the Kuhnian 'Gestalt Switch' (suddenly shifting the scientist from his familiar world-view/perceptual field to an altogether different one so that where he had seen a duck before he sees a swan now or just *nothing* at all), is traceable to ideological commitment and ideological variance in the present sense of these terms.

Secondly, there is a well-known argument of Kuhn, Feyerabend and others against Popper's falsificationist methodology of scientific revolutions by theory-dislodgement. As is borne out by actual historical case studies, mere observational testing of a given theory by means of refuting instances (= basic statements) may not suffice to overthrow it and to bring about scientific revolution. On the contrary, theory-falsification becomes possible only after potential alternatives become available, and not just by some simpler or abstract logical process of observational refutation. Traceable as it is to Popper's own original position (1934), this is an argument from *theoretical pluralism* as Feyerabend (1965) once chose to call it. Thus, what Kuhn and Feyerabend have stressed is this. Theory-dislodgement and scientific revolutions are matters of greater historical complexity than is seen by Popper. Test or no test, more than observational testing, it is the potential alternatives incompatible with and rival to the given paradigm, already reigning supreme in normal science, which play a decisive role in scientific revolutions.[63] Obviously, they can do so only if they are able to bring to surface, or highlight by sharpening, the difficulties and limitations of the disciplinary-matrix-driven puzzle-solving enterprise in which the *normal scientists* are engaged. Chances are that one of these potential alternatives takes over, dislodges the given paradigm and then completes the whole process of taking the scientific community from its state of crisis to a state of scientific revolution. But if this has been historically the kind of role the rival theories have played with respect to one another, how can the scientific community regard them as being incommensurable? Kuhn and Feyerabend have, of course, yet to explain this as a historical generalization, to which they look for support while arguing for more radical versions of the incommensurability thesis. I have argued before that appeal to the semantics of scientific terms is at best a poor strategy to derive incommensurability of theories/paradigms as a consequence, especially when methodological standards too are involved.

There is considerable truth in the historical generalization, just considered, that every *new* scientific theory, however revolutionary, is seen to throw fresh light on the older theory which it tends to supersede. If we go a step beyond this, we can make it a matter of methodological/epistemological requirement that it shed new light on its successes as well as failures. When, in a known case, the new theory actually does so, it can serve as a source of epistemic appraisal of the older theory. Thus, it will itself become one of the reasons why the scientists should prefer it to its predecessor. With this, I believe, any Kuhn-Feyerabend type doctrine of incommensurability is seriously undermined, not just exposed as a false historical generalization grounded in simplistic arguments from the semantics of meaning-variance from one theory to another.

6.5.2. *Philosophy as a Higher-Order Enterprise: Against the Under-Labourer Conception*

In the wake of the development in this century of different traditions of philosophical analysis, all wedded to *language* in its characteristic states of rule-governed creativity as their popular paradigm, it has become customary in English-speaking countries to characterize philosophy as a 'second-order' enterprise which inquires into different aspects of what are thought to constitute the 'first-order' disciplines in the different fields of science. This characterization is taken to imply that philosophy is best construed and cultivated as a meta-theory, expressible in a suitable meta-language, with the language of science as the object-language. Thus, as a meta-theory, philosophy is distinguished by its projected character as the logic of science. As a classic statement of this view consider what R. Carnap says:

> In our Vienna Circle discussions it had turned out that any attempt at formulating more precisely the philosophical problems in which we were interested ended up with problems of the logical analysis of language. Since in our view the issue in philosophical problems concerned the language, not the world, these problems should be reformulated, not in the object-language, but in the meta-language. Therefore it seemed to me that the development of a suitable meta-language would essentially contribute toward greater clarity in the formulation of philosophical problems and greater fruitfulness in their discussions.[64]

I still think that Carnap's thesis of the modes of speech – and, in this context, of the conventionality of language forms[65] – constitutes an unparalleled contribution to the methodology of philosophy conceived as a meta-theory committed to the logical (structural) analysis of the allegedly 'philosophically neutral' sciences.

Carnap's conception may be traced to Locke's 'under-labourer' conception which requires philosophy to clear the ground and remove "the rubbish that lies in the way of knowledge".[66] Its contemporary roots are, however, as deep as the assumption that there is a great divide between *language* and *the world of facts* such that a statement is always either *analytic* or *synthetic*. Commonly shared by many contemporary analytic philosophers with the exception of thinkers like Quine who attack it,[67] assumptions such as these have lent plausibility to the fashionable doctrine, originally the central tenet of logical positivism, that it is the sciences as first-order disciplines which search for knowledge of the world of facts. Philosophy as a second-order discipline deals only with the language of science, seeking to clarify its structure and meanings. Let us consider some of the implications of this conception as follows:

(a) To begin with, sciences are out there as philosophically neutral first-order disciplines. After a given science has acquired sufficient maturity, philosophy as a second-order discipline appears on the scene to inquire into the nature and structure of the first-order disciplines, or to clear the rubbish, if any, in their way.
(b) As a consequence of (a), philosophy is parasitic on science, with no genuine problems of its own, since all its problems are essentially rooted in the conceptual side of the sciences. Thus, if the language of science did not pose the second-order problems of clearing the rubbish in its way, we could dispense with philosophy. Therefore, what philosophy clearly lacks is the character of a sufficiently *autonomous* rational enterprise.
(c) As a consequence of (a) – (b), philosophy is better understood as an epi-phenomenon of the language of science, with the latter invariably *informing* the former but not *vice-versa*. Thus, science without philosophy is a meaningful and reasonable proposition. After all, what matters as regards human knowledge are the various sciences but not the philosophy of the sciences.

With consequences such as these, the under-labourer conception and its more refined contemporary versions have exercised considerable influence in recent years, particularly by encouraging various types of sceptical trends in philosophy and methodology of the sciences. Thus, it is held in various quarters, with T. S. Kuhn and Paul Feyerabend in the lead, that the sciences are self-sufficient. There is no basis in them for epistemology or methodology of science as a separate discipline. The roots of this kind of scepticism are traceable to those contemporary versions of the under-labourer conception which are widely embraced by analytic philosophers of different persuations. The harm, even if unintended, already done to philosophy by the over-enthusiasm of the 'scientifically minded' contemporary philosophers is considerable.

Now, it is necessary here to indicate how mistaken it is to think of science as a philosophically neutral first-order discipline with philosophy just appended to it as its epiphenomenon. I think that the best way of doing so is to argue for a different conception as a better alternative, which I propose to do in the concluding § 6.5.4. below. As I have argued before, science is inconceivable without its methodological assumptions and rules which *inform* its practice at various levels. And it is these which carry philosophical significance, making science philosophically non-neutral. Once they are recognized either as being followed or as being violated, in a given concrete situation, they will warrant explicit philosophical elucidation and criticism with a view to seeking better alternatives to them. Thus, to put it in a nutshell, of every science we can say that it is philosophically non-neutral in respect of its being constantly *informed* by the antecently determined methodological assumptions and rules. Instead of being just a hierarchically organized theoretical problem-solving enterprise, it is a far more complex enterprise comprising the *NFCPS* systems in the sense of my (1983). Among the rules (and search-cum-discovery procedures that accompany them) which make these systems, as subjects of scientific discovery and appraisal, possible we must recognize those methodological rules which are antecedently determined and uncritically accepted. The task of elucidating these along with other types of rule with a view to finding out better alternatives to them, far from being 'second-order', is a philosophical one. In the very process of inquiring into them, a contact is made between philosophy and scientific practice, particularly the search-cum-discovery procedures in the individual sciences. Thus the philosopher-methodologist's search for the best possible methodology for the sciences is bound to inform each and every science in ways which is crucial to *its* interactive organization as an enterprise comprising the *NFCPS* systems.

Rejection of the idea of an alleged philosophical neutrality of science removes the ground from under the currently re-emerging sceptical trends of thought concerning philosophy and epistemology in respect of their role in the sciences. But it also restores the basis in science for an objectivistic epistemology in which the methodology of science can be quite legitimately embedded. How important is it, then, to objectivistic epistemology to insist on the twin principles of demarcation between (a) science and non-science and (b) the psychology and logic of discovery?

6.5.3. *The Demarcation Problem: Empirical Science Viewed as a Game of Conjectures and Refutations*

The most significant attempt ever made to demarcate science from non-science is, of course, to be found in the following doctrine of Popper. In order to be scientific, a theory, considered as a conjunction, or structure, of statements, must be falsifiable.[68] It can be shown to be falsifiable, if we can specify the class of its *potential falsifiers*. Thus, we must reject a given theory as unscientific and unworthy of further investigation if it is *impossible* to specify the class of its potential falsifiers. At first sight, it may not strike us as to how this criterion commits an oversimplification in respect of the very structural identity of science as a whole and of objective knowledge systems in the individual sciences. In so far as it restricts *all* scientific statements to those which bear a certain logical relationship with a class of 'basic statements' as potential falsifiers, it presupposes a restriction on the epistemic-structural identity of these *systems* themselves. Thus it presupposes a statement-view of the epistemic-structural determinants, or codeterminants, of objective knowledge systems in the sciences, restricting them to statements that are capable of being true or false. As has been argued by me elsewhere (1983), such a view is an oversimplification of the structural identity of these systems. It must be admitted, of course, that the falsifiability criterion of demarcation carries an important philosophical insight to be considered presently. But it does not imply or commit one to a Popper-type falsificationist methodology of theory-appraisal/evaluation – a methodology which repeats the same oversimplification. Popper's mistake here is then two-fold, viz., (a) to think that the falsifiability criterion does imply or lead to the falsificationist methodology of theory-appraisal in science; and (b) to think that while it is *impossible* for us to *know* at any time whether a given theory is *true*, it is *possible* to know, on the basis of the rules of the falsificationist methodology, whether a given theory is *false*. All that the falsificationist methodology does is to evaluate a theory according to the seemingly simple logical law of *modus tollens* and establish it as falsified, if possible. Such a methodology, obviously, shares with the falsifiability principle of demarcation the presupposition that the epistemologically relevant structure of science is identical with a structure of theoretical statements that are capable of being true or false.

Now, the most serious difficulty which arises at this point has to do with the fact that it is open to doubt and controversy whether it is at all possible for us to *know* – of course, on the basis of refuting instances – that a given theory stands refuted. Very powerful arguments against the falsificationist methodology are traceable to Duhemian holistic thesis which says: it is not the individual scientific theory/hypothesis in isolation which is confirmable/disconfirmable but a whole 'group' of hypotheses.[69] What Duhem's own argument for this thesis does, in effect, is to change the unit of scientific/epistemic appraisal from an individual theory/hypothesis to a whole *theoretical system*.[70] But does Duhemian holism overcome the Popperian presupposition – the statement-view of the epistemic structure of science? The answer is in the negative. Instead, it assumes the same traditional conception of this structure. Howsoever complex a given theoretical system may turn out to be, it is still a system of statements that are capable of being true/false, confirmable/disconfirmable. We can reject even Duhemian holism as committing essentially the same kind of oversimplification as underlies the falsificationist methodology of Popper.

The pattern of the growth of knowledge that one can think of under the statement-view of the structure of science is the simple pattern of one set of such statements being constantly replaced by or reduced to the succeeding sets that possess comparatively greater empirical content and 'verisimilitude' in Popper's sense.[71] That is, as conceived by Popper, it is essentially a pattern of 'conjectures and refutations' – a pattern of science being in flux. It is clear that neither in the structure nor in the growth of scientific knowledge conceived in this manner, do *theoretical problems* of science play any significant part. But certainly there must be something seriously wrong with a theory of the structure and growth of scientific knowledge if it is *indifferent* to the objective character of a science both as a problem-posing enterprise and as a problem-solving enterprise – i.e., as an *NFCPS* system. Thus, of any view of science it can be said that the greater its indifference in this sense the more mistaken and oversimplified it is likely to be. In most cases of structural growth of knowledge in the sciences, it can be argued, *one and the same problem* may pass from one particular stage of problem-determination to another. But how do we account for this? Neither Kuhn nor Popper takes this into account. And yet this is one reason why a scientific revolution is a revolution at all. Are we to conclude, then, that the Popperian falsificationist methodology is incapable of accounting for methodological variance (by *CSD*-variance) which is the very essence of scientific revolutions?

6.5.4. *Methodological Aspects of Science As an Enterprise Comprising NFCPS Systems*

Every science must sooner or later explicitly ask:

(a) What *kind* of theory, with what sort of structure and interpretation, should it aim at?
(b) Given the kind of theory that it should aim at, what kind of methodology of epistemic appraisal should it adopt?

The question (a) raises a fundamental methodological$_1$ problem of theory-construction (and problem-determination) which can be solved by means of the appropriate kind of subject-specific rules of *method* and *strategy* which suit the given science. But the question (b) raises a different kind of problem relating to the best possible methodology of epistemic appraisal in the sciences. It can be solved by specifying the methodological$_2$ rules of epistemic appraisal which may stay invariant across the sciences. Accordingly, recalling our discussion in the §§ 6.3. and 6.4. above, we may distinguish methodological$_1$ rules of theory-construction and problem-determination in a science from its methodological$_2$ rules of theory-appraisal which may be common between it and other sciences. In other words, we may speak of the former as the methodological$_1$ rules of problem-solving and problem-determination and of the latter as the methodological$_2$ rules of epistemic appraisal of the theory-problem interactive systems in the sciences. In so far as it is necessary for a science to be *constituted* and *informed* by these two kinds of rules, its structural growth is linked with those interactions which trigger *methodological variance* in it from time to time. Interactive structural growth of objective knowledge systems in the sciences is thus inconceivable without methodological variance in the sense of variance of methodological$_1$ rules just alluded to. But methodological variance in this important sense is itself possible only if there are methodologically *invariant* theoretical universals, or structure-property correlations, in the background not just to delimit it but to make its epistemic appraisal possible. Theoretical universals have to do with the structure-property correlations in the objective knowledge systems (= theory-problem interactive systems) in the sciences, which can (a) stay invariant relative to any/all methodological variance a science may undergo and (b) in effect demarcate science from non-science. Thus, we can speak of theoretical universals as constituting a common factor of scientific invariance/scientific rationality, implying, among other things, fundamental constraints on all the sciences. At no time is it possible for us to claim that we have formalized all that goes to make scientific rationality in this important sense. But it is within its constraints alone, as these may be interpreted and investigated from time to time, that the individual sciences will follow permissible, though unpredictable, patterns of structural growth by methodological variance at various levels.

Now, what are the basic issues which underlie the problem of demarcation? I think that these relate to important methodological issues not only at the level of theory-construction but also at the level of theory-appraisal. The problem may accordingly be construed as one of determining, or identifying, those invariant structure-property correlations which alone can distinguish the objective knowledge systems in the sciences from those which are not such systems. Thus the whole problem arises as follows:

> What distinguishes the objective knowledge systems in the sciences from those which may not be such systems and yet which may find their way into science or into its very language?

This certainly calls for another look at Popper's falsifiability criterion of demarcation, seeking to distinguish the theoretical statements in a science from those which may not be scientific. I believe that its real philosophical significance can be brought out more clearly if seen in the light of epistemic structuralism, as I like to call it. Epistemic structuralism will be interpreted and discussed in Chapters 7 and 8 as a fundamental principle of structure-property correlations in the objective knowledge systems in the sciences. As an interesting variation on methodological structuralism, epistemic structuralism develops from the original insight (Pandit 1983) that the epistemic structures in a science, such as its *theories* and *problems*, are at the same time *developmental structures* of knowledge in that science. In order to make a clear sense of the fundamental concepts which are involved here, it is necessary to ask the following questions:

(a) What do 'epistemic structure' and 'developmental structure' signify in the present context?
(b) How can we test whether a given candidate is truly a developmental structure of objective knowledge in a science? Or, whether it is only a pseudo-developmental structure of objective knowledge?
(c) Whether the structural identity of the objective knowledge systems in the sciences is determined simply by the deproblematized epistemic structures of timeless *propositions* that are true? Or, whether it is more complex and type-distinct epistemic structures, *reproblematized*, which codetermine their structural identity?
(d) What principles of scientific change and interactive growth are at work in the very construction and development of the epistemic structural codeterminants of the objective knowledge systems in the sciences?

Once these questions (a)–(d) are answered, it will become clear that to recognize the insight underlying the falsifiability criterion of demarcation, it is not necessarily to commit ourselves to an oversimplification of the sort which Popper's falsificationist methodology of scientific change (= theory change) commits. On the contrary, it will be possible to incorporate this insight in deeper principles relating to theoretical universals, without any such commitment. If we want to find out about any member of the relevant classes of theoretical statements in a science whether or not it is a *developmental structure* of its objective knowledge systems, the best way to do so would be to subject it to a requirement as stringent and significant as Popper's falsifiability criterion.

From an objective epistemological point of view, a science is an enterprise comprising theory-problem interactive systems, as its objective knowledge systems, with theories and problems as their chief epistemic structural codeterminants. It is to its antecedently given methodological rules that it owes its *interactive* organization and growth. Two kinds of these rules were distinguished above: (1) the methodological$_1$ rules of theory-construction and problem-determination, which may also be called *the canons of scientific description* (= *CSD*) in a science; and (2) the methodological$_2$ rules of epistemic appraisal, or simply the canons of epistemic appraisal, which the sciences share in common. The procedures of physical problem-determination and theory-construction in classical Newtonian mechanics provide a good example of (1). Examples of (2) in the individual sciences, on the other hand, present a rich variety from the point of view of the historical development of the sciences. Thus, in this context, there have been instances of scientists in certain fields appraising their scientific generalizations with a view not to testing their explanatory or predictive power but their power for inductive systematization. This is not surprising if the very task of a science is wrongly assimilated to the taxonomic one of inductive systematization. No mature science will, however, postpone for long those explanatory tasks which await it in its field. Consider, for example, the case of linguistics. It had to come a long way before it could shed its (Bloomfieldian) taxonomic commitments and become a truly explanatory science. Very rich in its assumptions, now it uses its own procedures of theory-construction and problem-determination and its own rules of appraisal. Of course, it took the continuing work in and on the mentalistic linguistic theory of Noam Chomsky and others to achieve even this. However, all mature sciences must agree on a common methodological framework of epistemic appraisal for their respective theory-problem interactive systems. This is necessary, if scientific rationality on the one hand and the principles of variance and invariance in the sciences on the other are to make sense at all.

The methodological rules$_1$ of the first kind, the *CSD* in *a science*, determine not only *its* search-cum-discovery procedures but its very conceptual framework of problem-determination and theory-construction at a given time. But it is the methodological$_2$ rules of the second kind, the canons of epistemic appraisal, on which a *controlled growth* of this framework depends. The growth of the objective knowledge systems in the sciences is thus directly linked with methodological variance in the sense of *CSD*-variance. Without the latter the former is just inconceivable. But what kind of a causal dynamics does this, then, entail? What is it which necessitates methodological variance? And what is it which makes it possible? For an answer, we may turn to that aspect of scientific rationality which we can call *scientific objectivity*. This refers to scientific crises-situations of the type in which a science finds itself entangled in its own *CSD*. Scientific objectivity has, then, two aspects to it. One, which refers to a science in its crisis-state of rule-entanglement. Second, which refers to a science in its states of rule-governed creativity. In this second aspect, it is the *CSD* of a science which makes theory-problem interactions possible; and its rules of appraisal which regulate these interactions. Thus, it is a characteristic feature of scientific objectivity that a science is bound sooner or later to enter in states of rule-entanglement. These are the kind of crises-states which are causally relevant to rule-enriching creativity in the sciences. This is so precisely because these are the kind of states in which the theory-problem interactive systems in a science are just unable to cope with the complexity of the natural phenomena and the laws that govern them.

How are we to distinguish the crises-states of rule-entanglement in the sciences from those simple problem-situations in which a scientific theory may be faced with a logical difficulty either because of a contradiction in its basic assumptions or because of a conflict between it and empirical evidence resulting in its refutation? The same question can be asked about those crises-states of normal science in which, as Kuhn tells us, it is the puzzle-solving skill of the normal scientists themselves which is at test. It is not here possible for me to attempt a detailed answer to these questions. Obviously, there appear to be, then, three kinds of crises-states in a science. The one which I am concerned with relates to the theory-problem interactive systems in the sciences and their entire framework of methodological$_1$ rules (= its *CSD*). It is these which are in a crisis-state of rule-entanglement when they are quite erroneously extended to cope with *unanticipated* and *defying* complexities of natural phenomena and the laws which govern them. Every science is, of course, bound sooner or later to land itself in such highly complex states of rule-entanglement. Invariably, every such state is a result of an extended *rule-government*, in a newer problem area, in terms of the antecedently given *CSD* in a science. Unable to cope with newer and more complex problems, which are awaiting a clear and precise formulation, all that a science in a crisis-state of rule-entanglement can engage itself in are oversimplifications of its problems.[72] No *ad hoc* patch-work repair of its theory-problem interactive systems can be an adequate response to such crisis-states of rule-entanglement. On the contrary, when such states occur and come to light, they can trigger states of rule-enriching creativity so that a science may overcome its state of rule-entanglement. It should always be possible for the scientist, and the philosopher of science, to *monitor*, by setting up bold experiments, whether a science in a state of *extended rule-government* is really in a state of rule-entanglement. But actual monitoring of such states, and their correlations, need not wait for novel experimentation to take place. This is borne out amply by the example of Einstein's STR, by the very manner in which it made it possible for physicists and methodologists to overcome the crisis in physics in the last decades of 19th century[73] (for more details, turn to Chapter 9 below). When, using an alternative *CSD* quite commensurable with the one which may be superseded, a science actually overcomes its crises-states of rule-entanglement, we can say that a scientific revolution has taken place. Thus we may define a scientific revolution as *a state of scientific change which is a function of a science working its way backwards from its crisis-state of entanglement in its antecedently determined methodological$_1$ rules* (= its *CSD*) in order to lay down its *rules afresh by a process of rule-enriching creativity*.

Thus, a science can pass through an unending process of scientific change. The states of scientific change themselves range from states of rule-governed creativity, extended rule-government and unintended oversimplification to those of rule-entanglement, rule-enriching creativity and scientific revolution. In this context, states of knowledge change may be regarded as special states of scientific change, which call for a detailed scientific investigation. Knowledge change here refers to any change in the epistemic structural codeterminants of objective knowledge systems in the sciences. Thus, theory-change, and its correlated problem-change, in a science would be an example of knowledge change. Whole theory-problem interactive systems can change because of changes in the antecedently assumed methodological$_1$ rules of theory-construction and problem-determination in a science. Dynamics of knowledge change also involves interactions[74] between theories and problems as the developmental structures of objective knowledge systems in the sciences. Thus, we must carefully avoid all oversimplifications of the growth of scientific knowledge based on the traditional statement-view of the structural identity of these systems. This applies to the falsificationist methodology of scientific change as much as to other methodologies, since they are based on similar oversimplifications. Moreover, a re-interpretation of the falsifiability criterion of demarcation along the following lines is now in order. The relevant types of epistemic structures in the sciences are the developmental structures of their objective knowledge systems.[75] Relevant types of theoretical statements in the sciences can be tested to this effect by subjecting them to the demands of the falsifiability criterion as formulated by Popper. But, then, this is not the end of the matter. A different kind of test will be called for, if we want to test scientific problems as developmental structures of objective knowledge.

The whole idea of science as an enterprise comprising *NFCPS* systems, and of states of scientific change as developmental states of a science taking it in various directions, is conveyed more precisely, and in alternative ways, by the block diagram of Figure 6.1. and the tree diagram of Figure 6.2. Each of these may be taken as a formalizing model of science and scientific change. Each of them shows how theoretical universals, or structure-property correlations in the objective knowledge systems in the sciences, make epistemic appraisal at various levels possible. Epistemic appraisal, as shown here, is either appraisal of theory-problem interactive systems themselves or of methodological variance in the sense of *CSD*-variance in a science. Epistemic appraisal at all the relevant levels would be impossible without the theoretical universals. Theoretical universals as the methodologically invariant principles of epistemic appraisal in this sense were first proposed by me in my (1983).[76]

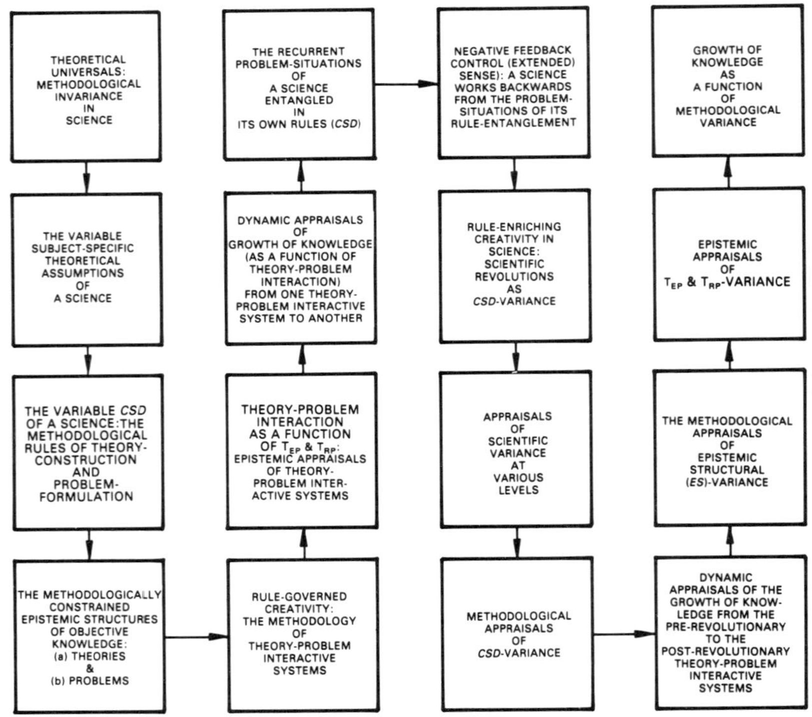

Fig. 6.1. A formalizing model of empirical science as an enterprise comprising *NFCPS* systems in the sense of Pandit (1983). Among its complexities captured in this model are its character not just as a problem-solving system but as a problem-posing system; the variable network of *CSDs* which anchor it realistically in its movement from problem-situation to problem-situation; and its methodologically invariant theoretical universals.

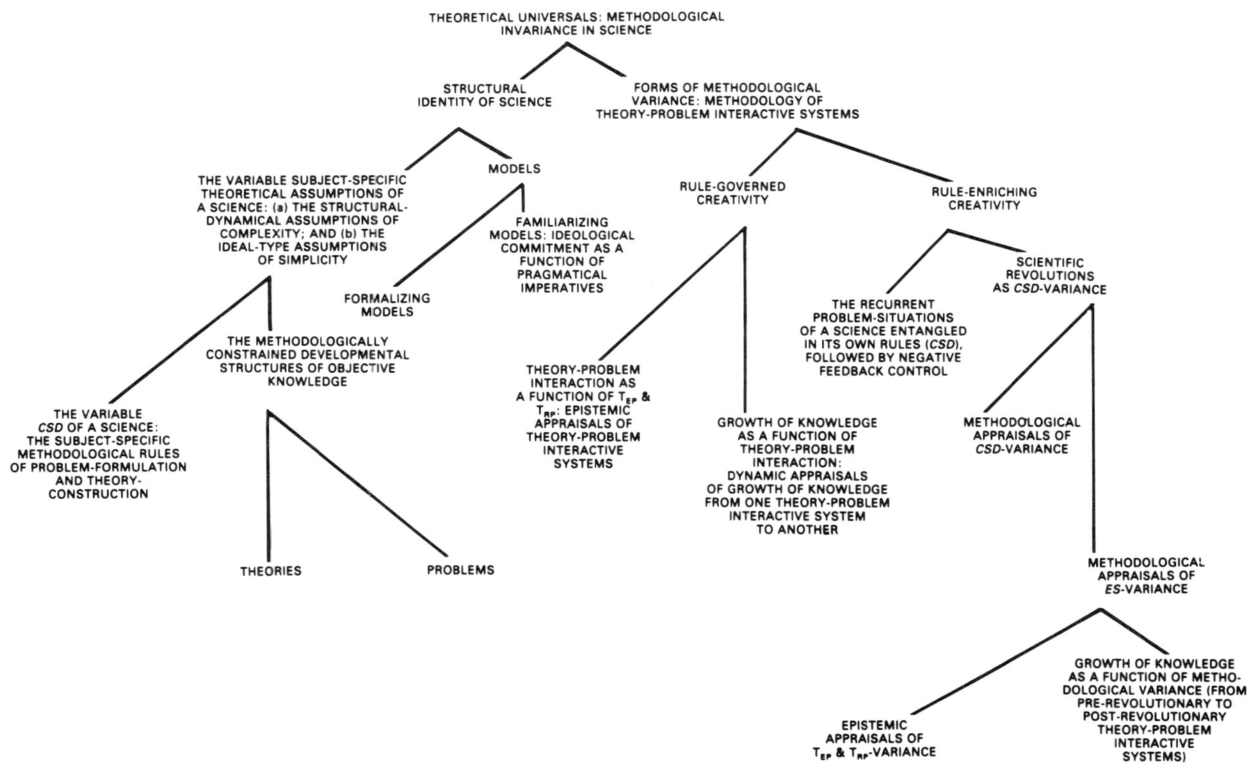

Fig. 6.2. A formalizing model of empirical science as an enterprise comprising *NFCPS* systems in the sense of Pandit (1983).

CHAPTER 7

FALSIFIABILITY AND METHODOLOGICAL INVARIANCE IN SCIENCE

> Nicht der Besitz der Wahrheit, sondern das erfolgreiche Ringen um sie macht das Glück des Forschers aus; denn alles Verweilen ermüdet und erschlafft auf die Dauer. Ein starkes, gesundes Leben gedeiht nur durch Arbeit und Fortschritt. Vom Relativen zum Absoluten.
>
> Gotthold Ephraim Lessing
> (as quoted by Max Planck 1949: p. 182)

7.1. THE PRINCIPLE OF FALSIFIABILITY

According to Popper, "*it must be possible for an empirical scientific system to be refuted by experience*".[1] Thus, "not the *verifiability* but the *falsifiability* of a system is to be taken as a criterion of demarcation".[2] There can be no doubt as to the genuineness of the problem of demarcation. But in order to understand the full philosophical significance of a methodological requirement as sweeping as this, we must ask for its possible philosophical justification as follows:

> What kind of assumptions/presuppositions underlie Popper's methodological thesis that it is not the possibility of its being verified as *true* but the possibility of its being falsified which makes a theory scientific?

I think that the whole thesis rests on the presupposition that at no point in the growth of our knowledge is it possible for us to know/verify that a theory T is true. But refutation of T is a permanent possibility. Thus, the presupposition in question is not, in its turn, a methodological requirement but an objectivistic epistemological thesis[3] concerning T. The two theses – one of them being methodological in nature and the other epistemological – should not be confused with each other, even if they may sound similar. The chief reason why I say that the one rests on the other is simply this. If we could not reasonably *say* of T what the epistemological thesis says *in effect*, the methodological rule of falsifiability would have no extra-methodological source of plausibility left whatever; it would be rendered arbitrary from an objectivistic epistemological point of view.[4] The epistemological thesis does not say that T can never be *true* or that it is logically impossible for it to be true. It would be absurd to imply any such thing, for it would amount to assimilating the synthetic statements of scientific theories to contradictions. The epistemological thesis says only that at no point or time in the growth of our objective scientific knowledge can we *know* (in the subjective sense of the term) that T is true, or claim to have proved that T is true. In other words, what it rules out as logically impossible is any knowledge/verification to the effect that T is true.

But how far is it philosophically plausible for one to maintain the epistemological thesis itself? Why is it that at no time/point in the growth of our objective knowledge can we claim to *know* that T – T being any one of our theories – is true/finally verified? In other words, what sense is one to make of the argument for falsifiability as a criterion of demarcation, which proceeds from considerations of the logical possibility of falsification on the one hand and those of the logical impossibility of verification on the other? This question assumes greater significance, given that the epistemological thesis rules out as logically impossible precisely what many philosophers of science – notably the logical empiricists – once upon a time prescribed as a criterion of demarcation between the *meaningful* scientific statements and those which are only pseudo-statements. No doubt, logical empiricism underwent recurring and significant changes at the hands of Rudolf Carnap, Herbert Feigl, Carl G. Hempel and others. But it continues to evoke memories of their past insistence on the verifiability criterion of meaningfulness. On the other hand, however, Popper himself appears to have anticipated similar questions. This is indicated by the fact that mainly two kinds of answer are discernible in his (1968), one of them involving a simplistic appeal to logic while the other appeals to methodological rules as *conventions*. Both of these arguments call for an examination as follows.

7.1.1. *The Argument from Logical Form*

While arguing for the falsifiability criterion of demarcation, Popper says: "My proposal is based upon an *asymmetry* between verifiability and falsifiability; an asymmetry which results from the logical form of universal statements".[5] Thus, it is simply a matter of the logical form of the *universal* statements of scientific theories/hypotheses that one can, with the help of the *modus tollendo tollens* of classical logic, deductively argue from the truth/falsity of *singular* statements (of observation) *not* to the truth but to the falsity of the former.[6] Such *modus tollens* arguments have the following structure: $[(t \rightarrow p) \cdot \bar{p}] \rightarrow \bar{t}$, or in words: "if p is derivable from t, and if p is false, then t also is false", where t is " . . . the system of statements which may consist of theories and initial conditions" and p is a basic statement in Popper's sense.[7]

The main steps in Popper's argument for the falsifiability criterion can be stated more clearly as follows. The falsifiability criterion as a methodological rule of demarcation is warranted by the epistemological consideration of what turns out to be an asymmetry between verifiability and falsifiability. And the epistemological consideration of such an asymmetry is itself warranted by deeper considerations of logical form of the universal statements of scientific theories/hypotheses. These statements are never *derivable* from any number of statements of singular form, but can be contradicted by singular statements whose negations are derivable from them.[8] It is on top of all this that Popper raises his controversial falsificationist methodology of scientific appraisal, i.e., the methodology of subjecting scientific theories/statements to severe tests with a view to refuting/falsifying them.[9] Among other things, such a methodology must appraise the severity of tests to which a theory has been subjected.[10]

Now, can we say that the application of *modus tollendo tollens* to empirical science – without which the epistemological asymmetry between verifiability and falsifiability would just be inconceivable – is made possible only by the *logical form* of the universal statements of scientific theories/hypotheses? An answer in the affirmative is implied by what Popper says explicitly in his argument quoted above. But what about those aspects of the exemplifications, if any, of the *modus tollens* in empirical science which are made possible by the singular statements of observation in their unique capacity either as the possible *initial conditions* or as the *potential falsifiers*? It is these statements which Popper designates as 'basic statements'.[11]

From the point of view of both the problems of demarcation and scientific appraisal of theories the class of basic statements is of unquestionable importance. Thus, on the one hand, they are supposed to play the most fundamental methodological role of *potential falsifiers* in deciding whether a theory/statement is to be called falsifiable, or scientific.[12] Of course, it is not necessary for one to accept/assume them as true/false while they perform the role of potential falsifiers. On the other hand, according to Popper's falsificationist methodology of scientific appraisal, it is always necessary that a set (proper sub-class) of basic statements is *accepted* as true/false in order to use it in the falsification of a theory which together with a set of (accepted) basic statements (of initial conditions) is under test. The crucial questions which arise here are: (1) How is one to decide the scientific status/falsifiability of the basic statements themselves in their methodologically important capacity as potential falsifiers/corroborators, unless they are to be accorded a *privileged* epistemic position in a somewhat logical positivist fashion? And, (2), in what significant sense can one speak of the acceptance of a basic statement as true/false so as to use it in the falsification/severe testing of a scientific theory? As regards question (1), Popper boldly admits that basic statements are no less subject to the falsifiability criterion than scientific theories and, consequently, embraces a doctrine of testability/falsifiability *ad infinitum*.[13] As regards the question (2), Popper has, I think, a mixed strategy to offer while he tries hard to avoid the logical empiricist-type trap of the basic (observation) statements as an *ultimate* test-basis and *foundation* of empirical knowledge. On the one hand, then, as a methodological strategy, we must *accept* (as true/false) those basic statements which *we decide to accept* if the deductive testing of our theories is to lead us anywhere.[14] That is, if the deductive testing of theories is to be possible at all and if it is to lead us anywhere, we must invariably resort to the methodological strategy of so deciding/agreeing on certain relevant basic statements as to accept them as true/false, if only for the time being. On the other hand, however, what must not be overlooked is the important *pragmatic* consideration "that the decision to accept a basic statement, and to be satisfied with it, is causally connected with our experiences – especially with our perceptual experiences".[15] In other words, what we must not overlook as a possible strategy is a (projected) *pragmatic theory* of basic statements which one could articulate along the lines and in the spirit of a Feyerabendian enterprise.[16]

By now it should be clear that not one but several assumptions enter into the Popperian methodological doctrine of the falsifiability of theories or even into what he regards as an asymmetry between falsifiability and verifiability. The falsifiability of theories is inconceivable without the class of possible basic statements; for it is the test of a theory's falsifiability that it should unambiguously divide this class into the two non-empty sub-classes of potential falsifiers and potential non-falsifiers.[17] Falsifiability of theories, whether in the sense of a methodological requirement or in the sense of Popper's thesis of asymmetry, must thus be premised on a number of factors and not just on the logical form of the universal statements of scientific hypotheses alone. Thus, among other things, it must be premised on the possibility that there is a class of statements of singular form such that every member of this class is (a) synthetic; (b) *capable* of being accepted, even if only as a matter of convention, either as true or as false; and (c) falsifiable in Popper's sense. The underlying assumption here is that there is a class of statements of singular form which are, in ultimate analysis, to be taken as falsifiable hypotheses but which remain essentially untouched by the asymmetry between falsifiability and verifiability. Thus, Popper's own position reduces to one which may be summed up by saying that if any class of synthetic statements (of universal form) is characterized by an asymmetry between falsifiability and verifiability then it is necessary that there is another class of synthetic statements (of singular form) which is not affected/characterized by such an asymmetry. Thus, it is not open to Popper to assert the thesis of an asymmetry between verifiability and falsifiability without presupposing a class of falsifiable statements (of singular form) which remains unaffected by this thesis, i.e., a class of basic statements with respect to whose members it makes sense to say that their truth/falsity is decidable/knowable by us in some appropriate sense of the term.

Our problem was to make sense of the Popperian assumption that the asymmetry between verifiability and falsifiability arises out of the logical form of the universal statements of (scientific) theories. The assumption as stated reduces to the claim that the falsifiability of theories in science is the sole function of their logical form. On such an assumption, the very possibility of exemplification of *modus tollendo tollens* in science would have to be attributed to the same logical form. On the contrary, however, our considerations above conclusively show that the falsifiability of theories presupposes not only their logical form but, among other things, also a class of falsifiable basic statements which are capable of various kinds of determinations at various levels. Thus the alleged asymmetry between verifiability and falsifiability and the very possibility of exemplification of *modus tollens* in science rest on the presupposition that there is a class of basic statements, each member of which has these characteristics, among others: (a) it is unaffected by the asymmetry between falsifiability and verifiability in Popper's sense; (b) it is falsifiable in Popper's sense and (c) it can be accepted as true/false and used in the *falsification* of a theory. It also turns out that no epistemological significance is to be attached to the assumption (c). For this assumption is seen to assume more and more the significance of a mixed strategy involving both methodological and pragmatical considerations. It remains to be seen how far (a), (b) and (c) are consistent with one another.

In the light of the considerations such as these, I think that Popper's argument from logical form loses the force of a purely logical argument. On close examination, it takes on a more complex form than one would have expected on the basis of the argument as formulated by Popper.[18] However, far from strengthening it, its complexity brings in an element of circularity in that the falsifiability of one particular class of statements (of scientific theories) is sought to be defined in terms of the concept of a falsifiable class of basic statements. What is in this context a more serious consequence is this. Popper's argument for the falsifiability criterion of demarcation proceeds from the assumption of an *asymmetry* between falsifiability and verifiability of theories, which is sought to be grounded in their logical form, but from which the equally falsifiable class of basic statements is essentially exempted. Above all, I think that Popper's claim that the epistemological asymmetry between verifiability and falsifiability "results from the logical form of universal statements"[19] must be rejected as being an oversimplification. And, in so far as Popper's falsificationist methodology of scientific appraisal presupposes his argument from logical form, the latter is open to all the consequences of the holistic onslaughts of the Duhemian argument,[20] which tends to blur, if not completely undermine, the alleged epistemological asymmetry between verifiability and falsifiability.

7.1.2. *The Argument from Methodological Conventionalism*

While viewing methodological rules as 'conventions', or as "the rules of the game of empirical science"[21], Popper simply lays it down as a methodological rule that "The game of science is, in principle, without end. He who decides one day that scientific statements do not call for any further test, and that they can be regarded as finally verified, retires from the game."[22] Construed as an argument from methodological conventionalism, this makes it, I think, a matter of methodological decision that at no time/point in the growth of scientific knowledge we can establish any of our theories as true. In other words, Popper seems to be arguing here for the epistemological asymmetry between verifiability and falsifiability by invoking a methodological rule concerning the 'game' of science. Thus, it is relevant to ask: Is there anything in Popper's view of the methodological rules as *conventions*, and particularly in his argument above, which could prevent an inductivist from proposing/sticking to the inductivist methodological rules for the 'game' of science? Or, more generally, are alternative methodological rules, which may be quite incompatible with the one invoked by Popper, unthinkable?

A closer look at Popper's argument would once again reveal that it is not as simple as its formulation would suggest. What Popper actually says in the present context implies at least more than one answer to the question above. Thus, in one of its aspects, the underlying pattern of Popper's argument is as follows. As a methodological rule, theories in order to be ranked as scientific must be *falsifiable*. The rationale for this methodological rule is the further methodological rule that the inductivist methodological rules must be avoided in playing the 'game' of science – particularly the inductivist requirement of the *finally verified theories in science*. Against such an argument it must be pointed out, however, that the type of considerations that are invoked for the falsifiability criterion cannot be invoked as a rationale for it. One does not propose and justify one's methodological rules by simply advising against/prohibiting the use of possible alternatives to them. Popper's methodological argument clearly seeks to prohibit the use of the inductivist methodological rules. He warns that one who *decides* to adopt such rules simply *retires* from the 'game' of science. Now, one must ask: Is it at all open to one to adopt such rules or those incompatible with the Popperian rules under the (Popperian) condition that the very *decision* to do so implies retirement from the 'game' of science? The answer is clearly in the negative. For the *methodological intolerance* this implies is in contradiction with Popper's methodological conventionalism which says that methodological rules are just conventions and a matter of decision.

It is here that the tension in Popper's methodology comes to surface. There is, on the one hand, his view, explicitly held by him, that methodological rules – in his case a whole hierarchy of them with the falsifiability criterion being at the apex – are to be treated as conventions and, therefore, as matters of decision. But, on the other hand, there is the view, *quite implicit in what he says*, that there are in science, or in scientific theory, certain inviolable universal features which lie beyond the limits of methodological variance due to the methodological decisions which the scientists may take from time to time. The former clearly leaves room for a principle of methodological tolerance with respect to alternative, possibly incompatible, methodological rules in the sciences. The latter implies not just a serious delimitation but a denial of methodological tolerance by admitting scientific invariance and the correlated methodological requirements which are just inviolable. This kind of tension is further aggravated by what I consider to be yet another serious consequence of Popper's somewhat stronger version of his methodological conventionalism. For example, consider what is being implied when Popper says: "What is to be called a 'science' and who is to be called a 'scientist' must always remain a matter of convention or decision".[23] Quite consistently with what Popper has said elsewhere, his declaration should leave no one in doubt as to its profound, though unintended, anarchist implications for the methodology of science.[24] Thus, it implies that, whatever criterion of demarcation we may propose, acceptance or rejection of the criterion is a matter of convention or decision. Should the whole scientific communities permanently divide into different and rigidly organized groups,[25] simultaneous emergence of and commitment to different, and even incompatible, criteria of demarcation will stabilize as a permanent possibility. It will no longer be, then, necessary for scientists, or for philosophers of science, to be equipped with a single, commonly agreed upon set of methodological rules at a given time. On the contrary, it should be easy to have any proposed criterion of demarcation knocked out. With the 'game$_1$' of science left to the mercy of the methodological decision makers for its very structuration and growth, the 'game$_2$' of methodological decision-making is bound to take on an autonomous though anarchist character. Outside this 'game$_2$' there is nothing for a criterion of demarcation to be grounded in. Ah! What an eternal 'game$_2$' of demarcation! Games, in order to be called 'games', must guarantee entertainment till eternity! And, in all probability, only anarchism is capable of doing that in the present context!

How are we now to resolve this tension, in Popper's methodological conventionalism, between a principle of methodological tolerance – which in its extreme form tends toward methodological anarchism – and that of methodological intolerance? The resolution of the tension is necessary if the falsifiability criterion is to be salvaged as a criterion – and not as *the* criterion – of demarcation with its philosophical significance not just a matter of methodological conventions/decisions. I think that a resolution in this direction is possible, provided we admit *scientific invariance* in the form of the universal features of scientific theories and problems – features which stay invariant relative to whatever variance the sciences may undergo from time to time. Of scientific invariance alone can it be significantly said that it is not just a matter of methodological conventions/decisions. I think that one of these universal features has to do with the very logical form of the universal statements of scientific theories. It is only on the assumption of scientific invariance in the sense to be considered below that one can rule out a certain methodological rule – such as the inductivist rules – as implying retirement from the 'game' of science (in Popper's sense). Let us, then, consider the very question of these universal features – which we may call the problem of *theoretical universals* – with a view to having another look at the falsifiability criterion of demarcation along with the epistemological thesis of an asymmetry between verifiability and falsifiability.

7.2. THEORETICAL UNIVERSALS: METHODOLOGICAL INVARIANCE IN SCIENCE

To return now to our original question above concerning a scientific theory T: What kind of philosophical justification, or rationale, can be offered for the epistemological thesis concerning T that there is an asymmetry between its verifiability and falsifiability? The question assumes considerable philosophical significance in the light of our conclusion above (§ 7.1.1.), against Popper's own argument, that the supposed asymmetry cannot be regarded simply a function of the *logical form* of the universal statements which may constitute T. It cannot be denied that the epistemological thesis must be understood in a sufficiently strong sense if it is to serve as a non-trivial presupposition of the methodological rule of falsifiability. Thus, it must be understood as asserting that (a) *it is logically impossible either to know or verify that T is true, even if it is really true, but (b) logically possible to know, or verify, that T is false*. Our original question can accordingly be restated as asking: What is the source of the plausibility of the epistemological theses (a) and (b)? What is this asymmetry itself a function of? We have already rejected Popper's answer to this question as an oversimplification. For a more adequate answer we cannot turn to traditional subjectivistic epistemology to trace it back to the knowing subject, its constitution, its powers of knowing or its imperfections. Nor can we turn to our *infinite capacity* to lay down rules including the Popperian rule not to play the 'game' of science by means of the inductivist type rules requiring verification instead of refutation. Thus, the logical impossibility of verification, in case T is true, cannot be assimilated, without absurdity and risk of a serious charge of psychologism,[26] to an impossibility of a psychological order. The same is true of the logical possibility of falsification, in case T is false. On the contrary, I think that for an adequate answer consistent with an objectivistic epistemology, we must turn to the sciences themselves, to their structural growth. This is necessary at least for two reasons. First, not only is the impossibility of verification of T of a logical order, the methodological rule of falsifiability of T cannot fall back on traditional subjectivistic epistemology. Secondly, this methodological rule cannot rest on the Popperian oversimplification making the epistemological asymmetry between the verifiability and falsifiability of T a function of its logical form.

But how are we, then, to understand from an objectivistic point of view the epistemological thesis concerning this asymmetry? The asymmetry is, I suggest, a function of 'theoretical universals' as those features of scientific *theories* and *problems* which stay *invariant* under all variance the sciences may undergo in the course of their historical development. As the invariant features of a science which undergoes scientific variance at different levels, theoretical universals can be tentatively identified as follows:

(a) All objective knowledge systems in a science, undergoing scientific change, show *structural invariances* giving them their essential structural identity as *theory-problem interactive systems*, $I(T_1, P_1)$, $I(P_2, T_2)$, $I(T_3, P_3)$, . . . and making their scientific study possible.
(b) Scientific theories and problems are the basic (type-distinct) *epistemic structural invariants* of objective knowledge systems, codetermining their structural identity.
(c) Scientific theories have the *logical form* of unrestricted universal statements; and scientific problems the logical form of either *What*-questions or *Whether*-questions or *Why*-questions.

(d) All scientific theories are essentially *simple* theories and all scientific problems essentially simple problems about different aspects of the world of an unknown ontic complexity.
(e) Given their structural identity as theory-problem interactive systems, objective knowledge systems in a science are not only characterized by their structural invariances whatever but by their controlled growth in which *theories* and *problems* appear as the *developmental structures* of knowledge.
(f) The growth of objective knowledge systems in a science is unpredictable but *rational*. The principles of variance and invariance which characterize this growth, particularly scientific revolutions, can be identified and formalized.

I believe that any formalization of objective knowledge systems, their structure and growth, would have to reflect as its basic (defining) properties what the principles (a)–(f) seek to capture. First of all, the property of being an *epistemic structural invariant* itself carries considerable epistemological ontological significance. Thus, by virtue of being epistemic structural invariants of objective knowledge systems in science, theories and problems codetermine the structural identity of such systems in a science. The principle which holds them as type-distinct epistemic structural invariants of objective knowledge systems in this sense may be called *epistemic structuralism*, subject to a widening of its scope later on. With this basic identification in the background, it is possible to move ahead in the task of identification of theoretical universals which form the subject-matter of objectivistic epistemology and the methodology of science. Each such identification should, of course, correspond in relevant ways to our best intuitions about objective knowledge systems here *reproblematized* as theory-problem interactive systems. Moreover, it should formalize our understanding of such systems so that their deproblematization by traditional epistemology is no longer taken for granted. As formalizations, such identifications should show how theories and problems, and the corresponding subject-specific methodological rules$_1$ of theory-construction and problem-formulation, come under fundamental methodological constraints, or search-cum-discovery procedures, which are always at work in a science.

Epistemic structuralism brings about a reproblematization of the objective knowledge systems in a science. It implies a clear structural identity for these systems in terms of certain types of epistemic structural invariances. It would be quite reasonable to go beyond epistemic structuralism, just defined, in order to explore scientifically interesting *structure-property correlations* which may be characteristic of theory-problem interactive systems in science. The question which we must, therefore, ask is this: Given the structural identity of the objective knowledge systems in a science as theory-problem interactive systems, $I(T_1, P_1)$, (P_2, T_2), $I(T_3, P_3)$, . . . what further properties could we expect them to exhibit? As a first step in answering this important question we might restate epistemic structuralism in a modified form as follows:

(g) For scientific *theories* and *problems* to play a decisive role of epistemic structural codetermination of the structural identity of the objective knowledge systems in a science, with properties causally relevant to systems growth, is for them to be the developmental structures of such systems.

It is now possible to have another look at Popper's general requirement for 'the game of science', that its statements must be falsifiable and so forth. It is now possible to argue for this requirement as a special case of epistemic structuralism. There could be no better epistemological ontological basis for a general methodological rule like that of falsifiability of a scientific theory than is provided by epistemic structuralism. The latter is itself a principle of invariance about scientific theories and problems as the developmental structures of objective knowledge systems in science. In one single sweep, Duhem too is taken care of. There cannot be, in my opinion, a better defence of some of those very ideas of Popper which are central to his formulation of objectivism and critical rationalism. The falsifiability of a theory as a mark of scientific rationality gets thus grounded, as it ought to, in scientific invariance (= theoretical universals). The characterization of scientific theories and problems as developmental structures of objective knowledge systems in a science implies a certain *symmetry*, as a methodologically invariant property, in the underlying causal dynamics these systems must undergo. The falsifiability requirement might thus be seen as that kind of requirement which enables us to distinguish theories which have this and other related properties from those which just don't have it. With a grounding in theoretical universals, as deeper as this, the falsifiability of a theory cannot be reduced to a convention or to a methodological proposal among other such proposals. There is more to it than just that, as we have just seen. We cannot do without it if we remain within the framework of objectivism proper consistent with methodological essentialism. What Popper has done, quite unsuccessfully though, I believe, is to combine objectivism in his sense with methodological conventionalism. But the two are quite incompatible with each other. (For further discussion on this point turn to Epilogue in Chapter 10 below.)

The principles (a), (b) and (e) together, or their reformulated simpler version (g), can thus be taken to formulate epistemic structuralism[27] as a fundamental principle of objectivistic epistemology, identifying *theories* and *problems* first as the epistemic structural codeterminants and then as the developmental structures of objective knowledge. Without the type-distinct epistemic structures of *theories* and *problems* playing such a crucial role in the very structural identity of objective scientific knowledge systems, we would never know what type of structure-property correlations we should look for in these systems as an ideal subject-matter of objectivistic epistemology. Thus epistemic structuralism itself generates deeper problems notably the problem of *theoretical universals*. This problem relates to the structure-property correlations in the objective knowledge systems of science on the assumption that epistemic structuralism, as formulated by (g), holds true of these systems. These systems must all of them, therefore, show methodologically invariant but dynamically relevant properties which are appropriately correlated to their structural identity as theory-problem interactive systems.

Epistemic structuralism paves the way for the remaining principles (c), (d) and (f), each of them addressing to the problem of theoretical universals. Thus the principle (c) captures the *logical forms* of theories and problems as forms of scientific invariance (= theoretical universals). In effect, it not only expresses one of the conditions of scientific appraisal (in Popper's sense) but determines an invariant feature in the functional aspects of both scientific theories and problems.[28] This naturally leads us to *theoretical simplicity*, as expressed by the principle (d) above, as a functional invariance in scientific theories and problems (a subject for detailed discussion in Chapter 8, § 8.1.1.). Every scientific theory, as also every scientific problem, is deeply involved in highly simplifying theoretical identifications a science employs in its search-cum-discovery procedures with which it investigates the entangled systems in the world. Where explanation of known laws of such systems is concerned, a theory demands antecedent theoretical identifications at appropriate levels before it can explain them. As such, every theory must employ theoretical assumptions and laws which may not only simplify but oversimplify their explananda in the relevant domain. Similarly, of every scientific problem it is true to say that it arises as a result of problem-determination either as a subject of scientific discovery or as a demand for a new theory to replace the background theory. Procedures of problem-determination must, therefore, employ the same type of simplifying, or oversimplifying, assumptions as are characteristically employed by the background theory. The principle (d) above might thus be taken to assert, or imply, that at no time in its historical development it is possible for us to single out as *scientific* a theory or a problem of which the following does not hold true: that it employs simplifying, and oversimplifying, assumptions ranging over the relevant domains in the highly entangled systems of the world.

It is only natural, then, to take the next step in the principle (e) above, which results in the identification of the *growth* of objective knowledge systems as a theoretical universal. There is thus a significant correlation between the structural invariance of *theories* and *problems*, the type-distinct epistemic structural codeterminants of the structural identity of these systems, and their *functional invariance* in their capacity as the developmental structures of objective knowledge. This implies, among other things, a constant *interaction* between theories and problems in a science. It also implies that the growth of objective knowledge systems, itself a case of structure-property correlations alluded to above, is a function of this interaction at various levels. The idea of a *holistic* interactive growth model[29] for knowledge is thus built into the principle (e) itself. On the other hand, however, a different principle of invariance at the level of the growth of knowledge is implied by the rationality principle (f) above. Being yet another case of structure-property correlations in this field, the details of this principle need to be worked out separately. It must be pointed out that it has no implication that the growth of knowledge follows a predictable course.

Far from being exhaustive of structure-property correlations in the field of objective knowledge, all the principles (a)–(g) above together imply the epistemological thesis of an asymmetry between the verifiability and falsifiability of a scientific theory T. Thus, at no point in the growth of knowledge is it possible for us to establish, or single out, T as true or as finally verified. For otherwise, T would satisfy neither epistemic structuralism nor an interactive model of the growth of objective knowledge in the specific sense under discussion here. On the contrary, if it could be singled out for final verification or verifiability, it would be an impediment in the way of objective knowledge and its growth. Thus three significant conclusions follow, among other things, from our discussion above:

(1) The Popperian arguments for the epistemological asymmetry between the verifiability and falsifiability of T – and, therefore, for the falsifiability criterion of demarcation – rest upon the oversimplifications of the kind considered above. Moreover, the very formulation of this criterion is, therefore, exposed as based on an oversimplified view of science.
(2) The epistemological thesis of such an asymmetry is better understood as expressing a principle of invariance at the level of the growth of scientific knowledge in the sense implied by the *basic* principles of invariance (a)–(g) above.
(3) What (1) and (2) together entail is a reformulation of the falsifiability criterion of demarcation such that it reflects the philosophically significant fact of its being warranted by (2) itself. Thus it should reflect the important fact that the insight underlying it has more to do with scientific invariance in its present sense than the Popperian methodological conventions pure and simple.

Tentatively speaking, the following reformulation, which is more comprehensive than the Popperian formulation, suggests itself: *The distinguishing mark of the relevant epistemic structural codeterminants of objective knowledge systems in a science is their character as developmental structures of these systems*. This reformulation has the additional advantage of ridding the criterion of any (Popperian) commitment to a statement view of the epistemic structural identity of such systems, as we shall see presently.

7.2.1. *Epistemic Structuralism and the Problem of Demarcating Science from Non-Science*

Epistemic structuralism, as formulated here and earlier in my (1983), says that the structural identity of objective knowledge systems in a science is codetermined by its *theories* and *problems*, these being the basic type-distinct *structural* invariants of all objective knowledge on the one hand and its developmental structures on the other. In so far as it is a basic principle of invariance valid for all science, demarcating it from non-science at the structural as well as developmental level, it serves as a first step to seeking further demarcations in terms of *theoretical universals*. Thus, we can raise questions of demarcation about scientific theories and problems or theory-problem interactive systems. As regards theories, we can still speak of their falsifiability as a property which distinguishes them from non-scientific theories. But this must be restated as follows: The distinguishing mark of a theory T, taken as a developmental structure of knowledge in its field, is its falsifiability. But what about the problems themselves as subjects of scientific discovery? One might simply say that they are developmental structures of knowledge in their field relative to exactly those theories which provide their realistic, empirically reliable and simple solutions. But this would not be adequate enough, particularly for those problem-situations in which no new falsifiable theory may be easily forthcoming. For this and similar reasons, I propose that we adopt as a criterion the following requirement: A problem is a developmental structure of knowledge in its field if and only if it is uniquely determined by a set of scientific theories and assumptions in that field. Thus, scientific problems as developmental structures of objective knowledge can be restricted to those the determination of which through various stages is the function of the resolving power of a scientific theory. Problems not so determined will be generally considered *ad hoc* and, therefore, as pseudo-developmental structures.

As a further condition on theories and problems considered as developmental structures of objective knowledge systems, the following holds true in terms of the general interaction theory set forth in my (1983). The *growth* of these systems, itself a function of interaction between these developmental structures (see Figure 7.1.), is a methodologically invariant property correlated with their structural identity as theory-problem interactive systems. Where such interaction and structure-property correlation are either absent or rendered impossible, there we can speak of neither the developmental structures nor the growth of objective knowledge systems.

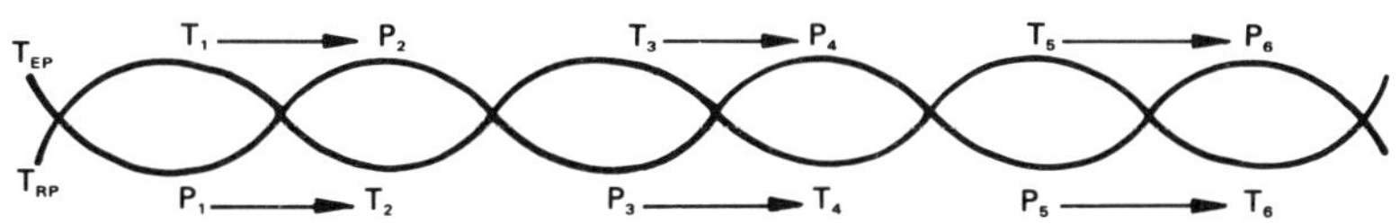

Fig. 7.1. The interactive model depicting the interactive pattern of the growth of scientific knowledge as a function of theory-problem interaction.

CHAPTER 8

THE METHODOLOGY OF THEORY-PROBLEM INTERACTIVE SYSTEMS

> Mechanics has the task of 'describing movements in nature completely and in the simplest manner possible'.
>
> G. Kirchhoff (1897: 1)

8.1. THE QUESTION OF THE NATURE OF A FALSIFIABLE THEORY

I agree with Popper that falsifiability of a theory is a mark of the theory's scientific character. But I disagree with him as regards the manner in which he formulates this principle and cannot accept the assumptions on which it is based. Particularly, there are two fundamental aspects to his formulation which must be examined:

(1) His formulation implies a certain universal restriction on the very structure/form of a scientific theory – a restriction which would according to him guarantee its falsifiability and thereby achieve the greater aim of demarcation between science and non-science.
(2) The only rationale, apart from the one in terms of its consequences, which Popper brings to bear on the falsifiability principle is offered essentially in terms of what is tantamount to a simplistic game-theoretic approach to science.

Both these aspects render Popper's basic approach highly unsatisfactory and unacceptable. But why?

While it may be clear that in order to guarantee its scientific character a theory must be falsifiable, it is not clear at all how falsifiability principle by itself can demarcate science as a *whole*[1] from non-science. In other words, and to put it in a nutshell, what this principle leaves essentially undemarcated are *scientific problems* themselves whose independent ontological status as epistemic structures cannot be denied. It is interesting to notice how even those philosophers who are currently busy arming themselves with a game-theoretic approach in the philosophy of language and science accord them a recognition to this effect.[2]

Again consider the important question which Popper never asks:[3]

Can falsifiability requirement itself be grounded in something more fundamental than a mere convention, where this 'something' may have to do with scientific invariance itself? I raise this question concerning the second aspect of the Popperian formulation, since the Popperian grounding of it is not just unsatisfactory but beset with serious difficulties. His grounding of it in a methodological convention is traceable, I think, to his methodological conventionalism, the view that what we call 'science' is a matter of convention or decision. On this view, then, *it is not even a matter of actual scientific practice, or its negative-feedback consequences whatever*. But, as we have noticed before, methodological conventionalism inevitably leads one to a position as extremely sceptical of scientific rationality as epistemological anarchism. The reason why this is so is this: It views empirical science as a 'game' among other games, and the logic of scientific discovery as *a set of rules* of the game among many such possible, even incompatible, sets. At any given time, then, it is not only possible but legitimate for a given science to have its rival theories in their field subjected to mutually incompatible sets of 'conventions' of scientific appraisal. With the consequent loss of rationality, *any* theory, even a false one, can easily gain 'scientific' respectability, come what may.

Now, it seems to me very important to ask: What, if any, is the nature of a falsifiable theory? Or, why is a falsifiable theory falsifiable? This is precisely the sort of question which the philosophers of science would not consider it worthwhile to ask. In any case, it is not permitted under the Popperian approach to science, approximating as it does to a game-theoretic approach. This is noteworthy because it is this approach itself which is responsible for bringing in the principle of falsifiability to demarcate science from non-science. Under this approach, all that one is permitted to ask concerning a (scientific) theory T, given this principle, is this: Is T falsifiable or not? This, together with an answer to it, will 'always' constitute the *first* move in applying the *rules* that one might lay down for playing the (Popperian) 'game' of science.

Now suppose, given a particular theory in any one of the sciences, we find, while making the 'first move' above, that the theory is a falsifiable one. Why can't we ask: Why is the theory falsifiable? What makes it falsifiable? That is, why can't we ask concerning any (scientific) theory found to fulfil the falsifiability requirement: Why is it falsifiable? Or, why does it satisfy such a requirement at all? In Popper, the only answer that one can discern is this: In the 'game' of science played according to the Popperian methodological rules there is, *conventionally speaking*, no move more basic than the one just considered, viz. a proposed theory T and the question concerning it: Is T falsifiable?

In my view, this whole approach to science, based as it is on methodological conventionalism, is essentially mistaken if only for the following reasons. First, as we noted earlier, methodological conventionalism itself is not an acceptable framework for a sound philosophy of science. Secondly, I believe that it is possible not only to raise but to answer a question more basic than the one permissible under the Popperian approach to the 'game' of science. The most fundamental methodological and epistemological aspects of science may, therefore, relate not to the falsifiability requirement as such but to deeper principles expressing what I have chosen to call theoretical universals.[4]

Popper links the falsifiability of a theory T to its logical form. On his view, then, falsifiability of T is a function of its logical form. Now, this is quite true, but not the whole truth. If there were nothing more to it than its logical form, the question concerning the extra-logical nature or structure of T itself would never arise as a fundamental question. The falsifiability requirement is after all a requirement which every theory, which is scientific, would satisfy. *But what is it precisely in the structure of a theory which enables it to satisfy such a requirement, when it does satisfy it*? This question, I believe, makes sense.

8.2. HOW SIMPLE IS THEORETICAL SIMPLICITY?

8.2.1. *Problems of Simplicity: Different Approaches*

It is generally recognized by both scientists and philosophers of science alike that (the parameter of) *simplicity* plays an important role not just in theory-construction but in the epistemic appraisal of theories in science. But there is not only a lack of agreement among philosophers as to how the simplicity of a theory might be defined and then employed for such purposes but considerable confusion between this concept itself and other notions that are frequently found at work in the philosophical discussions on science. Some of the philosophical accounts of simplicity even tend to make its role look more puzzling than it really is. This presently unsatisfactory state of discussion with respect to the *simplicity* of a scientific theory[5] is, I believe, traceable to the common dogmatic approach of many philosophers who take it either as the type of concept which is *reducible* to other methodological concepts (e.g. probability, falsifiability, etc.) or as of the type that relates to the *pragmatic* relations between a scientific theory and the community of its users/builders. And it is, I think, further complicated by the fact that while the instrumentalists have all along thought of *simplicity* as a criterion (Kriterium) for choice between any two empirically equally adequate (= observationally equivalent) theories/hypotheses, the realists look at the simpler of the two theories as being truer than the other. For the instrumentalist, and the conventionalist as well, the whole problem of the simplicity-ordering of scientific theories reduces to one of ordering the empirically (and hence explanatorily and predictively) equally adequate theories so that the most convenient ordering facilitating a choice between them is achieved. The underlying idea here is quite old.[6] It is, I think, the idea of equivalent descriptions, one of which is the simplest but without being *truer* than the other. One might quite plausibly think of a simpler presentation of mechanics by using vectors instead of coordinates as an exercise in descriptive simplicity in this sense.

For the *traditional* realist, on the other hand, the *degree* of simplicity of a theory, relative to its rivals, must be explicated in terms of its *likelihood to be true*. But there are several other versions of realism which do not bring in the notion of a theory's likelihood to be true so directly into its explication. As we would notice as we proceed to discuss them presently, they make the most interesting approaches from the side of realism.

It may already be somewhat clear from my preliminary remarks how our present understanding of the simplicity of theories in science has been shaped largely by our concepts or notions of inter-theory relations, theory-equivalence, empirical adequacy of a theory and so on, but all within the explanation-theoretic approach to science, namely: The function of a scientific theory is to explain natural phenomena at different levels of antecedently framed descriptions of these phenomena.

It will be one of my main assumptions in what follows that the *simplicity* of a scientific theory, in so far as it must enter in our epistemic appraisals of the explanatory and resolving powers of rival theories in a given field, has to do with the relationship between the theory and the world that it seeks to describe (or to predict). Assumptions such as this will, however, find their most natural expression in the *constraints* (1)–(4) that I shall presently propose for an explication of the simplicity of a scientific theory (henceforth theoretical simplicity or T-simplicity) to count as adequate. The T-simplicity of a theory, a legitimate parameter of its epistemic appraisal, is sometimes temptingly confused with its pragmatic simplicity (henceforth P-simplicity). To begin with, it is, therefore, necessary to make a few remarks concerning P-simplicity so as to distinguish it from the former. The P-simplicity of a theory can vary from person to person so that one and the same theory T_1, relative to another theory T_2 in the same field, is pragmatically simpler to one person but more complex to another, depending on one's degree of familiarity with the given theory and the questions one is *interested* in at a given time. Thus, a physicist may, in this sense, hold Newton's gravitation theory instead of Einstein's as pragmatically simpler. Again, in a somewhat different sense, an entire scientific community may find it less complicated, and hence pragmatically simpler, to make calculations for limited purposes in concrete situations (e.g. technological ventures like space probes) by means of the laws of not the successor theory T_2 but its predecessor T_1. Taken in this sense, then, the P-simplicity (or complexity) of a theory has to do with the applications of its laws in planning bold but relatively safer technological ventures. Thus, we can easily think of technological problem-situations of the sort in which it is simpler to make calculations and predictions about the relevant parameters not in terms of Einstein's STR or GTR but in those of Newton's laws of motion and his gravitation theory, where either theory could be used for the same purpose. In such problem-situations, Newton's theory might be described as pragmatically simpler than Einstein's. In this sense, then, the P-simplicity of a theory, relative to its rival, easily extends itself from situations of this type to those of transition from an earlier established theory to its successor, where the very advent of the newer theory is strongly resisted by the entire scientific community if only because of its own commitment to the established theory. Thus, it might be pragmatically simpler to believe in and to continue to operate with a familiar theory than a new or possibly revolutionary one.

All this entails, I think, what might be called an asymmetry between T-simplicity and P-simplicity in the following sense: Of any two (or more) rival theories, T_1 formulated at a time t $(=t_1)$ and T_2 formulated at a time t $(=t_2)$, it is always possible for T_1 to be pragmatically simpler, but theoretically more complex, than the successor theory T_2. Since science progresses in time from less simple to more simple (successor) theories, the P-simplicity of a theory cannot play any decisive role in the epistemic appraisal of scientific theories. But we don't have yet a generally acceptable definition of the T-simplicity of a scientific theory, the important role that it plays not just in theory-construction but in the epistemic appraisal of theories notwithstanding. My main aim, then, in what follows is three-fold, namely: (a) to propose a set of conditions for an explication of the T-simplicity of a scientific theory to count as adequate; (b) to briefly examine the received views on the nature of T-simplicity, in particular those that seek to reduce it to other methodologically privileged but derivative concepts; and (c) to propose, at least in clear outline, an alternative explication within the framework of the conditions to be set forth under (a).

In their search for *simple* theories, scientists have, I believe, always landed themselves in oversimplifications of one sort or other. The reason why this is so might be traced to the fact that, since both logical form and simplicity matter as important constraints on theory-construction in science, oversimplification is bound to be built into scientific theories and problems formulated, as they are, by means of universal statements of an unrestricted character and *what*-questions, *whether*-questions, and *why*-questions respectively. And if every simple theory is bound to have a certain degree of oversimplification built into it, as it were, then every simple theory will be falsifiable. Falsifiability of a theory might, therefore, be said to be not just a function of its logical form but a function of its simplicity. Greater, then, the degree of T-simplicity of a theory, greater will be its degree of falsifiability, and more easily testable it will be.

If this is so, it not only explains why a falsifiable theory is falsifiable but suggests that the T-simplicity of a theory might be essentially an irreducible property of that theory even if in a relative but important sense. And if this is so, how are we then to understand it? How irreducible and simple is the concept of T-simplicity? It might be argued that in so far as it is possible to talk of the simplicity of a hypothesis *h* as different from the simplicity of a theory T, there are considerable complexities in this concept. And it is true that some philosophers talk of hypotheses as being simple, while others talk of theories as being more or less simple. But I think that this is avoidable. It has nothing to do with the complexity, if any, in the concept of T-simplicity. The ultimate place of a hypothesis or a law is, I think, among the consequences of a comprehensive theory. Thus, in so far as a hypothesis is derivable from a theory as a whole – from its fundamental and auxiliary assumptions (= T · A) – it will, as a rule, inherit its degree of simplicity from that of the theory itself. For reasons such as this, it is better, I think, to work with a concept of T-simplicity which directly relates to a theory T_2 (=$\acute{T}$ · $\acute{A}$), given its rival T_1 (=T · A). But before embarking on the task of conceptualization of T-simplicity, one might ask a preliminary question, namely: What do we expect from T-simplicity once it is explicated? What constraints must be observed in explicating it? Is it possible to explicate T-simplicity in a manner which is neutral between the rival approaches of deductivism (Popper) and probabilism (Harold Jeffreys and others)? I think that these are important questions. But, I think, they do not anticipate any problem of justification with respect to a theory of T-simplicity. Any such theory needs no justification, except in the sense that it must prove its mettle in terms of its consequences within the larger methodological and epistemological setting for science.

As a first step, then, towards such a theory/explication, it is, I think, important that its adequacy be judged in terms of a set of conditions (or constraints) which I propose as follows:

(1) the condition of irreducibility;
(2) the condition of symmetry;
(3) the condition of dynamical relevance; and
(4) the condition of ontological complexity.

Each of these conditions will get clarified as we proceed to discuss the received views on the nature of T-simplicity. But let me begin by making a few preliminary remarks as regards the conditions (1)–(3). By the condition (1) of irreducibility I intend here to propose that the T-simplicity be understood as an irreducible concept in so far as the T-simplicity estimates themselves might be said to play an irreducible role in the epistemic appraisal of theories in science. The T-simplicity should, therefore, be viewed as a property (= *a theoretical universal*) that belongs to a scientific theory itself and has little or nothing to do with any methodological conventions/requirements that we might choose and vary from time to time as we want. Even though the T-simplicity, in this sense, refers to an inherent property of a scientific theory, it is erroneous to presume it to be some kind of a self-revealing property, as might be the case with aesthetic appreciation of simplicity and beauty in science and in nature. The condition (1), so understood, is entailed by my conjecture, proposed in my (1983) and to be discussed presently, that the concept of simplicity of a scientific theory, unlike that of falsifiability or probability, belongs to that class of concepts that offer themselves as the best candidates for the role of *theoretical universals*. An adequate explication of T-simplicity should, therefore, bring out its irreducible character in its role as a theoretical universal.

Consider now the conditions (2)–(3) of symmetry and dynamical relevance. In so far as the T-simplicity of a theory T_2, given its rival T_1, refers to one of its *inherent* properties, any approach that renders our estimates of its T-simplicity *variable*, in a manner which provides for setting aside our previous estimates of it, cannot be correct. That is, an explication of T-simplicity, in order to be adequate, should not have the consequence of rendering our T-simplicity estimates *asymmetric* with respect to the earlier and later performances of a scientific theory. It should, on the contrary, provide for their symmetry with respect to the performance of a theory at different times. By the condition (3) of dynamical relevance what I intend here to propose is this: A sound explication of T-simplicity should so enrich the epistemological and methodological frameworks of theory-appraisal and theory-choice in science as to receive adequate support from a sound theory of the structure and growth of scientific knowledge. An explication of T-simplicity will be, then, adequate in this sense if it has the following consequence: It is the simpler of any two or more rival theories in their field that can serve as a basis for the future development of scientific theory in that field. It is at this point, then, that such an explication will find its most natural justification in the support which it receives from a sound theory of the structure and growth of scientific knowledge.

I think that one immediate consequence of the conditions (1)–(3) above is this: they rule out exactly those very approaches to T-simplicity in which T-simplicity estimates are assimilated either to aesthetic appreciation of the elegance of a theory, or to its probability estimates, or to estimates of the degrees of falsifiability. That is, they rule out all approaches to simplicity that seek to reduce it to one or other of the various concepts of probability. This particular consequence will hold good, no matter whether one operates, for purposes of reduction, with prior-probability estimates or posterior-probability estimates, say within a Bayesian framework or some other. But, one might ask, what has led philosophers to define theoretical or nomic simplicity either in terms of falsifiability or in those of initial, prior-probabilities? The main reason for this sort of approach lies, I think, in the fact that it is generally, though wrongly, believed that the simplicity of a scientific law or theory, being just a matter of its initial plausibility, can be gauged well before its serious empirical and methodological investigation begins.[7] And it is accordingly thought to be either a matter of our aesthetic appreciation of a theory's elegance or, still more seriously, a matter that can be settled as simply as certain other parameters of the theory, such as its initial plausibility/probability or falsifiability. But this is done without ever realizing how this kind of approach must inevitably raise the question: How could then the T-simplicity estimates play as important a methodological role in theory-appraisal and theory-choice as they seem to do in actual scientific practice? The question 'How important really is their role in theory-appraisal?' might, of course, itself warrant further investigation in this context. This is one of the questions, then, that I shall be addressing as we proceed.

Consider, e.g., Harold Jeffreys's view of the simplicity of a scientific law, viz., his simplicity postulate.[8] The simplicity or complexity of a law is, says it, explicable in terms of the concept of its *initial* probability as follows: higher the simplicity of a law, higher will be its initial probability; and lower the simplicity of a law (or higher its complexity), lower will be its initial probability. The simplicity of a law, we are thus told, belongs to that order of either increasing or decreasing initial probabilities which can be calculated quite independently or in absence of any posterior-probabilistically relevant bodies of observational/experimental evidence. As Jeffreys puts it,

> . . . in the absence of observational evidence, the simpler law is the more probable and the initial probabilities can be placed in an order . . . the order of decreasing initial probabilities is that of increasing complexity.[9]

Now once the simplicity of a law is defined, in this way, as a parameter that has to do with its initial probabilities before it is subjected to real empirical-cum-methodological investigation, the whole concept of simplicity is, I think, deprived of its methodological role in the appraisal and selection of hypotheses or theories. An approach such as this obviously presupposes the framework of a probabilistic logic and inductivistic epistemology of the sciences in some interpretation or other. But the point to note is this: The simplicity estimates of laws or theories conceived as a function of their prior-probability estimates are bound to be non-subservient to their posterior-probability estimates taking relevant bodies of observational evidence into account. As is generally acknowledged, the two – the prior-probability estimates and posterior-probability estimates – will always tend to diverge *asymmetrically* as the observational evidence (for or against a theory) accummulates, except in the trivial-most, extreme cases where the prior-probability takes either of the two values 0 and 1. Simplicity of a theory as an irreducible parameter of scientific appraisal is thus simply made to lose its separate identity. On the other hand, moreover, not only are sharp differences over the initial probability estimates of one and the same set of laws/ theories possible under certain interpretations, it makes sense to speak of probability estimates of either kind as being revisable, depending on the variations in those parameters that are considered relevant for a probabilistic epistemology.[10]

Consider again how I. J. J. S. Good and A. F. McMichael make an interesting attempt, in their (1984: 120–127), to sharpen the distinction between the P-simplicity of a hypothesis and its T-simplicity. They approach this task in terms of the concept of probability within a 'Bayesian' framework as follows: The theoretical merit of the T-simplicity of a hypothesis "is achieved if we allow for the probabilities of hypotheses, without further explicit reference to simplicity."[11] In the same paper, however, the corresponding concept of the theoretical complexity of a hypothesis, abbreviated as $K_1(H)$, receives the following treatment: "$K_1(H)$ is related to the aesthetic appeal of H or to the brevity of its shortest statement when expressed in a language regarded as efficient in the area of interest."[12] Again, they observe: "$K_1(H)$ is perhaps roughly proportional to the 'cost' of *stating* H."[13] But what remains here unclear is how one might spell this out precisely, or relate such different conceptions with one another.

On the other hand, however, there have been several attempts to define the concept of T-simplicity somewhat non-probabilistically, notably by Nelson Goodman and Karl Popper.[14] I shall consider here briefly the approach of Popper (1968), who, unlike those who define it in terms of high probability, identifies it with falsifiability, logical improbability and the like. Thus, in his (1968), Popper proposes that the concept of T-simplicity, taken strictly as an epistemological concept, be equated with the degree of falsifiability of a theory.[15]

The degree of universality and of precision, [he argues] of a theory increases with its degree of falsifiability . . . Thus we may perhaps identify the *degree of strictness* of a theory – the degree, as it were, to which a theory imposes the rigor of law upon nature – with its degree of falsifiability; which shows that the latter does just what Schlick and Feigl expected the concept of simplicity to do.[16] Above all, our theory explains *why simplicity is*

> *so highly desirable*. To understand this there is no need for us to assume a 'principle of economy of thought' or anything of the kind. Simple statements, if knowledge is our object, are to be prized more highly than less simple ones *because they tell us more; because their empirical content is greater; and because they are better* testable.[17]

Again, in his (1972a), Popper suggests that the simplicity of a theory T has to do with its logical improbability instead of its probability.[18] Greater the degree of falsifiability of T, greater will be its degree of simplicity; but less probable will it be (where 'probable' means the 'prior-probability' that T is true irrespective of what empirical tests its predictions are subjected to). Whether we agree with Popper or not, it seems to me beyond doubt that science aims at theories of ever greater degrees of T-simplicity. But it is quite simplistic to suggest that the simpler theories are really those "which offer us the best chance to submit them to severe tests",[19] the ones with a higher degree of testability than the more complicated ones.

It is, in any case, erroneous, I think, to define T-simplicity in terms of the degree of falsifiability of a hypothesis or theory. There are at least two reasons why this is so. One of these has to do with my suggestion above that one might quite reasonably view the falsifiability of a theory as a *function* of its simplicity. As I shall argue subsequently, this does not entail that to know the degree of falsifiability of a theory is to know its degree of simplicity. This is then quite unlike Popper's view which does lead to such a consequence. Secondly, to reduce T-simplicity to the falsifiability of a theory is, I think, to deprive the simplicity estimates of their important and legitimate role in empirical and methodological appraisals of two or more rival candidate-theories. Certainly, while the falsifiability requirement might demarcate science from non-science, it is not clear how it might affect the scientist's appraisal of two or more scientific theories in a given field. T-simplicity estimates, on the other hand, not only can but do play an important role in theory-appraisal and theory-choice in science. To expect the falsifiability requirement in Popper's sense to both demarcate science from non-science and define the parameter of T-simplicity is, I think, to expect miracles from a methodological requirement as global as this. Or is T-simplicity itself a parameter so simple in character as to be decidable in terms of the falsifiability requirement? Popper and his followers might of course argue that the degrees of falsifiability of rival theories are precisely what one must resort to in those difficult situations of theory-choice where either (a) the two competing theories are equally corroborated as empirically adequate theories at a given time, or (b) they are equally falsified by negative evidence at a given time. This would of course mean assigning a special role to the simplicity estimates, conceived as the functions of degrees of falsifiability, in situations of type (a) and (b). But, I think, all this could not be regarded as an answer to the main question that we should answer, namely: How do the T-simplicity estimates play an irreducible role in theory-appraisal and theory-choice, not in this or that special situation but in the normal course where rival theories in their field are not related either by equality of corroboration or by equality of falsification? What sense does it make, moreover, to say or to imply that one might in the extreme situation of two equally falsified theories decide that one of them has higher degree of falsifiability and is therefore preferable to the other?

Both kinds of attempts, probabilistic and non-probabilistic, to define T-simplicity also suffer from serious common drawbacks. Thus there is further reason to conclude that it is fundamentally erroneous to reduce T-simplicity either to the high logical probability of a hypothesis/theory, or to its prior probability, or to its logical improbability/falsifiability. For either type of approach it is necessary to proceed from an unquestioned commitment to the deep-rooted traditional statement-view of the structural identity of science or of objective knowledge. Notice that T-simplicity has, under such type of approaches, nothing to do with scientific problems, since they do not figure in this structure as statements or hypotheses do. In what follows, I shall proceed on the opposite assumption to the effect that T-simplicity has to do with something inherent in *theories* and *problems* in science, but nothing at all with those methodological conventions that we may adopt and vary from time to time. It is important, I think, that we adopt a more careful approach to the concept of T-simplicity while guarding against the tendency to confusingly identify it with our favorite methodological conventions/requirements, no matter whether these are themselves cast in terms of the concept of falsifiability or that of probability. As a first step in this direction, I suggest that T-simplicity could be better understood as a *theoretical universal*, in the sense I use this term in my (1983) and in Chapters 6 and 7 above. Once it is understood in this sense, no attempt to explicate it can, I think, ignore the fundamental epistemological-ontological principle[20] that describes the structural identity of science as follows: The principal factors codetermining the epistemic structural identity of science are as abstract and type-distinct as *theories* and *problems* – including, of course, their dynamic interactions. With this principle in the background, one might then argue for a view of T-simplicity that relates the relevant parameters in scientific theories to those in scientific problems, in a manner that fulfills the conditions (1)–(4) above.

If we assume then the correctness of this principle (see § 8.3. below) – that says that the structural identity of science is codetermined by its *theories* and *problems* – together with an interactive model of the growth of scientific knowledge, first proposed in my (1983), the simplicity of a theory, relative to its rivals, is rendered definable in terms of a certain relationship between the theory and the set of problems (the question-set) that it can solve. The resulting relational concept of T-simplicity can then be seen to presuppose theory-comparison that centers around the sets of commonly shared problems, the Kuhn–Feyerabend incommensurability doctrine not-withstanding. Interestingly enough, Elliott Sober's (1975) attempt to define simplicity as a kind of informativeness is, I believe, an important step in this direction, since it is nearer to our requirement for treating theoretical universals than any other attempt I know of. Among the salient features of his theory of simplicity of hypotheses the following are noteworthy:

(A) . . . the simplicity of a hypothesis can be measured by attending to how well it answers certain kinds of questions . . . the more informative a hypothesis is in answering these questions, the simpler it is. The informativeness of hypotheses relative to questions is characterized by the amount of extra information they need to yield answers. The more additional information a hypothesis needs to answer a question, the less informative it is relative to that question.[21]

(B) . . . simplicity will be seen to be an important quality for hypotheses to have, because informativeness is a desideratum of knowledge claims, and simplicity is informativeness.[22]

The inverse relationship between the informativeness of a hypothesis and the content of its 'minimum extra information' (MEI set) relative to a question is then laid down by Sober as follows:

(1) $\mathrm{H} >_i \acute{\mathrm{H}}$ relative to Q iff $\acute{\mathrm{I}} >_c \mathrm{I}$ relative to Q,
or,
A hypothesis H is more informative than another, Ή, relative to a question Q just in case the MEI set Í of Ή is higher in content than the MEI set I of H.[23]

Sober then formulates his conjecture about simplicity thus:

(2) $H >_s \acute{H}$ [relative to Q] iff $H >_i \acute{H}$ [relative to Q]
or,
H is more informative than $\acute{H}$ in the sense of (1) just in case H is simpler than $\acute{H}$, where '$>_s$' means 'simpler than'.[24]

Combining (1) with (2), the result is the following definition of T-simplicity:

(3) $H >_s \acute{H}$ [relative to Q] iff $\acute{I} >_c I$ [relative to Q][25]

One might state the basic idea here more simply as follows: A theory T_2 is simpler than its rival, say T_1, if and only if, relative to a given question set Q, T_2 is more informative than T_1 in Sober's sense. [26] Or,

(4) $S(T_2) > S(T_1)$ [relative to Q] $\leftrightarrow$ $I(T_2) > I(T_1)$ [relative to Q].
Or,
the theory T_2 is more simple than T_1, given Q, if and only if T_2 is more informative than T_1, given Q.

How do we assess a theory of simplicity as informativeness in this sense? Elliot Sober, who develops it further in considerable detail,[27] thinks that any assessment must proceed in terms of the twin desiderata of 'support' and 'simplicity'.[28] What he has here in mind is this: that a theory of simplicity, like any other theory, needs justification chiefly of two kinds. In so far as it is expected to capture and systematize our pre-systematic intuitions about simplicity, a theory of simplicity must receive sufficient support from these intuitions to serve as an adequate explication. But in so far as it must also explain 'diverse phenomena' (here he has, I guess, hypotheses-choice in mind), a theory of simplicity must itself be 'simple'.[29] Earlier, we considered certain questions concerning the nature and role of a theory of simplicity, and of constraints on such a theory. I think that it is erroneous to so interpret these questions, as questions of justification for a proposed theory of simplicity, as to apply, in turn, some considerations of simplicity or other in actually justifying a given theory. But this is what one finds in Sober (1975) when he answers his own question as follows:

The theory that I have been elaborating and defending can be said to justify the use of simplicity in hypothesis choice on the grounds that informativeness is one of our goals in choosing hypotheses and, according to the theory, simplicity is informativeness. But what justifies the theory of simplicity? Since it is a *theory* like any other, its justification derives from its being sufficiently supported and simple.[30]

Our approach on the issues involved here is, however, different. The soundness of a theory of T-simplicity will ultimately have to be decided in terms of its ability to find its proper place within a properly formulated realistic theory of the growth of scientific knowledge and the methodology of scientific appraisal. For example, Sober's definition of simplicity could find its best justification not within a theory of questions as sets of natural predicates, in his sense (1975: 19–23), but within the framework of the methodology of theory-problem interactive systems in the sense proposed here and in my (1983). Such a definition, therefore, needs to be further explored as to its consequences for actual methodological or epistemic appraisal of scientific theories. The realism of the definition lies, I think, in its ability to project rival theories/hypotheses in their fields in intimate interaction with the antecedently formulated problems. It already presupposes, therefore, a problem-solving model of the growth of scientific knowledge. But the model thus presupposed is the typical model of the explanation-theorist. For this very reason Sober's definition turns out to be inadequate from the standpoint of the methodology of theory-problem interactive systems. For it is sufficient just to yield a comparative measure of the simplicity of rival theories T_1 and T_2 considered as rival *explanations*: Thus

(5) $S(T_2) > S(T_1)$

if and only if T_{EP} (T_2) employs less number of assumptions (Sober's MEI sets) than T_{EP} (T_1) does in explaining the same explananda e_1, e_2, . . . e_n, where S abbreviates the comparative predicate T-simplicity. It should be, however, possible to strengthen or complete this concept of T-simplicity S by means of what might be appropriately called counterpart T-simplicity $\acute{S}$, which has to do with a theory's relationship, of course relative to its rivals, not with the antecedently formulated problems it is intended to solve but with those other problems which it itself entails as theoretically novel problems (for discussion see Chapter 9). A definition of the counterpart T-simplicity $\acute{S}$ is necessary, I believe, if our talk of the simplicity of a theory is to be methodologically and epistemologically – and not just aesthetically – comprehensive enough to range over both the *explanatory* and the *resolving powers* of our scientific theories. I propose, therefore, the following definition, with the help of the abbreviations P_1, P_2, . . . P_n (for the newly identified problems) and A_1, A_2, . . . A_n (for additional assumptions). Thus, given any two rival theories T_1 and T_2 in their field,

(6) $\acute{S}(T_2) > \acute{S}(T_1)$

if and only if (i) T_{RP} (T_1) *and* T_{RP} (T_2) individually give rise to problems $P_1, P_2, \ldots P_n$ in a manner such that (ii) while T_{RP} (T_2) theoretically identifies $P_1, P_2, \ldots P_n$ *without* employing as many assumptions as $A_1, A_2, \ldots A_n$, that T_{RP} (T_1) needs if T_1 is to serve as a theoretical basis for either entailing or identifying (or formulating) the *same set* of problems. Or,

(7) $\acute{S}(T_2) > \acute{S}(T_1)$ iff
$T_{RP}\ (T_1/\ A_1, A_2, \ldots A_n) \rightarrow P_1, P_2, \ldots P_n$

&

$T_{RP}\ (T_2/\ A_1) \rightarrow P_1, P_2, \ldots P_n$

As I shall be arguing in Chapter 9, it is important to remember that if T_2 is to be not just *simpler* than T_1, in the sense just defined, viz.,

(8) $S(T_2) > S(T_1)\ \&\ \acute{S}(T_2) > \acute{S}(T_1)$

but really better in overall performance than T_1, then it is necessary, and perhaps also sufficient, that

(9) $T_{EP}\ (T_2) > T_{EP}\ (T_1)\ \&\ T_{RP}\ (T_2) > T_{RP}\ (T_1)$

The twin characterization of T-simplicity in (5) and (6) is very significant. It shows how irreducibly complex the concept of T-simplicity is. A careful application of the concept to actual cases of theory-appraisal and theory-choice in science should tell us how successful our proposed definition of simplicity is, especially in terms of the conditions (1)–(4) above.

Up to now I have been arguing as if it is entirely a sound approach to define T-simplicity as informativeness on the ground that science aims at the kinds of theories that are more and more informative. I shall now, however, restate the various positions with a view to arguing for a radically different conception of T-simplicity. It will turn out that even Elliot Sober's view of simplicity involves considerable oversimplification, as the other views do. The novelty of my own view will then lie in showing how complex the concept of T-simplicity really is. Let us start with the usual, and perhaps most dominant, probabilistic model (H. Jeffreys and others).

8.2.2. *The Probabilistic Model of Simplicity*

There are generally two different sorts of approach to defining T-simplicity in terms of 'probability', depending on how this notion is itself defined and sought to be related to other concepts such as 'evidential support', etc.. In the usual philosophical discussions, one notices either the Bayesian way or the more traditional, statistical approach[31] to dealing with these concepts, including the concept of T-simplicity. For Bayesians, a hypothesis or a theory T_2 is a simpler explanation than the theory T_1, given of course the explananda $e_1, e_2, \ldots e_n$, only if

(1) $p(T_2) > p(T_1)$
Or,
T_2 has higher probability than T_1, where 'p' ranges over 'initial' or prior probabilities – the initial plausibilities of theories themselves.

One might ask here: What is 'prior probability' itself? Salmon (1967: 121) has suggested that its clarification "be sought in the light of the interpretation of the probability concept in general". The precise interpretation of the prior probabilities (of hypotheses or theories) themselves, one notices therefore, differs between different schools such as 'Bayesians', objectivists and frequency theorists, favoring subjective, logical and 'the limit of the relative frequency' interpretations respectively. Thus, for the 'Bayesians' (or subjectivists) the prior probabilities of a hypothesis/theory are a matter of degrees of belief in its truth, which may vary from person to person according to their variable subjective contexts. (For a detailed lucid discussion see Salmon, 1967: 108–132). The basic idea underlying this approach, circulating in different versions, is this: A theory T_2 is simpler than another, T_1, only if T_2 is *initially* more likely to be true (its initial probability is higher) than T_1. That is, the simplicity of any theory T is a function of the degree of the initial probability that T is true:

(1a) $S(T_1) = p(T_1)$ &
$S(T_2) = p(T_2)$;
therefore $S(T_2) > S(T_1) \leftrightarrow p(T_2) > p(T_1)$

It would be, however, wrong to think that, if we took the posterior-probabilistically relevant observation statements O_1, O_2, into account, this kind of simplicity-ordering might expand itself as follows:

(1b) $S(T_1) = p(T/ O_1, O_2)$ &
$S(T_2) = p(T_2/ O_1, O_2)$;
therefore,
$S(T_2) > S(T_1) \leftrightarrow p(T_2/ O_1, O_2) > p(T_1/ O_1, O_2)$

It would be wrong to think so simply because it would amount to confusing the initial/prior probabilities of hypotheses with their posterior probabilities. So we have to be content with as simple a model as (1a) or (2) as follows.

There are, on the other hand, those who think (quite in conformity with the traditional statistical practice) that T_2 is a better and simpler explanation than T_1, given the explananda $e_1, e_2, \ldots e_n$ only if

(2) $p(e_1, \ldots / T_2) > p(e_1, \ldots / T_1)$
(T_2 confers higher probability p, or likelihood, on $e_1, e_2, \ldots e_n$ than T_1 does, where 'p' ranges over probabilities taken either as *apriori* or as a function of background inductive relations).

In each of these major approaches, one notices a variation on the same old inductivist theme, namely the attempt to link the simplicity of a theory to the likelihood of its truth: The simpler theory is more likely to be true.[32] The idea is very old. And it is easily traceable to the ontological assumptions about the so-called simplicity of nature.

8.2.3. *The Popperian Model of Simplicity*

As we have already noticed, the distinctive idea in this model is to characterize T-simplicity in terms of the (demarcative) requirement of falsifiability. That is, the simplicity of T is a function of its falsifiability, such that the following relations hold:

(3) $S(T_1) = F(T_1)$ &
$S(T_2) = F(T_2)$;
therefore, $S(T_2)>S(T_1)\leftrightarrow F(T_2) > F(T_1)$

The close similarity between these two types of models (one of high probability and the other of low probability) is striking. Since falsifiability, unlike falsification, does not depend on actual observational evidence, Popper's model will not lend itself to an expansion such as (3a) below:

(3a) $S(T_1) = F(T_1/ f_1, f_2, \ldots)$ &
$S(T_2) = F(T_2/ f_1, f_2, f_3, \ldots)$;
therefore, $S(T_2) > S(T_1) \leftrightarrow F(T_2/f_1, f_2, f_3, \ldots) > F(T_1/f_1, f_2, \ldots)$

where f_1, f_2, f_3 represent the falsifiers in terms of actual observational evidences.

I think that both the two types of models just restated give rise to the puzzling problems of how the parameter of T-simplicity can play any role in theory-choice if it is defined in either of these two ways. With B. van Fraassen (1980),[33] one can say, in the context of the probabilistic model, that the view that simpler theories are more likely to be true suggests strongly or implies that the world is more likely to be simple than complicated. But an implication such as this is quite absurd. Similarly, I think, Popper's attempt to define the simplicity of a theory in terms of its degree of falsifiability has the following absurd consequence. On Popper's view, of the two theories T_1 and T_2, the one which runs the higher risk of falsifiability, because it tells us a lot more about the world than its rival does, is the simpler one. Let T_2 be this theory. Suppose, our attempts to falsify T_2 are now successful. We have then an as yet unfalsified theory T_1 on the one hand, and a falsified theory T_2 on the other, the two having already passed through various stages of competition. Are we not entitled now to reject T_1 but accept T_2 because it is our simplest theory conceivable at the moment, if only because it is the easiest to refute? For on Popper's account, simplicity of a theory is a function not of its likelihood for truth but of its likelihood to be false.

8.2.4. *Elliot Sober's Model of Simplicity*

Let us now have a closer look at Sober's model of T-simplicity. As I understand it, his view that simplicity is a kind of question-relative informativeness can be restated as follows:

(4) $S(T_1) = I(T_1/Q_1, Q_2, \ldots)$ &
$S(T_2) = I(T_2/Q_1, Q_2, \ldots)$;
therefore, $S(T_2) > S(T_1) \leftrightarrow I(T_2/Q_1, Q_2, \ldots) > I(T_1/Q_1, Q_2, \ldots)$

This corresponds closely to Sober's own formulation (3) which we considered before, viz.

3. $H >_s H'$ [relative to Q] iff $I' >_c I$ [relative to Q].

As we have seen before, this definition is quite interesting. But it presupposes a supporting problem-solving model of the growth of scientific knowledge itself based on the view that the task of science is just an explanatory one. Although I think that this definition marks a significant step in our understanding of the simplicity ordering of scientific theories within these limitations, I find it, for various reasons, quite inadequate and deficient. This is not to deny that, if one were simply to work with this definition, one would find it full of far-reaching consequences. Yet I shall make it clear why I believe in a different kind of simplicity ordering of scientific theories.[34]

As a realist, Sober makes it obligatory on a *complete* realistic methodology (of theory-appraisal) that it provide a ground for our differential attributions of truth and falsity to theories that are being evaluated. Summing up his own view, he says,

> According to our explication of simplicity, a simplicity criterion distinguishes between observationally equivalent theories. According to our justification of simplicity, a difference in simplicity between two hypotheses counts as a reason for differential attributions of truth and falsity.[35]

This may be understood as an attempt to ground a realistic methodology, viewed as capable of differential attributions of truth and falsity, in a criterion of simplicity which is itself realistic in some sense, while assuming in common with the instrumentalists that the main aim of science is to achieve ever better explanatory/predictive hypotheses/theories. How might one justify such a grounding by Sober? And, more importantly, how is it different from those versions of realism in which the simplicity of a theory is explicated as its likelihood to be true? "Simplicity is a ground for attributions of truth and falsity", argues Sober,

> not because simple hypotheses are more likely to be true, but because our goals in accepting and rejecting hypotheses as true or false involve a commitment to constructing informative hypotheses.[36]

I will here not go into the further justification that Sober (1975: 173) tries to provide by pointing to "the deep affinities between simplicity and support", parallel to those between simplicity and realism just considered. Sober is, I think, basically right in so far as the latter affinity is concerned. But even here his commitment to a truth-theoretic approach to realism is as dogmatic as that of other thinkers. If I am right,[37] then an alternative account of this affinity between simplicity and scientific realism should be possible. In my account, which follows, the basic underlying idea of such an affinity will be captured and explored by arguing for an interplay between the methodological simplicity of a scientific theory and the ontological complexity of *the* world as a *whole*, taken not as an entity, or as a set of entities, among other entities but as an *invariant* reference frame for science.

8.2.5. *Re-ordering Theoretical Simplicity: Towards an Interaction-Theoretic Model*

I find B. van Fraassen's observation, alluded to above, interesting in so far as it seems to suggest a *reductio ad absurdum* type consequence against defining T-simplicity as a theory's greater likelihood to be true. As I have argued above, it will be an important step in explicating T-simplicity if we know beforehand how not to try to explicate it, and which false clues to avoid. The most popular idea that a theory T_2 is simpler only if it is more likely to be true, than T_1, is, I believe, one such false clue. But why? This has to do with the deeper question of what ontological assumptions is it most reasonable to entertain if science is to have the structural identity, codetermined by theories and problems in mutual interaction, that it has and if it is to aim at those theories, with explanatory power (T_{EP}) and resolving power (T_{RP}), that are increasingly simpler, increasingly *elegant* in the sense to be considered presently. Our scientific theories are, of course, destined to be more or less simple, i.e. they will always be capable of participating, with respect to one another, in an epistemologically-ontologically significant simplicity-ordering, so long as our search for them is based on the realistic assumption of the existence of *the* world not as an entity of direct reference, nor as a simple entity among other possible entities (worlds), but as a background ontological complexity of an unknown, if not unknowable, order that serves as an invariant constraining reference frame for science. If the world itself were not denumerably (or infinitely) and, hence, irreducibly complex, both in structure and causal dynamics, our theories would never be able to participate in such a simplicity-ordering. For it can be shown that if we assume that the world is as simple as may be conceivable to us on a given theory, the simplicity of a theory, given its likelihood to be true, will then reduce itself just to a *triviality* only, just to a movement in a circle which is empty: From the (non-vanishing) simplicity of a theory, that is likely to be true, to the (non-vanishing) simplicity of the world itself. This implies that, if T-simplicity is itself an irreducible parameter not just in theory-construction but in theory-appraisal and theory-choice, it is explicable only within the framework of general epistemological-ontological assumptions of *complexity* of a broadly realistic nature. This also implies that there exists an unrecognized but fundamental *asymmetry*, deep in the realistic framework of science, between the very complexity of *the* world as *its* invariant reference frame and its theories as increasingly *simple* theories about this world even if they may not refer to it directly.[38] It is quite reasonable, therefore, to introduce a general constraint (4) on an adequate explication of T-simplicity, in addition to the conditions (1–3) discussed before, using the idea of this *asymmetry* as follows:

(4) An explication of T-simplicity will be adequate only if, among other things, it does not reduce it to the kind of triviality which is destructive of the fundamental asymmetry between the complexity of *the* world as an invariant reference frame for science and our theories as simple theories about this world.

This suggests at least two things, viz., (a) That a really sound explication should preserve this asymmetry in a manner which is adequately reflected in every T-simplicity appraisal that we make; and (b) that it is not possible to fulfil (a) outside a suitably and newly elaborated scientific realistic framework.[39] The conditions (a–b) themselves codetermine then an important constraint of a scientific realistic nature on an adequate explication of T-simplicity, as will also be borne out by our subsequent discussion.

Is a *world* with an ontological structure of matter and electromagnetic radiation in mutual interaction, one might ask, *simpler* than one in which, besides these, gravitational radiation also enters into dynamical interactions of far-reaching astrophysical and cosmological significance? The answer really lies, one might argue, in those theories themselves that refer to these two worlds respectively, since it is these that we should compare with each other. If the simpler of the two theories, T_1 and T_2, entails, of course through those new problems that it itself entails, or through its novel predictions, that the world (say w_2) as describable by means of T_2 is more complex than the world (say w_1) as describable by T_1, this may be so because of an unrecognized compatibility between the ontological complexity of the world, or of different worlds corresponding to different theories, and the methodological simplicity of a theory which is intended to describe it. It is the simplicity of Einstein's GTR which demands, being greater than the simplicity of NGT (see Chapter 9, § 9.1), that we attend to the greater complexities in our universe than could be conceived of on the basis of NGT, NCT and Newton's three laws of motion. Greater the simplicity of a theory, one might therefore conclude, more likely is the ontological complexity in the structure and causal dynamics of the world that it predicts. Conversely, lesser its simplicity, less likely is the complexity in the structure of the world that it predicts. Thus, given any two theories T_1 and T_2 in their field, and given the two worlds w_1 and w_2 which they seek to describe respectively, we are led to think of their T-simplicity, and of the relative ontological complexity which they confer on w_1 and w_2, as constrained by the complexity of *the* world itself (see Figure 8.1. below). It can now be argued that science aims at those theories that are increasingly simpler, in an unconventional sense, such that the order of their T-simplicity in no way entails that the order of *nature* is simple. It will, on the contrary, always entail an ever-increasing complexity in the order of nature.[40]

A science takes its birth from general assumptions of interaction in which it remains embedded ever after.[41] These relate to our enigmatic (*geheimnisvoll*) interactions with *the world* as a complexity which, though unknown to us, constrains science, as its *invariant frame of reference*, in its search for simple theories. The assumption of global complexity that enters science at this point may be expressed by saying that the world is an infinitely (or denumerably) complex structural hierarchy of interactively entangled systems. The possibility of disentangling these in the form of simple equations describing their interactions or functional relationships is then what science explores. It seems that there is a definite relationship between (our) scientific knowledge of the world (= *W* in Figure 8.1. below) and our interactions with it, where the former constantly shapes the latter. As (our) scientific knowledge of *W* grows, our interactions with it become more and more enigmatic or problematic. It is then in this relationship that we may notice more clearly the kind of role that *W* plays. It appears here, as one might put it, as an invariant background of ontological complexity constraining all significant human interaction. Without *W*, then, there would be no general assumptions of interaction available to science that are insensitive to differences of subject-matter *between* or *within* the special sciences, nor any subject of objective knowledge for science to explore.

The basic idea underlying this may now be made more precise as follows. Presupposed as the invariant frame of reference for science, *W* is identifiable as a background ontological complexity that constrains science in its movements *backwards*, from problems to theories, and *forwards*, from theories to problems, in a manner which is subject to negative feedback control in the sense of (Pandit: 1983). It is, I believe, the function of a scientific theory not just to solve those antecedently determined problems that prompted its formulation but to raise new problems, or make novel predictions, and, therefore, to indicate how complex *W* is in those very aspects that might relevantly relate to its own domain of application. A theory which performs this job well will not only imply that *W*, as a subject of objective knowledge, is more complex than we may have, to begin with, assumed it to be but confirm *W* in its role as a constraining background complexity or as the invariant reference frame for science. But a theory can do this, so might one argue, only if science already assumes not that *W* is a simple entity among other possible entities (= worlds) but that it is a complexity of an order, or of an indefinite number of orders, whose nature as a whole we want to understand but can never know completely. Science can, therefore, progress not *cumulatively* from the older theory to the new one but *interactively* from theory-problem interactive systems to theory-problem interactive systems (Pandit: 1983). The development of problems plays thus an important role in the structural growth of objective knowledge. If we look at a theory not just as an explanatory structure but as a problem-posing structure, we might say that objective knowledge grows from less simple theories to more simple ones in a manner consistent with the principle that holds *W* as an invariant reference frame for science. That a theory T_2 is simpler than another theory, T_1, is not to say, in this context, that *W* is simpler on T_2 than on T_1. Quite the contrary. Without *W* as an invariant background of ontological complexity, there would be nothing for our theories to be simple theories about. *W* as an invariant constraint on science requires, therefore, that with every new theory newer ontological complexities of *W* come to be identified such that the enigmatic character of our interactions with it does not diminish to a point where there are no new problems to look for newer theories with newer ontologies. It is, then, the simplicity of a theory which guarantees its falsifiability against the invariant background of *W* which is not of our creation but with which our interactions are in certain crucial ways controllable. A theory judged, against this background and against its rival, as simpler should, therefore, itself entail (the empirical possibility) that it is not the last theory in its field. That is, its simplicity should not have the consequence that *W* is simpler ontologically than one may have assumed it to be on earlier theories. Instead of destroying the ontological complexity of *W* as an invariant reference frame for science, it should conserve it such that the following principle, which may be called the principle of conservation of ontological complexity of *W* – henceforth c(*W*) – constrains the methodological simplicity-appraisals of theories as follows (see Figure 8.1 below):

(a) The simpler theory T_2 satisfies c(*W*) better than the less simple theory T_1 only if the following holds true of them:
$C(w_2)/\, T_2 > C(w_1)/\, T_1$
Or, the simpler theory T_2 satisfies c(*W*) better than the less simple theory T_1 only if the ontic complexity of w_2, given T_2, is greater than that of w_1, given T_1.

But how is then the simplicity of a theory itself to be understood, since it is no longer necessary, in this context, to appeal to its likelihood to be true? In the light of our discussion above, in particular the assumption that the epistemic-structural identity of a science is codetermined by *theories* and *problems* in mutual interaction, we can say that science aims at the type of theories that are not only of ever greater explanatory power (T_{EP}) but of ever greater resolving power (T_{RP}).[42] While with the former a theory solves those antecedently determined problems that prompt its formulation, with the latter it determines novel problems and makes novel predictions, opening a way to novel experimental possibilities to test them. The rational scientific appraisal of a theory T_2, given its rival T_1, will accordingly involve the following two-step appraisal:

$$T_{EP}(T_2) > T_{EP}(T_1)$$
$$\&$$
$$T_{RP}(T_2) > T_{RP}(T_1)$$

Such appraisals are possible only if the postulates and auxiliary assumptions of T_2 are less in number and more unified than those of T_1 in respect of both these two functions. It is then here that the simplicity

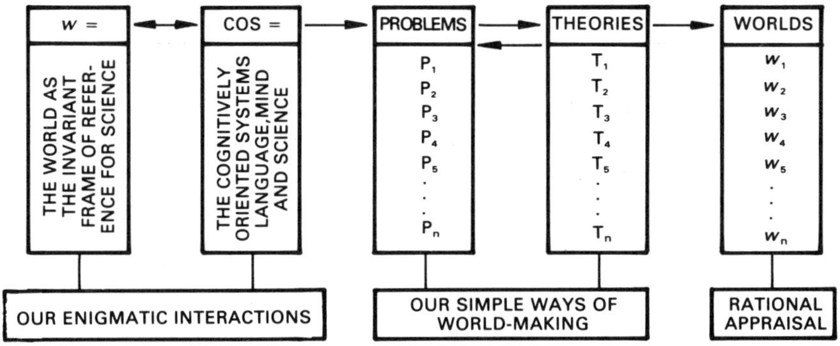

Fig. 8.1. A formalizing model of our *ways* of world-making as constrained by *W* which is the invariant reference frame for science. From a scientific realistic point of view, our enigmatic interactions with *w* are conceivable here in a manner akin to Plato's allegory of the cave. For a scientific realist the real challenge is, therefore, not just to explain the possibility of the nexus: $\langle P_1, \ldots, P_n \rangle$ $\langle T_1, \ldots, T_n \rangle \rightarrow \langle W_1, \ldots, W_n \rangle$, but to explain non-trivial overlap-relations that bind them together in rational epistemic appraisal. The most plausible way conceivable to do this is to recognize their common anchorage in *W*: while being anchored in *W*, they are our only telescopic windows opening towards *W*. There can be, therefore, no ways of world-making without *W* as their invariant background.

of a theory, relevant to the version of scientific realism I am arguing for, lies:

$$S(T_2) > S(T_1) \leftrightarrow T_{EP} \cdot T_{RP}(T_2) > T_{EP} \cdot T_{RP}(T_1)$$

As a consequence for scientific realism, it is now possible, on this definition of simplicity, to look at T_2 as *better* than T_1, in their field, not because it is more likely to be true or truthlike but because its framework is (in a sense different from that of the received view) *simpler* and, therefore, preferable as a basis for the future development of science in its field. What is most important here is this. With T_2, which employs more unified and less number of assumptions than T_1, it is possible to *say* more as well as to *ask* more questions about *W* than with T_1.

Where in science might one then look for the best examples of two (kinds of) theories T_2 and T_1 such that: $S(T_2) > S(T_1)$, in the sense that the ontology of T_2 is more complex than that of T_1? I think that classical mechanics (CM) and quantum mechanics (QM) provide a good example, the fact that they are not direct rivals in a common domain not-withstanding. Or, one may even consider QM and its possible non-local hidden-variable theoretic reformulations. In any case, the case of QM is interesting in that its formalism – which preceded its interpretation – has been free from the kind of controversy which its interpretation has evoked ever since its creation in the period 1925–1927 in the work of Heisenberg, de Broglie, Erwin Schrödinger and Max Born. If at any time there have been proposals to modify the formalism itself or to seek an alternative to it, these have been prompted by the interpretational problems themselves. That many of these problems arise due to the unexpected and unknown ontological complexities in the quantum domain is brought home to us by so many *Gedankenexperimente* concerning quantum phenomena, notably those of the Einstein-Podolsky-Rosen[43] (EPR) type that have recently culminated in those of the Bell type whose realized versions have more recently been performed.[44] These may be taken to confirm that since, in certain special cases, the predictions of QM violate Bell inequality,[45] they carry the significance of ontological complexities of a kind that we still do not understand fully. The ontological complexities of the quantum world, given what QM predicts, are then clearly of an order which one could not foresee just on the basis of its past successes in dealing with the experimentally established phenomena. They are also of an order which would have remained undetected if no attempt had ever been made to use its own framework to pose novel problems. And that is exactly what the *Gedankenexperimente* of EPR type have done to QM. What can we now conclude from this? The methodological simplicity of a scientific theory resides no doubt in the structure of its basic postulates and auxiliary assumptions with which it not only deals with the given set of antecedently known problems but raises more, possibly novel, problems than it can solve. The latter function of raising novel problems corresponds to novel predictions and novel experimental possibilities (including *Gedankenexperimente*) which a *simple* theory may itself entail. The ontological complexity inherent in a theory may be, therefore, not only compatible but intimately connected with its methodological simplicity. It is possible that a methodologically simple theory is bound to be ontologically complex. QM, when compared to either CM or to its possible hidden-variable theoretic reformulations, seems to illustrate the point.[46]

Let us now elaborate a little more the basic idea of T-simplicity which we have developed in the preceding discussion. It has been one of our important assumptions that the epistemic structural identity of the objective knowledge systems in the sciences is codetermined by their *theories* and *problems* in mutual interaction. Thus, a science is interested in developing those theories which possess both T_{EP} and T_{RP} in ever-increasing degrees. With this assumption in the background, one might still like to ask: Why worry then at all about T-simplicity appraisals? Why should such appraisals at all be thought to play an *irreducible* role in scientific appraisal, as is also suggested by the conditions (= constraints) of adequacy (1)–(4) laid down by us above? My answer to this question, which clarifies further the nature of these very conditions themselves, is as follows. Only T-simplicity appraisals, in the sense to be explicated presently, could tell us, I think, whether any two or more rival theories, T_1 and T_2, are really related in the following manner:

$$T_{EP} \cdot T_{RP}(T_2) > T_{EP} \cdot T_{RP}(T_1)$$

such that T_2, and not T_1, serves as an empirically reliable theoretical basis for the future development of objective knowledge systems in the field of our scientific interest. In what follows, it will be made clear why our simplicity-ordering of scientific theories and problems should be expected to reflect dynamic interactions between theories and problems at different levels of problem-determination and problem-solving in the sciences. The simplicity of a theory will, then, turn out to be essentially a function not of its informativeness in Sober's sense (1975) but of certain kinds of *asymmetries* of information, as they might be called. A scientific theory generates these asymmetries (i) *within* itself and (ii) between itself and its rivals at various levels. Thus, we can say that a theory T_2 is methodologically simpler than another, T_1, if and only if the explanatory power and resolving power of T_2 are greater than those of T_1. Or,

(b) $S(T_1) = f\{[T_{EP}(T_1)/e_1, e_2, e_3]\ \&$
$\qquad [T_{RP}(T_1)/P_1, P_2, P_3)]\}\ \&$
$S(T_2) = f\{[T_{EP}(T_2)/e_1, e_2, e_3]\ \&$
$\qquad [T_{RP}(T_2)/P_1, P_2, P_3]\}$
Therefore, $S(T_2) > S(T_1) \leftrightarrow T_{EP} \cdot T_{RP}(T_2) > T_{EP} \cdot T_{RP}(T_1)$

But when does a theory T_2 have greater T_{EP} and T_{RP} than that of T_1? The answer is as follows:

(c) $T_{EP} \cdot T_{RP}(T_2) > T_{EP} \cdot T_{RP}(T_1) \leftrightarrow$
c_1 $T_{EP}(T_2/A_1) \vdash e_1, e_2, e_3, \ldots$ &
c_2 $T_{EP}(T_1/A_1, A_2, A_3,) \vdash e_1, e_2, e_3, \ldots$ &
c_3 $T_{RP}(T_2/A_1) \rightarrow P_1, P_2, P_3, \ldots$ &
c_4 $T_{RP}(T_1/A_1, A_2, A_3) \rightarrow P_1, P_2, P_3, \ldots$

T-simplicity defined in this way can be seen clearly as the kind of parameter that implies asymmetrical ordering of the information *operation* on theories. Simplicity does not, therefore, imply just informativeness of hypotheses, as is suggested by Sober. That is, it does not simply depend on how informative a hypothesis is relative to an antecedently determined set of questions. On the contrary, it implies what might be called informational asymmetries of two chief kinds that are always intertwined with each other. These may now be described as follows (d)–(e):

(d) As is exhibited by (c_1, c_3) and (c_2, c_4) above, one and the same theory is informationally asymmetrical with itself by virtue of its being a source of *solutions* and *problems* at one and the same time. Information gained at one end (of T_{EP}) implies a search for information at the other end (of T_{RP}):
$T_{EP}(T) \vdash e_1, e_2, e_3$
$T_{RP}(T) \rightarrow P_1, P_2, P_3$

(e) Asymmetries of the second kind generate asymmetrically ordered pairs such as (T_1, T_2) in (b) above. Information *ordered* in a *new* theory T_2 implies a *re-ordering* of information available in a previous theory T_1:
$S(T_2) > S(T_1) \leftrightarrow T_{EP} \cdot T_{RP}(T_2) > T_{EP} \cdot T_{RP}(T_1)$.

The twin-asymmetry (d) and (e) is implied by, I think, the one already considered above under the adequacy condition (4) and the consequent ontological *constraint* (a). The two are thus deeply related, since the former must obtain if the latter is to feature in the simplicity-ordering of our scientific theories. It is clear that while my theory of simplicity as a complex parameter could have far-reaching consequences for the methodology of theory-appraisal and theory-choice, the theory is itself a simple consequence of the theory of the structure and growth of scientific knowledge set forth in my (1983).

But now the question might be raised: What are the distinctive advantages of our explication of T-simplicity as formulated by (e) above, especially when (a) enters as an essential constraint in it? (a) looks counter-intuitive, especially because it is customary to expect a simpler model of reality if the theory that accounts for it is simpler than its rivals. In this way, we are accustomed to link the idea of the progressively simpler theories, T_1, T_2, T_3, . . . with the idea of the progressively simpler models or worlds w_1, w_2, w_3, . . . satisfying these theories respectively. But if we ask why this is so, we have to answer by appealing to some principle of the simplicity of the world (or nature). In this way, we will have completed the very empty circle in which our customary thinking is deeply caught, viz., our theories must be progressively simpler because nature is simple; and we can know how simple nature really is if our theories are progressively simpler. A *trivialization* of theoretical simplicity more perfect than this could not be imagined. What is the way out of this trivialization? All I can think of is (a) above as a first step, and then the formulation (e) of our proposed explication itself. The idea underlying (a) above is, I think, at least partly illustrated by the fact, which is characteristic of the natural sciences such as physics, that the essentially simple equations of a theory can have complicated solutions. Consider in this context the recent idea of the spontaneous symmetry breaking in the field of elementary particle physics. As Louis Michel (1981: 21) puts it: "The natural occurrence of spontaneous symmetry breaking suggests that physical laws might have more symmetry than physical states". "The reason for suspecting", as Steven Weinberg (1979: 55–56) points out,

> that the substratum underlying elementary particle physics, the field theoretic substratum, is governed by symmetry, has to do with a phenomenon which has become known by the rubric 'broken symmetry'. This is the idea that it is possible for the fundamental field equations to have a symmetrical form and yet for the solutions of these field equations not to share that symmetry.

A well known example in elementary-particle physics is provided by the equations of the quantum theory, as a theory of weak and electromagnetic interactions. These equations are here known to possess a local gauge symmetry which the observed particles, such as electrons, do not display in their masses. While there may be not just one type but different types of symmetry constraining scientific theories in fundamental ways, what is most significant methodologically in the present context is this. The very assumption that the symmetry principles, of a mathematical group-theoretic nature, must govern the form of the field equations in a fundamental sense enables the elementary particle physicist to proceed by a method of *working backwards* from the antecedently determined problems, he is seeking to solve, to their simplest possible resolution (explanation). The case of the Maxwell field equations of electromagnetism, that are known to obey a gauge-symmetry, illustrates the point, in the words of Steven Weinberg (1980: 115):

> assuming this gauge-symmetry principle, one can deduce all the properties of electromagnetism, including Maxwell's equations and the fact that the mass of the photon is zero.

Behind the search for a unified field theory, that began with Einstein's unsuccessful attempt to unify the Maxwell field equations of electromagnetism with those of gravitation, lies then the same assumption that with such a theory, with properties of gauge-symmetry built into its basic equations, it might be possible to *explain* the four different laws of force known to govern the weak, the electromagnetic, the strong and the gravitational interactions respectively as consequences of spontaneous symmetry-breaking (= as aspects of the same underlying field). It is in this sense, I think, that Weinberg and Abdus Salam suggested (independently of each other) in 1967 that the weak and electromagnetic interactions are governed by a broken gauge-symmetry group (see S. Weinberg, 1980: 110–119). But what is likely to be overlooked in this context is this. The method of working backwards from a complex of known problems to their simplest possible explanations will, as a rule, provide a setting in which a science moves quite in the opposite direction. As I have argued before, this is the method of working forwards from the proposed theoretical explanations (= new theories) to novel problems (or novel predictions), or to significant reformulations of the inaccurately formulated old problems. If a grand unified theory in physics is discovered one day, the theory itself will never entail, if it is really a simple theory, that it is the last word on the fundamental laws of physics. What it will entail, on the contrary, will be a new beginning with new questions.

There are then at least two chief advantages of explicating T-simplicity as in (e), over and above those that have to do with the adequacy conditions (1)–(4) above. First, only the simpler of any two theories, in the present sense of this term, could be said to achieve what might be called *maximal nomic specificity*, involving causally relevant properties and connections, as regards its T_{EP}, thereby reducing *possible ad hoc*/arbitrary elements in its basic assumptions to the minimum.

Simpler a theory is, in our sense, lesser are the chances of our finding in its basic assumptions complicating factors of an *ad hoc* or arbitrary nature. Secondly, only of the simpler of the two or more rival theories, in our sense, could it legitimately be said that it (or its T_{RP}) might serve as a better basis for the future development of scientific theory in its field. Thus, simpler a theory is, in our sense of the term, greater is the possibility of our finding in its structure and implications a basis for the future development of scientific knowledge in its field.

8.3. METHODOLOGICAL IMPLICATIONS OF EPISTEMIC STRUCTURALISM

Our attempt to answer the question concerning the nature of a falsifiable *theory*, we have raised at the very outset, might be said to shift its fundamental aspects beyond methodological conventionalism, demanding a conceptualization which is not only deeper but realistic in nature. One of the principles, epistemic structuralism,[47] with which this task might be accomplished, will be a subject of discussion in § 8.6. below. This principle addresses itself to possible types of *structure-property correlations* as a characteristic feature of all growing objective knowledge systems in the sciences and, therefore, as a subject for further investigation. It constrains these *systems* in a general but significant manner, since it says that the epistemic structural codeterminants of their structural identity – viz., *theories* and *problems* in their many-faceted interactions in the individual sciences – are at the same time their developmental structures.

If empirical science aims at ever growing systems of objective knowledge, as I believe that it does, then it could not conceivably do so without this deeper principle holding true. Moreover, it can now be seen clearly how one might answer the question above concerning the nature of a falsifiable theory mainly in terms of this very principle as follows:

> A falsifiable theory is falsifiable in so far as it is a developmental structure of objective knowledge; and to test its falsifiability is not just to make a move in the Popperian 'game' of science but to test its fundamental character as such a structure.

With this, the falsifiability requirement, as formulated by Popper, stands exposed as to its fundamental inadequacy: *it leaves scientific problems essentially, if not entirely, undemarcated*. (Further discussion on this point follows in § 8.6. below).

Moreover, no attempt to demarcate science from non-science just structurally can succeed. That a science is not simply a structure of some sort – generally viewed as a hierarchical one of either reductive or deductive relations between statements/theories – but a growing system makes necessary the kind of demarcation which is both structural and dynamical, i.e., which satisfies both synchronic and diachronic constraints of science. Proceeding from the broadly conceived framework of methodological essentialism as it does, epistemic structuralism demarcates science in two steps as follows: (1) The structural identity of science is codetermined by its methodologically constrained type-distinct epistemic structures as abstract as *theories* and *problems* in their mutual relations of interaction. And (2) these epistemic structural codeterminants of objective knowledge systems in the sciences are correlated with the holistic properties of such systems, namely their development and growth in time. Thus, they may also be called developmental structures of knowledge (see Figure 7.1).

Epistemic structuralism does not imply that these epistemic structures (= developmental structures) are available to the scientist each in its individual capacity and in isolation from one another. On the contrary, it implies that a given theory will be available to the scientist always in relation to a problem. Similarly, a given problem will be available to him in relation to a theory. Details of this implication were discussed by us before in Chapter 3 above, showing how it goes a long way taking care of Quinean scepticism concerning propositions. Now it is methodologically interesting to consider how the strict non-availability of theories and problems in the present sense radically changes our conception of *background knowledge* from a Popperian statement-view to a non-Popperian view. For Popper, 'background knowledge' denotes "all those things which we accept (tentatively) as unproblematic while we are testing the theory",[48] such as initial conditions, auxiliary theories, and so forth. This definition, accepted as it is by the followers of Popper, tells us, I think, more about the Popperian *attitude* to *background knowledge*, when taken in relation to a theory under test, but less about what *it* may amount to structurally. However, a homogeneity assumption stemming from the statement-view of science is, as it were, built into this conception by allowing background knowledge to be either true or false. The (Popperian) theory under test and the 'background theory' are thus governed by mutual consistency. On the contrary, what epistemic structuralism implies is the conception of background knowledge as a theory-problem interactive system in the background: A methodologically constrained system of theories and problems in interaction, and not just of theories alone. And it is this which explains both the possibility and necessity of epistemic appraisal in science along the following lines (to be discussed in detail in Chapter 9): $T_{RP}(T_2) > T_{RP}(T_1)$. More precisely, epistemic structuralism says that there must be methodologically significant *structure-property correlations* across theory-problem interactive systems of the type

$$I(T_1, P_1), I(P_2, T_2), I(T_3, P_3), \ldots$$

which have remained uninvestigated so far. It implies, therefore, that there is no such thing as (background) knowledge deproblematized. Any such idea, on the contrary, is just a myth. Thus, epistemic structuralism implies a *reproblematization* of all knowledge, both in respect of its structuration and that of its growth (Pandit 1989).

8.4. WHAT IS WRONG WITH THE RECEIVED VIEWS ON THE METHODOLOGY OF SCIENCE?

If the different approaches to the problem of demarcating science from non-science suffer from one kind of oversimplification or the other – e.g., science is another name for consensus, or it involves falsifiability, or it is characterized by verifiability, or it yields justified true belief – this is because of the oversimplifications committed at deeper levels that have received very little attention. It is true that philosophers of science and the practising scientist alike remain committed to the oversimple doctrine that science aims at truth. Moreover, the structural identity of its objective knowledge systems is thought to lie in the hierarchical structures of *type-identical* theoretical statements, or empirical generalizations, capable of being true or false. The rest of their structure, if any, is of epistemological interest only in so far as it is subservient to the attainment of its aim just alluded to. Given this kind of basic commitment, the whole problem of demarcation is accordingly seen as one of distinguishing the scientific statements/beliefs from those that are not scientific. At a higher level – that of scientific appraisal – the problem is consequently seen as one of appraisal of a theory's explanatory power.

The result is hardly surprising: most of the alternative methodologies of scientific appraisal proposed one after another are only variations on the methodology of T_{EP}-evaluation. This is especially true of inductivism and probabilism and of Popper's falsificationist methodology of theory-evaluation.

On the other hand, however, the recent sceptical approaches to the philosophy of science – e.g. those of T. S. Kuhn and Paul K. Feyerabend – tend to generate the impression that the very enterprise of the methodology of science is illegitimate in so far as it (allegedly) involves the 'philosophically prescriptive' exercise of legislating for science. This is exactly what Feyerabend intends to convey more persuasively through his anarchist slogan – "anything goes" – that he himself claims to "prescribe" as a medicine for epistemology and philosophy of science.[49]

I don't think that philosophers of science have ever undertaken to legislate for science in a manner suggested by Feyerabend.[50] Even if there may be some truth in this allegation – which does not apply to scientists such as P. Duhem, E. Mach, A. Einstein and a few others – one must whole-heartedly agree with Feyerabend on the main point that philosophical legislation for science from outside may not be necessary to the kind of enterprise called philosophy of science. What a philosopher of science can legitimately do, however, is to show (a) how every science legislates for itself in a manner which is reflected in its historical development as a relatively autonomous self-regulating theory-problem interactive enterprise and (b) how it does so very often in ways which are of common philosophical significance to *all* science, Feyerabend's epistemological anarchism not-withstanding.

Philosophy of science has recently tended more and more to degenerate into a game of witty strategies which are being employed by the specialists just to win over one another. We are hardly surprised to notice this in the presence of recent moves among philosophical circles suggesting, on the one hand, that science be viewed as a game among games and scientific revolutions as akin to political revolutions, or even religious conversions, and philosophy, on the other hand, as some sort of a super-game. This is true not only of the assumptions underlying Kuhn's conception of science but also of those underlying Popper's conception. What remains puzzling, however, is this. Along with his gametheoretic conception of science as being analogous to chess, Popper has brought in a scientific theory T's *verisimilitude* not only as a parameter of theory-appraisal but as a strategic means of achieving the ultimate aim of science. But for such a concept there is hardly any need or justification in science so conceived or even in a strictly falsificationist methodology of T_{EP}-evaluation.

Looked at from another side, however, commitment to T's verisimilitude seems to function as a source of compulsion for Popper to remain committed to a falsificationist methodology of T_{EP}-evaluation. There is no other way to explain the poverty of the Popperian methodology when seen against the background of his otherwise considerably rich epistemological ontology. Given an ontology as rich as Popper's, or as I have been here proposing, the question that needs our attention is this:

> Can we conceive of science, or what it aims at, without the concept of truth as a regulative ideal (in Popper's sense)?[51] And of the methodology of epistemic appraisal in science without the concept of verisimilitude, or without being committed just to T_{EP}-evaluation?

A realistic answer in the affirmative is possible, provided we conceive of science as an enterprise of theory-problem interactive systems in my sense. So conceived, it becomes possible to recognize how the concept of truth, or verisimilitude, ceases to play any role (regulative or otherwise) in the methodology of epistemic appraisal in science (a point I discuss in detail in Chapter 10 below).

In this context, an important role is played by the relationship between epistemological ontology and the methodology of science. The former is presupposed by the latter in so far as it concerns itself with fundamental questions of the structural identity of objective knowledge systems in science and the nature of correlations obtaining between their structure and their holistic properties such as their growth and rationality. Considerations such as these have led me elsewhere[52] to lay down the following principle of embedded methodology for judging any proposed methodology of science as adequate or acceptable:

> The methodological models of epistemic appraisal in a science must be properly embedded in sound models of the epistemic structures of its knowledge systems and of their growth.[53]

If a proposed methodology of science fails to satisfy this principle, it will be unacceptable. All methodologies which do not satisfy it for one reason or other are bound to lead to consequences that cannot be accepted. Such methodologies, it will be seen, either proceed from epistemological frameworks which are inherently inadequate (such as inductivistic empiricism) or hardly have any such framework to proceed from (as is the case with the methodological positions of Kuhn and Feyerabend). The poverty of a methodology in respect of its epistemological framework may be shown by those of its consequences which spell nothing but scepticism for the methodological enterprise proper. In this sense, so poor is the epistemological basis of the Kuhnian, Lakatosian and Feyerabendian conceptions of science and its method that

(a) for Kuhn there is no escape from relativism as regards the question of evaluation of a methodology of science. Thus no such evaluation makes sense, science being according to him a disciplinary-matrix driven puzzle-solving enterprise, undergoing states of normal science interrupted by those of revolutionary science;
(b) for Lakatos it is the history of science, with more and more external history becoming internal history, which is the *basis* of evaluation of alternative methodologies – a view beset with its own great difficulties;
(c) for Feyerabend one's very task will be misconceived if one tries to bind science to any fixed, universal methodological rules.

According to Feyerabend (1978a),

> Science is an essentially anarchistic enterprise: theoretical anarchism is more humanitarian and more likely to encourage progress than its law-and-order alternatives.[54]

> Indeed [Feyerabend observes] one of the most striking features of recent discussions in the history and philosophy of science is the realization that events and developments, such as the invention of atomism in antiquity, the Copernican Revolution, the rise of modern atomism (kinetic theory; dispersion theory; stereochemistry; quantum theory), the gradual emergence of the wave theory of light, occurred only because some thinkers either *decided* not to be bound by certain 'obvious' methodological rules, or because they *unwittingly broke* them.[55]

> More specifically [he continues], one can show the following: given any rule, however 'fundamental' or 'necessary' for science, there are always circumstances when it is advisable not only to ignore the rule, but to adopt its opposite.[56]

This last remark is quite noteworthy in the present context. What it advocates is not the idea of science, or scientific progress, without methodological rules. It advocates the opposite idea that this ('liberal') *practice* of solving difficult scientific problems by (a) breaking the antecedently given rules and (b) adopting entirely new ones is "both reasonable and *absolutely necessary* for the growth of knowledge".[57] Interpreted in this manner, it would seem fairly reasonable to accept Feyerabend's anarchist view of science. It would not be correct to look at scientific progress just as a function of the violation of the antecedently given methodological rules, nor at a given case of violation just as a violation pure and simple. For every violation of a rule brings in its fold some different, even opposite, rule.

From this one might easily conclude that even the possibility of violating a given rule on a particular occasion must rest on a different, and more fundamental, set of rules of a particular scientific activity. If the growth of knowledge systems in a science is viewed as a function of *methodological variance* in its field in the sense of its constant movement from one set of rules (and strategies) to another, one must sooner or later face the following question: What is the detailed nature of those principles which underlie methodological variance in a science in this sense? It is at this point that Feyerabend's anarchism has so far escaped scrutiny.

Methodological rules of subject-specific relevance to a science – and it is these which Feyerabend has in mind most of the time, given his examples – are no doubt always present there and meant not only to be followed but also to be broken in times of crises. And new rules must be adopted or discerned as having been adopted in their place. But can this whole process be entirely arbitrary or random? An anarchist like Feyerabend would reply in the affirmative, suggesting that science itself is either an aimless enterprise or a random activity. For it is this question itself which Feyerabend's slogan "*anything goes*" not only anticipates but seeks to answer, if only anarchistically. This slogan, says Feyerabend (1978b), "is *not* the one and only 'principle' of a new methodology, recommended by me. It is the only way in which those firmly committed to universal standards and wishing to understand history in their terms can describe my account of traditions and research practices . . . If this account is correct then all a *rationalist* can say about science (and about any other interesting activity) is: anything goes".[58] Feyerabend has elaborated this point further as follows:

But 'anything goes' does not express any conviction of mine, it is jocular summary of the predicament of the rationalist: if you want universal standards, I say, if you cannot live without principles that hold independently of situation, shape of world, exigencies of research, temperamental peculiarities, then I can give you such a principle. It will be empty, useless, and petty ridiculous – but it will be a 'principle'. It will be the 'principle' 'anything goes'.[59]

It is, of course, clear that anarchism in this sense advocates radical scepticism on fundamental questions of method. Thus, the very question whether there are any subject-neutral abstract principles true of all science does not arise at all. For while it is being admitted that scientific progress in a science involves methodological variance, the problem of what makes methodological variance itself, and appraisals thereof, possible is simply dismissed as a pseudo-problem.

We have seen how Feyerabend grants to each and every science its freedom to frame those rules which it must violate from time to time in concrete problem-situations. What he has failed to recognize is this: Over and above the subject-specific methodological rules that may be present in a science, there are subject-neutral methodological rules which are valid for all science. It is the rules of this latter, more abstract, category which he has sought to rule out by formulating his slogan: "Anything goes". The only explanation possible for his position, which argues quite *naively* against *common* methodological rules in science from the fact that every individual science violates its own subject-specific rules from time to time while making progress, is then this. What Feyerabend's conception of science together with his scepticism concerning universal methodological rules lacks is an appropriate grounding in a sound epistemological framework, thereby violating the requirement alluded to above. Thus consider his argument:

Every science must sooner or later violate its given rules and adopt their opposites. Hence there are no permanent, abstract rules not to be violated as required by rationalists.[60]

Does this argument threaten the existence of universal rules of method at all the relevant levels? The argument is mistaken, since Feyerabend argues here essentially from a true premiss to a false conclusion. There are at least three reasons why it must be rejected. First, methodological rules in a science need not invariably be assumed as *given* explicitly. On the contrary, many of them are learnt by example mostly from historical scientific practice. In the history of physics, for example, it has often been the case that the physicists did not clearly realize in which rules (and strategies evolved therefrom) their enterprise had been entangled until the violation of those rules took place before them in the proposal of a revolutionary theory going beyond the known explanatory tasks of their own time. Secondly, violation of non-linguistic methodological rules makes sense only in the context of their normative universality. No doubt, they will always be man-made. But they are not conventions, in Popper's sense, made just to be violated. On the contrary, their essential abstract character remains a subject for further investigation. Thirdly, and consequently, from the fact that every science *must* make, or has made, progress not by a dogmatic adherence to its *subject-specific* rules but by violating them – that it is these which Feyerabend has in his mind is what I infer from his examples which are themselves subject-specific – what one can conclude generally is the following:

> *As a rule*, scientific change in and across the individual sciences is possible by violating their subject-specific methodological rules of theory-construction and problem-formulation from time to time and adopting their opposites or incompatibles.

But before they are violated, they must be *developed* to the fullest level possible and then *tested* by means of *strategies* appropriate to them in respective fields (Pandit: 1983). This clearly implies that scientific change involves causal dynamics responsible for methodological variance in the sense of Chapter 6 above. There arises, therefore, the most significant problem: What is the nature of those principles which make methodological variance in science possible? Are there deeper principles without which it would be just impossible to rationally appraise the states of scientific change taking a science in a certain direction? But it is not possible to deal with such questions without looking for deeper principles relating to what I have chosen to call theoretical universals.[61] This may also explain how owing to epistemological poverty one may easily land oneself in scepticism concerning method and resort to slogans like "Anything goes".

How seriously should we take, then, Feyerabend's (1978b: 188) claim that the slogan sums up the predicament of the rationalist who searches for (subject-neutral) universal methodological rules/principles for science? It implies that rationalism, or essentialism in its present sense, must invariably end in anarchism. But, as we have just seen, this is exactly the point one cannot reach from the main premiss of his argument. On the contrary, I think that I have already shown in Chapter 5 above that it is not rationalism (nor essentialism) but methodological conventionalism in Popper's sense which must end in anarchism when it takes its extreme form.

8.5. THEORETICAL UNIVERSALS AND THE PRINCIPLE OF THE RESOLVING POWER OF A SCIENTIFIC THEORY

Henry Margenau (1982) singles out two features of scientific method in the context of the problem of reductionism/reducibility:

> one is inherent in the entire scientific enterprise and is so clearly understood by its workers that it often escapes discussion. It is the temporary character of scientific truth, its continual need of refinement, extension, and at times rejection followed by new research. . . .[62]

I think that Margenau's point can be understood as relating to *statements* of scientific theories, their very pervasive features when considered as epistemic structures of science. These pervasive features have, I think, essentially to do with their character as developmental structures of objective knowledge. It is this feature of a scientific theory – i.e. the feature of being an epistemic and a developmental structure of objective knowledge – among other things, which I call a *theoretical universal*; and it is this which, I believe, is intended to be focused here.

The second feature, on the other hand, is described by him as having to do with what he calls the metaphysical principle of "extensibility of constructs":

> It is usually employed as a criterion for eliminating useless theories; it is an elaboration of Occam's razor and demands that of two theories, both of which explain a certain range of observations (P-facts), the one with the wider range is to be accepted. This may mean that, with a certain reinterpretation of constructs two theories covering different fields merge into a single, larger one, or it may mean that one will be rejected, or, less frequently, that deeper understanding reveals two theories to be one. . . .[63]

Margenau's description of the second feature, while again quite significant in itself, puts a lot of ambiguity into what he calls the metaphysical principle of "extensibility of constructs". Moreover, it is not at all clear why such a principle be dubbed "metaphysical". The underlying point will however become clearer if we speak, instead, of what I have elsewhere[64] called the principle of (ever-increasing) resolving power (T_{RP}) of a scientific theory.

Once it is recognized, as Margenau does, that one of the universal features of a scientific theory has to do with its T_{RP}, say in the sense of my (1983), it is possible to work out a rationale for epistemic appraisal of theories on the basis of their respective resolving powers. It would then follow as a general methodological rule that a theory T_2 is to be preferred to a given theory T_1 relative to a given problem-domain on the basis of the following proposal:

$$T_{RP}(T_2) > T_{RP}(T_1)$$

But the concept of T_{RP} is not as simple as that. It raises issues that go beyond our traditional conceptions of the methodology of science, or of scientific appraisals. The concept of T_{RP} is definable as a theory's power to give rise to significant changes in problem-formulation at various levels of what I regard essentially as processes of theory-problem interaction. Thus only if T_2 is *able* to trigger greater or powerful changes in problem-formulation by means of what it carries in its conceptual train can we say: $T_{RP}(T_2) > T_{RP}(T_1)$. And this implies that T_2 carries newer penetrating concepts of greater theoretical depth[65] in its conceptual train/network. Resolving power, however, belongs to a theory as a whole, not to the concepts themselves which it might *appear* as introducing in their individual capacity.

Thus it is important to note that any appraisal to the effect that $T_{RP}(T_2) > T_{RP}(T_1)$ will entail methodological appraisals at the deeper level of problem-formulation, problems being subject to appraisal as to their epistemic *bases* in theories. Problem-formulation being a function of T_{RP}, like theories problems are, in my view, open to methodological appraisals along the following lines:

(1) Is there a set of *testable* theoretical assumptions such that the formulation (f_1) of a given problem (P) – $f_1(P)$ – could be traced to it, and only to it, as its theoretical basis?
(2) What is the detailed structure of the *theoretical basis* of $f_1(P)$?
(3) How does $f_1(P)$ compare with an alternative formulation $f_2(P)$, with an alternative theoretical basis of its own?
(4) Whether problem P_1 is $T_{RP}(T_1)$-determined? Or, is it just *ad hoc*? Whether P_1 could be reformulated as P_2 in terms of $T_{RP}(T_2)$? And so forth.

I believe that the methodological appraisal of problems in the proposed sense would make it less tempting for the scientist to resort to *ad hoc hypotheses*,[66] particularly when he finds himself placed in most difficult problem-situations. In the course of its historical development, such hypotheses are known to have arisen in science. And, in their turn, they have often given rise to problems incapable of interacting with any theory. Such problems are better recognized as *pseudo-developmental* structures of knowledge, being just incapable of interacting with any theory at any level of problem-formulation or problem-determination. This shows how methodologically powerful the correlation between the epistemic structures of objective knowledge systems and their developmental structures is in science. In this sense, epistemic structuralism implies *growth* itself as a methodologically invariant property of such systems. As an invariant property of theories and problems in mutual interaction, it formalizes the property of being a developmental structure of knowledge as a *theoretical universal.*

The structure-property correlations of the type under discussion can also appear in the form of methodologically invariant constraints on essentially variable methodological rules of problem-formulation and theory-construction in a science. Thus they are implied by more broadly conceived principle of methodological essentialism[67] as follows:

(a) A science is always interested in problems as a subject of scientific discovery in their own right. These include particularly those problems which can be formulated by *what*-questions, *whether*-questions and *why*-questions. Their possible solutions can always be formulated as unrestricted universal hypotheses and theories.

(b) *Its* theory-problem interactive systems, and the search-cum-discovery procedures, acquire deep methodological significance as and when they undergo methodological variance in the sense of Chapter 6 above.

Theoretical universals in their role as the methodologically invariant properties of theory-problem interactive systems in a science thus assume deep significance in this context. The requirement that the scientific problems share common logical forms may at first sight appear as of little methodological significance. But any doubt can be easily dispelled if we remember that methodological constraints common to the sciences can be seen at work in their respective subject-specific rules of problem-formulation and theory-construction. For example, their search-cum-discovery procedures are always constrained by the ideal-type assumptions of simplicity and the structural-dynamical assumptions of complexity they choose to employ from time to time. Thus principles of invariance operate at a fundamental methodological level, constraining the *types* of subject-specific problems and theories which the sciences can legitimately raise and develop.

According to the present view, then, *problems* and *theories* in a science are methodologically constrained type-distinct epistemic structural codeterminants of its objective knowledge systems. Theory-problem interactive systems in a science, as we have chosen to call them, are essentially linguistically neutral. But they are embedded in language itself considered as a *C-H* system, since they can be formulated, with lesser or greater accuracy, within alternative language frameworks. In this sense, all languages generally may be considered as offering alternative ways of formulating the methodologically constrained theoretical identifications of natural phenomena, as also problems and theories concerning them.

8.6. THE METHODOLOGY OF THEORY-PROBLEM INTERACTIVE SYSTEMS

It is quite reasonable to proceed further in our discussion by making an assumption of complexity concerning the universe as a whole, which is as general as follows. We live in a universe of diverse kinds of interactions between diverse kinds of systems which may be deterministic or indeterministic in nature. These systems are all, in their turn, *interactive systems* of different orders of organization. In this context, it is important to understand clearly the kind of role *observation* and *experimentation* play in the individual sciences. They serve, in my opinion, to regulate the processes of the growth of the objective knowledge systems. They do so by opening up before the scientist *newer* fields of interaction that every powerful scientific theory predicts. Thus, when an astronomer 'discovers' a new star-system already predicted by a theory, he really discovers a new field of human interaction and also of theory-building. But what about a *sudden* 'discovery' of such a system, hitherto quite unexpected/unpredicted on the basis of a known theory? Such instances in science can be reasonably interpreted as cases of *stumbling on a new complexity* in our universe. It is only the relevant background systems of knowledge (= theory-problem interactive systems) which enable us to recognize them as extending further not only the fields of significant human interaction but also the subjects of objective knowledge.

Against this general interaction-theoretic ontic assumption of complexity, let us recall epistemic structuralism, the ideologically neutral epistemological-ontological principle I first set forth in my (1983) as a basis for the methodology of science. At the time of formulating this principle I was perhaps far from being clear about its precise content and significance. As a result, I did not set forth its precise formulation. But I think that a precise formulation is not only desirable but necessary if we are to understand it in its full significance. In the brief discussion in § 8.3. above, we have seen how this principle is concerned with the structure-property correlations in the field of the growth of the objective knowledge systems in the sciences. Thus it attributes a certain fine structure, *epistemic structural complexity*, to all such systems and, then, *correlates* it with holistic properties of causal dynamics, growth and rationality which these systems exhibit in the course of the historical development of each science. As we proceed further in our discussion, it will turn out that this seemingly simple principle is quite complex, being a conjunction of at least three postulates or assumptions (A)–(C), that make it what it really is. The postulate (A) is really the fine-structure assumption that all objective knowledge systems possess a well-identifiable structural complexity, which has remained uninvestigated so far. The postulate (B) provides for *theories* and *problems* as type-distinct epistemic structural codeterminants of this (synchronic) structural complexity in objective knowledge systems. And the postulate (C) correlates the structural complexity of these systems thus identified with their holistic properties of causal dynamics, growth and rationality as follows:

(A) *The fine-structure assumption*:
Objective knowledge systems in science possess a well-identifiable structural complexity implying an essential structural identity for these systems in which epistemic structural invariances play an important role.

(B) *The assumption concerning the nature of the synchronic structural complexity of objective knowledge systems*:
Scientific theories and problems are the type-distinct epistemic structural codeterminants of this structural complexity implying that the objective knowledge systems in science have the structural identity of theory-problem interactive systems.

(C) *The dynamical assumption of the structure-property correlations*:
Objective knowledge systems in the sciences show significant structure-property correlations which make these systems highly dynamical interactive systems and their individual epistemic structures highly developmental structures of knowledge with properties causally relevant to *its* growth. The entire causal dynamics of the growth of objective knowledge systems depends on which type of structure-property correlations they actually show.

In my (1983), I then went on to building up what might appropriately be called a *causal dynamics* of the growth of objective knowledge systems in science. Thus I tried to work out a theory of how objective knowledge systems grow as complex theory-problem interactive systems in which theories and problems appear as the chief interacting elements. In this model, important role is played by the methodological concepts of *problem-determination*, or problem-formulation, in a science as a function of the *resolving power* of a theory (= T_{RP}) on the one hand and *problem-solving* as a function of its explanatory power (= T_{EP}) on the other. In Chapter 9 below, I shall state my theses in this field in greater detail. On a closer look, most of these will look like theorems based on the postulates (A)–(C) of epistemic structuralism. Some of the most important among these might be restated more precisely as follows (i)–(v):

(i) The dynamics of the *growth* of knowledge, a function of theory-problem interaction, involves important principles of methodological variance.
(ii) Of every scientific theory T, it is true to say that if it possesses explanatory power (= T_{EP}), then it also possesses resolving power (= T_{RP}), but not *vice versa*.
(iii) Of any two (or more) rival scientific theories T_1 and T_2, it will be true that if T_2 has excess explanatory power over that of T_1, then it has also excess resolving power over T_1, but not *vice versa*.
(iv) The growth of scientific knowledge follows a dynamical path characterized by the dualism of *working forwards from theories to problems* such that
$T_{RP}(T_2) > T_{RP}(T_1)$
and working backwards from problems to theories such that
$T_{EP}(T_2) > T_{EP}(T_1)$.

The kind of variance in epistemic structures (= ES-variance) being implied here is known to take place in the individual sciences in the form of *correlated* theory-change and problem-change on the one hand and *CSD*-variance as a more radical form of methodological variance on the other. Thus, we can say that

(v) growth of knowledge is a function of theory-problem interaction at various levels of methodological variance, viz. (a) at the level of scientific revolutions, themselves a function of *CSD*-variance – i.e. a function of our working our way backwards from problem-situations of rule-entanglement to rule-enriching creativity; and (b) at the level of ES-variance as triggered by *CSD*-variance.

Given, then, our characterization of the growth of scientific knowledge and its causal dynamics (i)–(v), one might ask whether the resulting model satisfies the adequacy conditions, formulated in my (1983), on a model of the growth of scientific knowledge. For example, one might like to know whether my model is really an *interactive* model of the sort which relates the growth of the objective knowledge systems in the sciences to the relevant parameters of methodological, or scientific, variance, thereby depicting it as a function of multi-variable interaction. I think that if we look at science, or its objective knowledge systems, through this model, it will turn out to be the kind of enterprise comprising the *NFCPS* systems at work in all the various sciences. And if this is really so, then the model is an acceptable one in the sense of the adequacy conditions just alluded to.

In Chapter 9 below, I shall extend further some of my arguments, developed in my (1983), with a view to showing in greater detail how our models of the epistemic structural (and of course developmental) codeterminants of objective knowledge systems in science must be taken as *embedding models* if they are to yield or generate any significant methodological models of epistemic appraisal with a sound epistemological-ontological basis. Thus, applying this consideration in the present case, the growth of scientific knowledge will be illustrated and gauged in any instance of theory-problem interaction involving two theories T_1 and T_2 by means of appraisals of the following type:

$$T_{EP}(T_2) > T_{EP}(T_1)$$
$$\textit{and}$$
$$T_{RP}(T_2) > T_{RP}(T_1)$$

This implies clearly that this growth is a function not only of *theory-change* but also of correlated *problem-change* in a science in the following sense. Relative to a background theory T_1, a new theory T_2 will constitute a really better theory and a (progressive) step forward in the growth of knowledge only if T_2 not only explains more than T_1 does but *determines* greater number of *newer problems* than T_1 does. But how is, then, correlated theory-change and problem-change in this important methodological sense possible? What makes it possible at various levels? The answer lies in (v) above. In other words, these constitute what I have designated as ES-variance which is itself shown to be a function of methodological variance at various levels including *CSD*-variance.

It also implies, then, an important methodological role for search-cum-discovery procedures with which a science can aim at theories of type T_2 with ever greater explanatory and resolving powers instead of those which may be described as being more or less *truthlike* (in Popper's sense). For it seems to me quite clear that the aim of getting at theories with ever greater explanatory power is neither achievable nor feasible without, at the same time, aiming at theories of ever greater resolving power. And this seems all the more appropriate within an objectivist-fallibilist as well as realistic-interactionist framework *without* any commitment either to conventionalism in any form or to *truth* as a regulative ideal. The two kinds of aims are thus closely intertwined with each other. If a science is to aim at theories of ever greater explanatory depth to solve the antecedently determined problems, it is necessary that it employs appropriate search-cum-discovery procedures and ensures a recurrent process of self-correction, as also a correlated process of theory-change and problem-change. But such procedures require really not just formulation but discovery of newer and deeper problems which are beyond the scope of the given theory. Once these problems are scientifically determined and formulated, they will expose the given theory's inadequacies and oversimplifying assumptions. And the problems of ever increasing depth in this sense would be forthcoming only as a function of what I have chosen to call the *resolving power* of a scientific theory.

The methodology of theory-problem interactive systems,[68] then, picks up neither the Popperian individual theories nor Lakatosian research programmes (= *series* of propositions as Lakatos calls them) as the units of rational epistemic appraisal in science. On the contrary, without resorting to any strategy of changing the very subject of such appraisal in science, it identifies *it* with the theory-problem interactive systems themselves. This identification must be seen as a consequence of the postulates (A)–(C) of epistemic structuralism already considered above.[69] They enable us clearly to identify *theories* and *problems* as those type-distinct epistemic structures which codetermine the structural identity of a science. It is here that non-trivially significant *structure-property correlations* in the growing systems of knowledge assume great importance, with *theories* and *problems* appearing as their structural invariants on the one hand and developmental structures on the other. They assume special methodological significance indicating universal constraints on the structure, growth and appraisal of scientific knowledge systems. We should, therefore, look for such correlations in every theory-problem interactive system

$$I(T_1, P_1), I(P_2, T_2), I(T_3, P_3), \ldots$$

that we can identify in a science. For example, the universal property of being a *growing knowledge system* – $G(K_s)$ – can itself be shown to be correlated with the very structural identity of a science as codetermined by its *theories* and *problems* in their many-faceted interactions. Consider, again, the kinds of constraint in terms of the methodology of scientific appraisal, which have to do with structural properties as simple as a theory's *non-ad hoc-ness*. The kind of constraint I am here hinting at may be expressed as follows. In science it is the theory-problem interactive systems themselves, instead of either the individual theories or series of theories in the sense of Lakatos,[70] which we must subject to epistemic appraisal. It may be quite useful to invoke the familiar criteria like independent testability to find out how *ad hoc* or *non-ad hoc* a newly introduced theory/hypothesis really is. But it is the scientific problems themselves as a subject of scientific discovery and, therefore, as developmental structures of the growing knowledge systems which can be shown to play a vital role in the scientific procedures of theory-replacement and theory-modification which are *non-ad hoc*. Scientific theories too, in their turn, play a vital role in all those scientific procedures of development of problems, whether by discovery or by reformulation, which are *non-ad hoc*. And yet problems do not figure anywhere in the Popperian or Lakatosian methodology of scientific appraisal either as a subject of scientific discovery or as developmental structures of growing knowledge systems.

The postulates (A)–(C) of epistemic structuralism also entail a set of adequacy conditions at a more fundamental level of the demarcation problem itself as follows:

(a) A demarcation criterion must be *comprehensive* enough to cover all the type-distinct epistemic structural codeterminants of objective knowledge systems in science; and
(b) it must be consistent with the structure-property correlations

which characterize these systems, particularly where their structure is correlated with properties like growth and rationality.[71]

Since it is quite reasonable to identify *theories* and *problems* as the epistemic structural codeterminants of objective knowledge systems in science, an adequate demarcation criterion must be such as defines the scientific character not only of theories but also of problems. It becomes now possible to criticize Popper's falsifiability principle for falling short of this very requirement. The same is true of Lakatos's methodology of scientific research programmes.
The adequacy conditions (a)–(b) above together entail then

(c) A demarcation criterion must demarcate both theories and problems as being scientific, keeping in view their character as developmental structures of objective knowledge systems. And, therefore, it must not admit as scientific anything which may be a pseudo-developmental *ad hoc* structure.

The condition (c), in its turn, entails the following constraints on the methodology of epistemic appraisal in science: (i) the appraisal must be addressed to interactions between *theories* and *problems* as the developmental structures of objective knowledge systems in a science; and (ii) *it* must be ideologically and historically neutral. Thus it must be quite independent of any considerations of one's (rational or non-rational) belief in specific theories and problems or in their historical aspects where their origin is concerned. Thus, if we recognize *theories* and *problems* as the epistemic structural (and developmental) codeterminants of objective knowledge systems in science, their epistemic appraisal will address itself to their interactions and to dynamic replacement of one particular theory-problem interactive system by another such system. And the appraisal will be neutral, ideologically as well as historically: whether anyone believes in a given system or not and the question of *who* discovered, or invented, it and *how* will all be irrelevant to such appraisal.

This particular consequence is of great significance, particularly in view of the following fact. Many philosophers, including the epistemological anarchist Paul Feyerabend, look at science as a family of beliefs among other similar families such as religion, political ideology, magic, witchcraft and ritual belief and so forth. Consider, again, the Kuhnian pragmatical-sociological conception of science, in effect bifurcating it into the steady-state phases of 'normal science' on the one hand and the destabilizing but shortlived states of 'revolutionary science' on the other. Wolfgang Stegmüller's (1976) reconstruction of it as a pragmatical non-statement view of the structure of science is also noteworthy in this context. The concept of a *scientific community* and that of *belief-formation* in such a community are, interestingly enough, invariably presupposed here as primitive concepts. Many others, on the other hand, are inclined to take a less extreme view of science. They simply look at it as an enterprise comprising systems of *rational belief*. It is hardly surprising that much misunderstanding concerning the epistemological and methodological aspects of science stems from this sort of simplistic subjectivistic approach.[72]

CHAPTER 9

THE RESOLVING POWER OF A SCIENTIFIC THEORY AS A BASIS OF ITS EPISTEMIC APPRAISAL

> The one principle – of designing experiments – is to design them so that they give an answer.
>
> David Keilin [As quoted in C. H. Waddington (1977)].

> In my judgement, theory has a double role to play in astronomy; the common one of providing interpretations for observed phenomena; and the uncommon one of providing for astronomy the kind of basis which experiments provide for physics. The latter role is largely unrecognized and largely not practised.
>
> S. Chandrasekhar (1974)

9.1. METHODOLOGICAL VARIANCE: FROM NEWTONIAN TO EINSTEINIAN THEORY-PROBLEM INTERACTIVE SYSTEMS

What does then, one might now ask, empirical science aim at? An adequate answer to this question will depend on how we view a scientific theory, in its nature and function, in particular in its relations of interaction with scientific problems themselves. The contemporary debates about the nature of quantum mechanics as a physical theory, that originate from Albert Einstein's penetrating criticisms of it,[1] are, I believe, quite significant and instructive in this context. I shall refrain from going here into the philosophically interesting aspects of this unended debate about quantum mechanics, as I intend to do so elsewhere. It should suffice here to point out that one of the assumptions on which Einstein's criticism is based relates to his idea of a *complete* physical theory.[2] The questions about quantum mechanics that Einstein asked himself and tried to answer quite naturally raise then questions of a similar nature about scientific theory in general, such as the following: When is a scientific theory a *complete* theory? Or, what makes a theory a complete scientific theory? I think that one could deal with the subject-specific question of the completeness of a physical theory such as quantum mechanics in a philosophically much better way if one first dealt with the more general question of the completeness of a scientific theory. But, unfortunately, the latter question has either remained unasked or it has received no attention at all. What I propose to do in what follows, as regards this question and the one concerning the aim of science, is this: I shall briefly indicate my own answer to both of them before we proceed further with the main subject of our discussion.

I think that scientific knowledge, in different fields, has already grown sufficiently enough for us to notice clearly how its structure is codetermined by theories and problems together, instead of by theories alone. A scientific theory T is definable, from my point of view, as (methodologically) *complete* if and only if it has the following major *properties*:

(1) The basic postulates, and auxiliary assumptions, of T entail, of course within the possibilities of experimental error, solutions to most of the problems in the *set* of antecedently determined problems that clearly fall in its field, provided that in the process of doing this (i) it resorts to no *ad hoc* assumptions and (ii) it receives greater empirical support than what it would have received if it merely reproduced the empirical success of those theories which either precede it or are its rivals in its field.
(2) The basic postulates of T entail *novel* problems, most of which have the logical form of *whether*-questions, such that T's framework itself is able to suggest novel experiments to find accurate solutions – a clear 'yes' or 'no' – to them (of course within the limits of experimental error).
(3) And, while T may not offer solutions to all those problems that it inherits or not suggest novel and bold experimental arrangements to test the *theoretical bases* of all those novel problems which it itself entails, it should provide significant and harmonious reformulations of those problems to which it is incapable of offering direct solutions.

In this sense, the completeness of a scientific theory has to do not with its truth or likelihood to be true, but with its logical or theoretical simplicity in the sense of Chapter 8 above. This explains, in a nutshell, the entire point that underlies the general methodological rule, I have proposed in my (1983), that every good scientific theory T must be susceptible to critical epistemic appraisal of its explanatory power, $T_{EP}(T)$, on the one hand and its resolving power, $T_{RP}(T)$, on the other. As is implied by the postulates of epistemic structuralism (see Chapter 8,

§ 8.6.), every individual empirical science can aim at theories of ever greater T_{EP} and T_{RP}. If we assume a strong correlation, in the objective knowledge systems in the sciences, between their epistemic structural codeterminants and their holistic properties of development and growth, with both *theories* and *problems* in science being accorded the epistemological-ontological status of such structures, we can say this: There must be not one but two aspects to every scientific theory T as a science moves on from one problem-situation to another, viz., (1) $T_{EP}(T)$, with which it solves *antecedently known/determined* problems, and (2) $T_{RP}(T)$, with which it *determines* more problems than it can, or is intended to, solve. This will hold good even if we may view its predictive power not as an independent aspect in itself but as that which can, in principle, collapse into $T_{EP}(T)$, a successful prediction being always viewed as that which collapses into the explanatory power of a given theory.

Since neither of these two concepts – $T_{EP}(T)$ and $T_{RP}(T)$ – is absolute in character, a theory will have greater resolving power than its rival only in so far as it is able to shift the scene of problem-formulation from the field of the rival theory to its own, thereby raising more theoretical problems than it can solve or cope with. And only when a theory is capable of doing so can it serve as a developmental structure of objective knowledge. An *ad hoc* 'hypothesis' does, I believe, precisely the opposite: it renders a developmental structure – a given theory caught in the spiral of a difficult problem-situation – as a pseudo-developmental structure.

Which is, then, the more fundamental aspect of a scientific theory, its resolving power or its explanatory power? This question is already anticipated, in my (1983: 98–101), in the form of the very suggestion that a theory or an assumption with T_{RP} but without T_{EP} is conceivable, while a theory with T_{EP} but without T_{RP} is not. I still think that this is essentially correct. It is, then, the resolving power of a scientific theory which is its most fundamental aspect, guaranteeing its character as a developmental structure of objective knowledge. Moreover, the explanatory power of a theory presupposes its resolving power. Hence, relative to a theory T_1, another theory T_2 will have (greater) T_{EP} only if $T_{RP}(T_2) > T_{RP}(T_1)$, but not *vice versa*. This implies the concept of $T_{EP}(T_2)$ as methodologically parasitic upon $T_{RP}(T_2)$. This should not be really surprising once it is recognized that T_2, relative to T_1, derives its T_{EP} from all those theoretical identifications which it makes through its basic assumptions. Thus, e.g., T_{EP}(GTR) depends on such theoretical identifications as are implied by the novel physical assumption that the presence of matter in space gives rise to a gravitational field thereby altering the Euclidean character of space into the non-Euclidean by causing material bodies moving in it to follow a curvilinear path. The kind of novel theoretical identifications built into the fundamental principles of GTR that I have here in mind is aptly described by John Stachel (1986b) as follows: "Even more novel than the introduction of a non-flat space-time structure into physics was the 'dynamising' of the structure: the metric tensor not only acts upon matter, but is acted upon by it".[3] It is then these and related theoretical identifications in which resides T_{RP}(GTR) $\rightarrow$ T_{EP}(GTR). That this is so can be gauged by a consideration of the methodologically significant *scene* of physical problem-formulation which the GTR shifts to its own field as follows:

(1) *The classical (Newtonian) problem*: What are the action-at-distance forces with instantaneous effects on material bodies moving in a non-inertial/non-Galilean accelerated frame of reference?

It is the problem underlying this *what*-question, with a propositional/methodological basis of its own, which the Newtonian gravitation theory (NGT) seeks to solve, bringing in its T_{RP} newer theoretical identifications. The question assumes, among other things, as its propositional basis the proposition that there are action-at-distance forces of the required type that are susceptible to exact theoretical identification. However, this question, along with its propositional basis, is rendered *unnecessary* and *replaceable* by a different set, provided we assume, among other things, the presence of *physical fields* of a continuous type where previously we assumed the existence of action-at-distance forces. But we must then additionally assume – as Einstein's special theory of relativity (STR)[4] does as an answer to the question whether our universe is a local or a non-local one, i.e. "whether there is a limiting velocity for physical interactions/signals in the universe?" – that the velocity of light, c, is the limiting velocity prohibiting velocity $v > c$. The result is precisely what one gets at in the Einsteinian reformulation of the classical (Newtonian) physical problems as follows:

(1a) What kind of structure and dynamical laws must be invoked to describe and explain the (complexities of the) behavior of a material body under the influence of a gravitional field?[5]

Given the assumption, e.g. in Einstein's STR, that with respect to the Galilean reference-body *K*, a ray of light always traverses a rectilinear path with the velocity *c*, the GTR entails the problem of curvilinearity of light rays in gravitational fields as follows:

(2) T_{RP}(GTR) → "Whether light rays are bent when passing in the vicinity of a strong gravitational field, and, if yes, with what measurable effect?"

It is not only of historical interest but of deep methodological interest to go into the whole question of development of this problem, from Newton to Einstein, since the T_{RP}(NGT) together with the resolving power of Newton's corpuscular theory of light (NCT), set forth in his *Opticks*, entails the same problem. Several aspects of theory-problem interaction, particularly in the field of problem-formulation, as a function of the resolving power of a theory are discernible in the Newtonian formulation (= determination) of it on the one hand and in the Einsteinian reformulation (= transformation) of it on the other. Significant differences in formulation of the *same* underlying problem, or of the corresponding *novel* prediction, can be seen here as a function of the differences in the resolving powers of the respective theoretical frameworks employed for this purpose. This is not to suggest, however, that the development of the problem of the bending of light rays through the attraction of a celestial body is no more complex than that. Quite the contrary. Its development is, I believe, an interesting case of methodological variance in classical physics covering a single problem but with several phases to it, *notably* the Newtonian, the Soldnerian and the Einsteinian, as we might call them. I shall here refrain from going into those phases of it that seem likely candidates for its development between the last two phases.

As early as 1801, Johan Georg von Soldner[6] (Erhard Scheibe drew my attention to Soldner's paper of 1801) formulated (= modified) the problem of the gravitational bending of a light ray, first formulated by Newton, as a problem of (calculating) *its deviation from its motion along a straight line through the attraction of a celestial body which it passes close by*. The formulation obviously *assumes* not just the Newtonian gravitation as an action-at-a-distance force but the optical law of motion of a light ray along a straight line together with the corpuscular theoretic assumption (see Jaki, 1978: 948) concerning a light ray that it consists of streams of corpuscles with rest mass. This is how a problem entailed by the T_{RP}(NGT.NCT), which Newton himself raises in the *Queries* (1–31) in the concluding Book three of his *Opticks*, receives its first decisively accurate formulation within the Newtonian framework. Remarkable for their range and logical form as classical examples of very general *whether*-questions arising as a result of the resolving power of physical theory, Newton introduces these *Queries* thus: "I shall conclude with proposing only some Queries, in order to a farther research to be made by others". In order to be sure that in Query 1, quoted below, Newton has not just reflection, refraction and diffraction but gravitation in mind, it may be helpful if we consider it together with his last Query 31 (pp. 375–376) which asks: "Have not the small Particles of Bodies certain Powers, Virtues, or Forces, by which they act at a distance, not only upon the Rays of Light for reflecting, refracting, and inflecting them, but also upon one another for producing a great Part of the Phaenomena of Nature? For it is well known, that Bodies act one upon another by the Attractions of Gravity, Magnetism, and Electricity; and these Instances shew the Tenor and Course of Nature, and make it not probable but that there may be more attractive Powers than these . . .". Let us consider now the Query 1 itself:

(2a) Do not Bodies act upon Light at a distance and by their action bend its Rays; and is not this action (*caeteris paribus*) strongest at the least distance?[7]

Now, what are the important methodological aspects of the problem of the gravitational bending of a light ray as Newton raises it as a problem among other problems? I think that these come to surface in Soldner's treatment of it (see Jaki, 1978: 939–948) which may be described as follows:

(A) Soldner *assumes* an affirmative answer to the Newtonian universal *whether*-question (= Query 1), which amounts to implicitly accepting the validity of the universal novel prediction of the gravitational bending of light rays without first subjecting it to the test of experimental confirmation.

(B) The behaviour of a light ray under gravitation thus predicted is considered as a *deviation* from an assumed (optical) law of its motion along a straight line.

(C) Deviations of this kind are then identified as those "which are considerable but, being combined with magnitudes one has not yet succeeded in clearly identifying, escape the observer" (p. 939), as opposed to those "which are so small that it is difficult to decide whether they are true deviations or errors of observation". (p. 939)

(D) The task which Soldner then sets himself is one of calculating the variable amounts of deviation in different cases depending on the celestial bodies involved and given certain "general properties of curved lines".

(E) Since the amounts of deviation calculated in different cases – the Newton value 0″.84 in the case of starlight passing close by the sun – turn out to be too small for (our) observations, Soldner concludes that "Therefore it is clear that nothing makes it necessary, at least in the present state of practical astronomy, that one should take into account the perturbation of light rays by attracting celestial bodies". (p. 948)

(F) Yet, not deterred by the result that all perturbations, calculated on the basis of theory, are unobservable, Soldner considers it worthwhile and important to know, as he puts it, "what is presented by theory".

Keeping all these features (A)–(F) in view, it is clear that Soldner is not so much concerned with the experimental possibilities of confirming a novel universal prediction as with the development of the problem of the gravitational bending of light rays as a quantitative problem of calculating the values of deviation from an assumed optical law in selected cases, of course assuming that the novel prediction itself is true. Notice that those (I have here Popper and his followers in mind) who think of novel predictions on the model of 'novel facts', that can be verified, would find such an enterprise not just uninteresting but degenerating and, therefore, disappointing if only because of its failure to produce such 'facts'.

The question which can now be asked is as follows: Do these early developments (as anticipations of the later ones) alter the nature of the underlying problem, or of the novel prediction of the gravitational bending of light rays, when considered in the framework of Einstein's GTR with its entirely new set of assumptions making newer theoretical identifications possible where there were either none or just an insufficient set of identifications available to Newton and to Soldner? If we recall how the difference in the basic assumptions of two rival theories in a common domain can lead to significant difference in formulation of one and the same problem, our answer to this question should be in the negative. The problem of bending of light rays, which the T_{RP}(NGT.NCT) entails as a (complicating) problem of deviation from an assumed law of their motion, is in a more simpler manner entailed by the T_{RP}(GTR) as a problem of the behavior of light particles as test particles in a gravitational field. Notice that for Newton, as for Soldner, the only way to incorporate light (as a stream of particles that have rest mass) into a problem of gravitation is to look at the whole problem as a two-body problem with one of them (= the center of mass) heavier than the other. Thus the assumption that light consists of particles with rest mass turns out to be crucial to calculations of the Newton value of the bending effect within the Newtonian framework. On the other hand, no such assumption is necessary for the GTR to predict or permit calculations of the Einstein value of the bending effect within its field-theoretic framework. For under the GTR, which is a 'generally covariant' theory of four-dimensional space-time structure, all particles, including those of light, are now seen as traversing geodesics in curved space-time (= the straight lines of a curved manifold or free-fall trajectories in space-time) but no more as being deflected from straight trajectories by the gravitational action-at-a-distance of a heavier center of mass. The prediction of the gravitational bending of a light ray stays, therefore, as a *novel prediction* under each of these frameworks, although in one as a new hypothesis concerning the behaviour of light with respect to the attracting celestial bodies, if only as involving a deviation from an assumed optical law of its motion, but in the other as a new hypothesis, in its own right, based on the postulates of GTR and other assumptions (e.g., that light, which has no rest mass, consists of a stream of wave fronts and so forth). The only way then to derive the gravitational bending of light rays from the NGT as a novel prediction in exactly the same sense as just considered would be to *reformulate* it, say as NGT́, so as to render it beyond recognition. And this vindicates our idea of the methodological variance as it underlines the development of essentially one and the same problem from Newton through Soldner to Einstein. Thus, in this case, as in many others, we find a truly remarkable example of how of any two theories T_1 and T_2 we can say that $T_{RP}(T_1) < T_{RP}(T_2)$, provided that T_1 (e.g. NGT) only inaccurately *determines* a theoretically significant novel problem formulated by a *whether*-question which later on undergoes more accurate reformulation in terms of the resolving power of T_2 (e.g. GTR). For we notice here quite clearly how T_{RP}(GTR) raises such problems whose solutions entail discoveries of hitherto unknown phenomena (e.g. bending of light rays by the gravitational field), thereby opening up before our view newer areas of theory-problem interaction. Of the T_{RP}(STR . GTR) we can then say, among other things, that it provides a much simpler and deeper theoretical basis for reformulation as well as solution of a problem which was significantly determined/developed for the first time by the T_{RP}(NGT . NCT).

Newton's gravitation theory, together with his corpuscular theory of light as a stream of particles, might then be said to predict the bending of star-light under the gravitational pull of celestial bodies in the sense of his Queries alluded to above. The novel prediction, if we may call it so, is clearly a function of the T_{RP}(NGT . NCT), irrespective of the fact that Newton merely expresses it as a general Query among other Queries. The Newton value (= 0″.87) of the amount of bending in the case of star-light grazing the sun, which he himself never computed, is however incorrect. Einstein's GTR makes far more accurate predictions to the same effect. It predicts specific values of the effect in the case of star-light bending while it passes in the vicinity of a gravitational field as strong as that of the sun. The experimental confirmation of the predicted value (= 1″.74) confirms that the T_{RP}(GTR) > T_{RP}(NGT . NCT). For the concept of the bending of star-light or of rays of light in general is logically not a component part of T_{EP}(NGT . NCT). On the contrary, it is a part of its T_{RP}. And it stays as a part of its T_{RP} even with GTR appearing on the scene. For this is allowed by [T_{EP}(GTR) > T_{EP}(NGT . NCT)] implying that T_{RP}(GTR . STR) > T_{RP}(NGT . NCT).

Let us consider some other similar cases:

(3) As a result of the T_{RP}(GTR), one cannot but notice the (non-Newtonian) physical problem of the structure of the gravitational field itself assuming significance as a problem essentially beyond the limits of the T_{RP}(NGT); and the theoretical problem of the equality of gravitational and inertial mass as only inaccurately determined by Galileo's law of free fall postulating uniform acceleration for bodies falling under the sole influence of the force of gravity – something unobtainable, in a simple and straightforward manner, for a falling body on assuming CM together with NGT.

(4) The general problem of the curvilinearity of the path of a

body moving in a non-Galilean accelerated frame-of-reference such as the sun's gravitational field: This problem – which can again be formulated as a set of *whether*-questions – arises as a result of Einstein's principle of equality of gravitational and inertial mass, itself a principle with great T_{RP}, taken together with the assumption of the gravitational field as determined by the curvature of the metric, or as that which the presence of matter in space itself sets up causing space to curve and giving rise to non-Euclidean space.

(5) The problem of redshifts[8] due to gravitational potential: This problem arises out of the assumption in the GTR that the gravitational field affects the measuring rods and clocks attached to non-Galilean accelerated frames of reference. The problem arises especially as one concerning the nature of the *predicted* gravitational redshifts in the spectra of distant stars and is currently a fundamental problem with implications for relativistic cosmology.[9]

Thus, there are powerful concepts, such as those of the gravitational field, space-time curvature, equality of gravitational and inertial mass and so forth, which lie at the very basis of the T_{RP}(GTR). The more specific consequences flowing from them relate to the curvilinearity of light rays in the gravitational field, perihelion of Mercury and gravitational redshift in the spectra of distant stars. They speak of its rich predictive power (in the specific sense to be discussed in § 9.2. below) which, in the light of experimental corroboration, confirms T_{RP}(GTR), as compared with T_{RP}(NGT). Corroboration of its predictions can thus be said to have corroborated not T_{EP}(GTR) but only T_{RP}(GTR) > T_{RP}(NGT). More generally, a theory T_2, given that $T_{RP}(T_2) > T_{RP}(T_1)$, will, as a rule, provide a theoretical framework powerful enough to determine hitherto unknown *whether*-questions based on newer theoretical identifications. From the point of view of testability, therefore, the most decisive aspect of a scientific theory is its T_{RP}. It is by virtue of its T_{RP} that a theory can determine *problems*, hitherto unknown and unexplored, which, in their turn, lead to the designing of novel experiments with results decisive enough to solve those problems.

There are, however, many instances in science where a (new) theory has been subjected to use for asking *whether*-questions of the type for which there is no realistic theoretical basis in the theory itself. This has happened in particular fields of scientific research under the type of strategies which scientists themselves have employed in order to bring a new theory and its domain under the familiarizing models associated with the theories in the background. For example, this is true of Maxwell's equations for electromagnetism utilizing Faraday's idea of the *field*[10] but subjected to the strategy of a familiarizing model during the last decades of the nineteenth century:

> Is there some medium, called ether, through which electromagnetic waves must travel, just as sound travels in the molecular medium of air?

The Michelson-Morley experiment of 1887[11] was designed precisely to resolve doubts as to the existence of such a medium. The experiment had, however, negative results. As a consequence of a negative feedback from that experiment, it was the strategy of the familiarizing mechanical models of the ether which became a casuality, and not Maxwell's equations themselves. This shows how under the negative feedback from experiments it is possible to drop a set of *whether*-questions as not belonging to the resolving power of the theory wrongly used as their theoretical basis. But this is something which it is difficult to realize in absence of a theory T_2 with $T_{RP} > T_{RP}(T_1)$. Thus, it took the development of the STR to realize that the *whether*-questions such as the one above did not belong to the field of the resolving power of Maxwell's theory. *Ad hoc* hypotheses such as the one invoking the familiarizing model of light as a vibration of the luminiferous ether are sooner or later exposed as hypotheses prompted by pseudo-developmental strategies rendering a theory in difficulty as a pseudo-developmental structure.

Assuming, as GTR does, that gravity is basically a field characteristic of matter and that the principle of equality of effects of inertia (inertial mass) and gravity (gravitational mass) must hold true, T_{RP}(GTR) can be gauged from its predictions relating to phenomena as diverse as the velocity of propagation of gravitational processes, the deflection of light rays in a gravitational field, and redshift of spectral lines in such a field (i.e., effect of gravity on the frequency of light).

On the other hand, however, we may look at the T_{EP}(GTR) as its power to explain the anamalous motion of the perihelion of Mercury in addition to all that Newtonian dynamics already explains. As is clear from the discussion above, a really good and powerful theory must explain what its predecessor explains and fails to explain. But its unique character must be found in its T_{RP}. Thus, it must also serve as a powerful scientific framework for the determination of *whether*-questions and *why*-questions as important steps in the search-cum-discovery procedures in its field. It is these, in their turn, which will result in discovery of hitherto unknown effects of the laws of nature, or of invariance and symmetry principles in respect of them, and of newer areas of theory-problem interaction. A really powerful scientific theory must, therefore, not only be able to incorporate the older theories as special cases, shedding fresh light on their successes and failures, but must shift the scene of problem-determination from their field to its own. Only then can it serve as a unique basis for an interactive growth of objective knowledge systems in its own field. I have here in mind particularly that aspect of a forward-looking scientific theory which enables the scientist and the methodologist to think of the future development of theories and problems in mutual interaction instead of their development in the past.

A closer look at the scientific *whether*-questions is called for, since they are among the most important types of questions which depend, for their very determination and formulation, on the resolving power of theories in the background. According to Tondl's (1973)[12] analysis:

> In these and similar questions the starting basis is a sentence in respect of which the pronouncement of the *whether*-question has the aim of attaining a decision as to the semantic characterization of that sentence, namely a decision as to its truth or falsehood.

Tondl offers a more precise semantical characterization with the help of the *whether*-question operator '(?)' and the symbol '$S_i(i \leq n)$' for the sentence-basis of the question:[13]

$$(?)(S_i)$$

or

Is S_i true or false or whether, for some reason, some other decision must be given as to the semantic character of S_i?

Tondl argues that in the very asking of

$$(?)(S_i)$$

what one is assuming, besides its sentence-basis S_1, S_2, S_3, are certain semantically relevant relations between S_1, S_2, and S_3 "which are fixed beforehand" and expressed by meaning postulates.[14] If we let P stand for these meaning postulates, the *whether*-question above can then be represented schematically as

$$(?)(S_1)\ [S_1, S_2, P].$$ [15]

where P has the form $S_1 \equiv \sim S_2$ and $S_2 \equiv \sim S_1$, and "the expression in the square brackets fixes the information which is assumed when the question is uttered".[16] On this analysis, then, one should be able to decide which sentence occurs in the actual formulation of the question and which occurs as part of the information assumed.[17]

When a theoretical framework gives rise to subject-specific problems formulated by *whether*-questions it paves the way for scientific discovery on the basis of the postulated or *assumed* theoretical identifications. These questions are generally of a universal character:

> Is gravity a field characteristic of matter?
> or
> $(?)(S_1)[S_1, S_2, P]$, where
> S_1 represents: "Gravity is a field characteristic of matter."
> and
> S_2 represents: "Gravity is not such a characteristic."

It is *whether*-questions such as these which are *methodologically* most important to science. That this is so should become clear from the following examples from physics:

(a) Are light rays bent by gravity in the vicinity of a gravitational field such as that of the sun?
or
Do light rays always follow a straight trajectory, or move along a straight line?
(b) Is light a stream of particles moving according to mechanical laws?
(c) Is light a vibration, a wave, propagated in some medium?
(d) Does ether exist as a medium through which electromagnetic waves must travel?
(e) Is gravity a force that acts-at-a-distance?
(f) Can a body move at a velocity $v > c$?
(g) Can signals be sent at a velocity $v > c$?
(h) Are there galaxies which are receding from our galaxy at great speed?
(i) Do black holes exist?

There can hardly be any doubt, then, that in the methodological context of theoretical problem-solving situations in science it is quite useful to distinguish between two kinds of questions: (1) explanation-seeking *why*-questions and (2) what might be called 'discovery-oriented' *whether*-questions such as the ones cited above. It is proper to give this name to the latter type of questions, since the theoretical problem-solving context, which gives rise to such questions, will as a rule send scientists in search of appropriate kinds of *novel* experiments to determine the predicted value of the effects described or the theoretical identifications made in the assumed sentence-basis of such questions.

This is not to imply that attempts to solve problems underlying *whether*-questions will invariably land one in the field of scientific discovery. There are greater complexities to the asking and answering of such questions in science and hence to the problems themselves. However, the following characteristics can be said to be true of *whether*-questions such as those cited above:

(1) It is always the theoretical systems/assumptions in the background which give rise to *whether*-questions along with their sentence-bases sets and meaning postulates. Thus *whether*-questions presuppose theoretical systems with T_{RP} without which the former would not only remain unasked but be just inconceivable.

(2) A precise answer to a *whether*-question in the sense of (1) entails novel experiments/observations whose design and accuracy will considerably be determined by the assumed theoretical background and experimental techniques based on this background.

(3) When a *whether*-question is prompted by and pursued within the framework of a new theory making newer theoretical identifications while replacing the older ones (e.g., the assumption in GTR that gravity is a field characteristic of matter which replaced the Newtonian assumption that gravity is an action-at-a-distance force whose influence is transmitted instantaneously across empty space) a decisive answer in terms of experiments/observations will constitute a discovery if the results happen to be positive.

(4) Whenever an answer to a *whether*-question prompted by T_2 as a rival to T_1 assumes, on the basis of experiments, the significance of a discovery/successful prediction, its real methodological significance will consist in its being a step towards the confirmation of $T_{RP}(T_2) > T_{RP}(T_1) \equiv_{DF}$ [excess $T_{RP}(T_2)$ over $T_{RP}(T_1)$].

(5) Theoretical problem-solving does not come to a standstill or a steady state with the making of a discovery, whether negative or positive in effect (consider, e.g., the discovery that ether does not exist as a negative result of the Michelson-Morley experiment, or the discovery that light rays bend while passing near sun's gravitational field). It is, on the contrary, accelerated. That is, an answer to a *whether*-question, in the negative or affirmative, will open up new problem-areas as areas of theory-problem interaction – newer subjects of objective knowledge.

(6) The above features (1)–(5) of problem-determination by means of *whether*-questions may be further explored in the form of a question: why is it that the confirmation or refutation of Newtonian gravitation theory does not rest on whether light rays do or do not bend under the force of gravity? And why is it that the results of the 1919 eclipse expedition confirming that light rays do bend when passing in the vicinity of a gravitational field as strong as that of the sun are taken to confirm experimentally Einstein's GTR? Has this anything to do with $T_{EP}(GTR) > T_{EP}(NGT)$ or with $T_{RP}(GTR) > T_{RP}(NGT)$?

What Newton's gravitation theory assumes is this: that gravitation is an action-at-a-distance force with which one body *instantaneously*, though admittedly mysteriously, interacts with another (at a distance) according to the law:

$$F = \frac{Gm_1m_2}{r^2}$$

(where G is a constant, known as the gravitational constant).

Now, when compared with Einstein's GTR, we can draw the following significant conclusion as part of their methodological appraisal:

$$T_{RP}(NGT) < T_{RP}(GTR)$$

GTR and its fundamental assumptions arise essentially in response to the methodologically revolutionary *whether*-question:

Is gravitation a field-characteristic of matter?

What – if not the Newtonian framework – made it possible to raise such a revolutionary question? Obviously, it was the theoretical innovation as revolutionary in significance as the introduction of the concept of the field in the problem areas of electricity and magnetism first by Faraday and then by Maxwell, itself demanded by an earlier innovation in the form of Huygens' wave theory of light as opposed to Newton's corpuscular theory. The unifying concept of the electromagnetic field made it possible to describe the highly complex phenomena of electricity and magnetism in a language violating the subject-specific rules of the Newtonian framework, viz., the framework of material particles moving under action-at-a-distance forces.

Einstein's answer, in the affirmative, to the *whether*-question 'Is gravitation a field characteristic of matter?' naturally gave rise to further questions such as this:

> In case it is, how will gravitation be related to space and time?[18] And how are gravitational and inertial masses related?

It is through this question and hence through the assumption that gravitation is a field characteristic of matter that Einstein arrives at another highly unifying assumption at once relating gravitation and thereby matter to space and time as follows:

> As a field characteristic of matter gravitation affects the space-time of any given region so as to modify its geometry from Euclidean to non-Euclidian one.

This speaks of T_{RP}(GTR) > T_{RP}(NGT) as a remarkable methodological advance in the sense of *methodological variance* making it possible for physics to come out of a difficult problem-situation of rule-entanglement by resort to a process of rule-enrichment discernible in STR and GTR The T_{RP}(GTR)-determined prediction of the non-Euclidean nature of space can be visualized and exemplified in terms of a thought experiment about the sum of the angles of a triangle ABC, which is formed by three different tracks of light rays AB, BC, CA around the sun, as differing from 180°. This is so precisely because the lines AB, BC, CA would bend under the effect of sun's strong gravitational field.

The 1919 eclipse expedition did indeed measure the actual value of the predicted effect, confirming the non-Euclidian nature of space and thereby the excess T_{RP}(GTR) over T_{RP}(NGT). The prediction or the theoretical identification of another kind of effect called gravitational redshift relates to another aspect of the T_{RP}(GTR):

> That the effect of gravitation is felt not only in space but also in time.[19]

This effect is demonstrated by measurable changes in the frequencies of spectral lines, "because these reflect the time changes in the atomic systems at the source".[20]

In a somewhat *trivial* sense (for details see §9.2. below), for any theory T_1 to explain the natural course of phenomena $E_1, E_2, \ldots E_n$ is to render their time evolution predictable, whether probabilistically or deterministically. Thus for Newtonian mechanics, taken together with Newton's gravitation theory, to explain why all planets move in elliptic orbits around the sun is to render their motion, at any time in future, predictable according to the following pattern:

$$(T_1, 0_1, 0_2 \ldots 0_n) \vdash E_1, E_2, \ldots E_n$$

where $0_1, 0_2, \ldots 0_n$ are a finite set of observation sentences describing the relevant initial conditions. But there is a non-trivial sense of prediction such that a theory may predict hitherto unknown aspects of reality in its field, as prospective new problem areas, which fall outside its *intended* scope of explanation. But this raises the question: Can a theory truly predict what it either cannot or is not intended to explain? Or, must a theory T_1 explain what it may predict by way of hitherto unknown effects? The correct answer is that it may, but it need not. First of all, it might be argued that there is no reason to require that, as a rule, T_1 should, among other things, explain what is not intended to be included in its explanandum at the very outset. So what we cannot require is that T_1 explain what did not form a part of the problem-situation prompting its construction. Since what T_1 may non-trivially predict as a hitherto unknown law of nature (e.g., that light rays bend by such and such amount when passing near a strong gravitational field) does not form a part of its explanandum, it may, but need not, explain the consequences of its own novel predictions. Moreover, it is through its novel predictions rather than its explanations that its chances of refutation are enhanced. What we can require, then, is not that T_1 explain the consequences of its own novel predictions but that it should explain *more than* what constitute its *intended* explananda $E_1, E_2, \ldots E_n$. If, by way of novel predictions, T_1 predicts $P_1, P_2, \ldots P_n$, which are not part of $E_1, E_2, \ldots E_n$, then T_1 may, but need not, explain $P_1, P_2, \ldots P_n$ even when there is confirmatory evidence in favour of them all.

We will, as a rule, need another more comprehensive but methodologically simpler theory T_2 such that the following holds:

$$(T_2, 0_1, 0_2, \ldots 0_n) \vdash (T_1, P_1, P_2, \ldots P_n),$$
$$\text{since } T_{EP}(T_2) > T_{EP}(T_1)$$

If it is correct to say that T_1 may, but need not, explain $P_1, P_2, \ldots P_n$ when there is experimental confirmation that they are true – i.e. when the predicted laws/effects are discovered – then it is important to ask: What does the discovery or confirmation of the predicted value of a certain hitherto unknown effect (such as the bending of light rays passing in the vicinity of sun's gravitational field) actually confirm, the $T_{EP}(T_1)$ or the $T_{RP}(T_1)$? What our previous discussion of NGT and GTR shows is this. Whatever is attributable to the $T_{RP}(T_1)$ may gradually come to be attributable to the $T_{EP}(T_2)$. The bending of light rays through the attraction of celestial bodies is, as we have seen, predictable, if only inaccurately, given the T_{RP} (NGT. NCT),[21] as is indicated by Newton's Queries 1 and 31 in his *Opticks*. But it cannot be said to be explained by the T_{EP}(NGT.NCT), if only because the latter is constrained, not to do so, by a number of assumptions that enter the very process of derivation of the novel prediction. These assumptions include the following assumption. Light, which consists of particles with rest mass, follows the trajectories of particles. The bending of light rays through the attraction of celestial bodies is, as we have seen before, predictable on the T_{RP}(GTR . NCT) just as a deviation from the assumed optical law of their motion along a straight line.

What did the confirmations of the predicted value of the deflection of light rays in the vicinity of sun's gravitational field – as a result of a series of test photographs (in January 1919) at Greenwich observatory showing the Hyades against a reference frame of other stars and of the solar eclipse observations (on 29th May the same year) in Principe and at Sobral[22] – confirm, T_{EP}(GTR) or T_{RP}(GTR)? These results showed the value of the deflection (at the edge of the sun) as lying between 0.87″ and 1′74″. The findings were variously hailed as the discovery of "a whole continent of new scientific ideas", as "the greatest discovery in connection with gravitation since Newton enunciated his principles".[23]

The aim of the 1919 eclipse expedition was to test Einstein's GTR in terms of an experimental arrangement with a resolving power flowing, as it were, directly from T_{RP}(GTR) and described by Einstein himself in a very decisive manner.[24] But which of its aspects, one must ask? What

GTR predicts is this, in Einstein's[25] words: "*that, in general, rays of light are propagated curvilinearly in gravitational fields*". This obviously raises a host of *whether*-questions among which the following is of crucial methodological significance, if only as a radical reformulation of the corresponding Newtonian question already considered above:

> Do rays of light traverse a path in gravitational fields that is similar to that of a body projected through a gravitational field?

One is reminded here of Newtonian mechanics and Newton's law of gravitation which brought terrestrial and heavenly bodies in motion under one and the same *set* of laws as a solution of a host of problems expressible by means of *whether*-questions. In Newton's system of laws, the law of inertia appears as an ideal-type assumption of simplicity of the natural order, with a resolving power rich enough to raise *whether*-questions to which Newton's law of gravitation (NGT) was an answer. I think that the methodological scene remains *essentially* the same in the case of STR and GTR. Now, one of the fundamental assumptions of STR is known as the law of the constancy of the velocity (and rectilinearity of the path, as one might add) of light in *vacuo*. It is interesting to consider how a positive solution of the theoretical problem raised by the *whether*-question above, as did indeed become possible, draws clear limits to the validity of STR and thereby to T_{RP}(STR). When considered in the light of GTR, STR turns out to be analogous to an ideal-type assumption of simplicity of considerable resolving power, a special case of more complicated laws embracing wider complexities of natural phenomena. In this role, however, it can be seen as making a theory, as powerful as GTR, possible as a solution of highly complex problems. A curvature of rays of light, similar in nature to that of the path of a body projected through a gravitational field, "can only take place when the velocity of propagation of light varies with position".[26] Thus, we can say with Einstein that the results of STR "hold only so long as we are able to disregard the influences of gravitational fields on the phenomena (e.g. of light)".[27] I think that successful predictions of the type exemplified by GTR's prediction of the bending of light rays confirm not $T_{EP}(T_2) > T_{EP}(T_1)$ but $T_{RP}(T_2) > T_{RP}(T_1)$. In the present case, therefore, we may say that the experimentally confirmed bending of light rays by gravitational fields confirms the validity of the deeper concept of the curvature of space and hence $T_{RP}(\text{GTR}) > T_{RP}(\text{NGT})$. That $T_{RP}(\text{GTR}) >$

T_{RP}(NGT) is what is more powerfully revealed in all the three successful predictions from GTR relating to (a) the motion of the perihelion of Mercury; (b) deflection of light rays by a gravitational field; and (c) displacement of spectral lines towards the red called redshift.[28]

What is the methodological significance of all this? It seems to me that Einstein himself paid little attention to the far-reaching methodological consequences of the kind of epistemic appraisal under consideration: T_{RP}(GTR) > T_{RP}(NGT). All the three successful predictions are looked at by him (1960: 123–124) as an experimental confirmation of the 'truth' of GTR, owing mainly to his general belief that a theory is what "finds the justification for its existence in the fact that it correlates a large number of single observations", and that "it is just here that the "truth" of the theory lies".[29]

On the contrary, however, I believe that what a successful *novel* prediction confirms is not a truth-appraisal of a theory T_1 or T_2 but an epistemic appraisal of the form: $T_{RP}(T_2) > T_{RP}(T_1)$, implying, among other things, that T_2 is capable of (i) reformulating antecedently known problems more accurately and (ii) raising new problems inconceivable under the $T_{RP}(T_1)$. In a nutshell, what it implies is this. T_2 is so powerful methodologically and conceptually that it shifts the methodological scene of problem-formulation from the field of T_1 to its own field thereby also raising more problems than it can solve (consider Newton's Query 1 in *Opticks* in this context). This is what is exactly true of the appraisal: T_{RP}(GTR) > T_{RP}(NGT). Here one has only to consider the range of the problems that GTR raises. From a methodological point of view, therefore, what science can aim at is, I believe, growth of objective knowledge in the form of theories of ever greater T_{RP} and T_{EP}, capable of giving rise to ever newer theory-problem interactions.

But what would have the scientists said had the results of the eclipse expeditions prompted by Einstein's own description of an experimental arrangement to test the bending of light rays in gravitational fields been negative? Did the validity of GTR depend crucially on the character of these results? In one important sense it did; and this has to do, as we shall see in the following discussion, not with T_{EP}(GTR) but with T_{RP}(GTR). For it will become clear as we proceed that all the *novel* predictions of GTR are a function of T_{RP}(GTR).

Let us consider the negative discovery relating to the existence of 'ether', which resulted from the famous 1887 Michelson-Morley experiment. Among other things, this experiment was prompted by the following *whether*-question:

> Whether the Earth's passage through the 'ether' has any measurable effect on the speed of light?
>
> or
>
> Whether the speed of light relative to an observer/ experimental arrangement moving (together with Earth) through the ether is different in different directions?
>
> or
>
> Does light propagate differently in different reference frames that move in different directions relative to the ether?

The problem thus formulated arose during the last decades of the 19th century against the background of a complex system of Newtonian rules of problem-formulation and theory-construction (the classical *CSD*)[30] on the one hand and the newly emerged theories grappling with problems of electromagnetic phenomena and wave-optics on the other. How are we to evaluate, retrospectively, not the historical but the methodological significance of this problem and of the Michelson–Morley experiment itself? The latter was designed chiefly to test the ether hypothesis, which seemed necessary if the wave theory of light was to hold ground against its only rival – the corpuscular theory? The difficulties inherent in our question are complicated by the fact that while the nullresult of the Michelson–Morley experiment went decisively against the ether hypothesis, physicists generally found it difficult to give up their belief that the postulated ether existed. Many alternative interpretations of the result suggested themselves in quick succession in a climate of general dogmatic commitment to the system of theories and problems in the background, the classical *CSD* and the ether hypothesis being notable as part of this background. Michelson himself is said to have called his experiment a 'failure', the results being so *unexpected* according to the then prevalent theories/ assumptions.[31] Above all, the physicists George Fitzgerald and Hendrick Lorentz shifted their attention away from the original ether hypothesis and other assumptions, if only to find a satisfactory interpretation of the null result and to show how it did not contradict the hypothesis of an ether *at rest*.

Ever since Einstein formulated his STR in 1905, historians and philosophers of science have been debating the question of possible influence of this experiment on the genesis of STR,[32] with Einstein himself nowhere recognizing any such influence. The whole controversy arising out of this, much of which was addressed to Einstein himself while he was alive, has, in my opinion, only distorted the methodological situation as it arose in the last decades of 19th-century subsequent to the formulation by Maxwell of his theory of the electromagnetic field. In the process, the methodological aspects of the situation have got mixed up with its historical aspects.[33] If we consider closely the whole set of general *assumptions* that are employed, however implicitly, in the formulation of the problem above, it will become clear that they are traceable not to the resolving power implicit in Maxwell's theory of the electromagnetic field but to the classical *CSD* – i.e., to the classical (Newtonian) rules of problem-formulation, theory-construction and model-building. As regards Maxwell's theory, what we have is the following, among other things:

The (experimentally confirmed) theoretical identification to the effect that light itself is an electromagnetic disturbance,that it consists of *electromagnetic waves* which propagate at a speed of c = 300,000 km./sec. in *vacuo* – a consequence of great methodological significance for wave optics, having to do with the resolving power of Maxwell's theory of the electromagnetic field.[34]

But as regards the classical *CSD*, the following assumptions that are traceable to it are noteworthy:

(a) All physical phenomena, no matter how complex, must be explained in terms of material particles either attracting or repelling each other according to action-at-distance force laws.
(b) Any attempted physical description of newly encountered phenomena tending to defy the constraints flowing from (a) must be subjected to the ultimate test of intelligibility and admissibility in terms of mechanical models.[35]
(c) Ether conceived as a subtle, all-pervasive though invisible medium for light waves to propagate must exist if the wave-properties of light are to be fully explored and explained – an assumption traceable to the work of the wave-theorist Huygens and subsequently to that of Young and Fresnel.
(d) The Galilean principle of relativity (Galilean transformation) that all inertial reference frames are equivalent in the sense that all laws of mechanics are identical in them: that all mechanical laws,

including the theorem of the addition of velocities, are invariant relative to Galilean transformation.

(e) When applied to the propagation of light in empty space, (d) yields the assumption, quite consistent with classical mechanics but incompatible with the STR, that the velocity of propagation of light must depend on the velocity of the body (the source) emitting the light – i.e. on its direction 'in space'.

The ether hypothesis itself is seen here appearing as an assumption in the formulation of the problem relating to the time-difference in the speed of light in different reference frames in motion relative to ether. But is there any necessary *conceptual* or *methodological* connection between this hypothesis and Maxwell's theory of the electromagnetic field or the theory that light itself consists of electromagnetic waves? I think that the answer is in the negative, the actual historical aspects of the problem-situation not-withstanding.[36] For if there were any such connection, these theories too would have to appear as necessary assumptions in the formulation of the whole problem to which the Michelson–Morley experiment was addressed.

If this is correct, it follows that this problem was not *determined* by the resolving power of Maxwell's theory of the electromagnetic field. And that, on the contrary, it arose as a consequence of a dogmatic extension of the classical *CSD* – the assumptions (a)–(e) – to newer problem-areas or to newer theoretical identifications such as those in wave optics. So dogmatic was the 19th century physicist's commitment to the classical *CSD* – its rules and strategies – that, newer theoretical identifications not-withstanding, it seemed perfectly rational not only to insist on the ether hypothesis but also to predict a positive outcome for the Michelson–Morley experiment. What is interesting is the fact that, among other physicists deeply committed to the classical *CSD*, Maxwell himself believed in the ether hypothesis.[37] It is interesting for two reasons. First, it was Maxwell's own field equations of electromagnetic phenomena which introduced the conceptual innovation of the order of the field concept – a methodologically significant aspect of the physical problem-situation of which Einstein later on, unlike his contemporaries, showed an exceptionally unique perceptive understanding. Secondly, no sooner than the ether hypothesis was pursued more and more vigorously it came under a host of constraints of the classical *CSD*. Among other things, the postulated ether had to have all the necessary mechanical properties as a medium for light waves to propagate at a speed whose value c was already determined in 1849, even if its density was supposed to be unconventionally small and its very postulation to entail a denial of empty space.

Thus, on the one hand, the real significance of the ether hypothesis lay in the underlying attempt to subject the physical problems of wave optics and electromagnetism to the requirements of classical *CSD*, its mechanical models, etc. – something whose accomplishment depended on the ether hypothesis turning out to be true. On the other hand, however, with the ether hypothesis once invoked, if only to explain the already known properties of light, and with the necessary mechanical properties attributed to it, the postulated ether could also be seen, once its existence was proved, in another role. It could serve as a basis for distinguishing *reference frames* at rest relative to it and hence as privileged for purposes of measurement, formulation of laws of nature, etc., *and* those in motion relative to it and hence more complicated for such purposes. But all this would clearly be in violation of the Galilean principle of relativity (Galilean transformation), since mechanical laws in the privileged frame of reference would be different from those in the rest of the frames.

The tremendous shock with which the negative outcome of the Michelson-Morley experiment was received is itself indicative of the vast complexity in the physical problem-situation as it had developed during the last decades of 19th century. It was not so much a situation of tension between the old and the new as it was one of *rule-entanglement* in the Newtonian rules of problem-formulation and theory-construction and in contradictions resulting from their extension to newer theoretical identifications. That is, it was the typical situation of applying very simple rules – the classical *CSD* – to complex problems arising out of newer theoretical identifications of optical and electromagnetic phenomena.

What was it, then, exactly which the Michelson-Morley experiment sought to test? Not Maxwell's theory of the electromagnetic field. Nor the wave theory of light. It was the ether hypothesis itself, and of course the classical *CSD* in so far as it prompted the invoking of this hypothesis to fit the wave theory of light into Newtonian framework. This is, then, another way of admitting that the ether hypothesis was after all an *ad hoc* hypothesis, invented not to save some individual theory in difficulty but to have the domain of much simpler mechanical laws – of the classical *CSD* – extended to complex problems of electromagnetism and wave optics.

The best way to have the ether hypothesis tested was to have an experimental arrangement to detect and measure the effects of Earth's motion through the postulated ether on the speed of light – i.e. the time-differences in its speed in different directions.[38] The Michelson-Morley experiment,[39] which was designed in 1887 to do this, produced to everyone's disappointment the null result: Earth's velocity relative to the ether frame was undetectable, with the velocity of light turning out to be the same in all directions.[40] The awkward implication of the null result suggesting that the Earth was always at rest relative to the ether frame could not be accepted for obvious reasons.[41] The problem-situation was however considerably complicated by the Fitzgerald-Lorentz contraction hypothesis seeking to interpret the null result with a view to rescuing the ether hypothesis.[42]

What is remarkable is the fact that Einstein did not fall a victim to prevalent dogmatism that characterized the physicist's approach to the classical *CSD* and to the Michelson-Morley experiment. Einstein took note of the complication introduced by Fitzgerald-Lorentz contraction hypothesis, perceived the problem-situation thus complicated as ripe for *methodological variance* (*CSD*-variance),[43] and formulated his STR in 1905. With the STR appearing on the scene, the methodological situation changed radically. It became possible not only to lay the ether hypothesis finally to rest but to show how unrelated it was to the resolving power of Maxwell–Lorentz theory of electromagnetic phenomena.[44] While the hypothesis as a sentence-basis of the *whether*-questions above has no roots in this theory, appearances to the contrary are traceable to the dominant strategy of mechanical models in classical physics. Exposing all interpretations of the null result as *ad hoc* defence-strategies to rescue the ether hypothesis, the STR not only led to Fitzgerald–Lorentz contraction of moving bodies as its special consequence but implied methodologically significant modifications of the contraction hypothesis itself.[45]

Thus, if physics can be said to have ever landed itself in a methodologically significant problem-situation of rule-entanglement it was, I think, during the last decades of the 19th century. For what one finds here is the typical situation in which even the newer theoretical identifications of complex electromagnetic and optical phenomena are subjected to the requirements of a framework as oversimple as the classical

CSD, if only because of a deep ideological commitment to it. What suffered a serious, though temporary, distortion in the process was the important methodological aspect of this problem-situation, viz., the great and unique resolving power of Maxwell's theory of the electromagnetic field leading to newer theoretical identifications (e.g. of light as an electromagnetic wave). It is this which deserved a closer and more perceptive attention than it did receive from the physicists of that period. All historical evidence, including Einstein's non-acknowledgement of any influence of Michelson-Morley experiment on his 1905 paper on the STR, points unambiguously to the fact that only Einstein, then working as a clerk in a Swiss patent office, had the correct insight into the newly unfolded complexities of the electromagnetic and optical phenomena; with others groping completely in the dark under the burden of a strong ideological commitment to the classical *CSD*. So strong was their ideological commitment to this framework, its ultimate perfection and validity, that even the best conceivable *monitoring* of the complexities of the problem-situation – the null result of the Michelson-Morley experiment itself – could not shake it off. Instead of *working backwards* to have some of the basic assumptions in classical physics subjected to a review and a revision, the physicists generally only attended to the null result, trying one explanation after another with a view to saving the ether hypothesis and its embedding framework in the classical *CSD*.

What Michelson-Morley experiment with its nullresult did *in effect* was to *monitor*, as it were, how an *extended rule-government* had landed classical physics – its entire theory-problem interactive systems – in a state of *rule-entanglement* which went unrecognized long after the experiment gave its negative verdict. Einstein's great contribution lay, I believe, here: in perceiving the situation precisely, but independently of this experiment and without having to wait for any such experiment, as (1) one of rule-entanglement and (2) sufficiently ripe for methodological variance – i.e. for bringing in newer subject-specific rules to overcome (1).[46] Experiment or no experiment, the only way out for physics caught in such a state, faced with newer theoretical identifications far more complex than could be expected under the classical *CSD*, was to have its rules of problem-formulation and theory-construction revised or *enriched* in an appropriate manner. Einstein's STR, among other works of his, did precisely that in terms of its two fundamental postulates (1)–(2) as follows:

(1) *The special principle of the physical relativity of all uniform motion or of equivalence of all inertial reference frames*: In Einstein's (1905: 891) famous words, ". . . for every reference system in which the laws of mechanics are valid, the laws of electrodynamics and optics are also valid".[47] In other words, "General laws of nature are co-variant with respect to Lorentz transformations".[48] Or, simply put, "the laws of physics take the same form in all inertial frames".[49]

(2) *The law of transmission of light with a constant velocity c that is independent of the state of motion of the source of light*: In the words of Einstein (1905: 891–892), ". . . light is always propagated in empty space with a definite velocity V which is independent of the state of motion of the emitting body".[50] In other words, "the velocity of transmission of light *in vacuo* is equal to a constant *c*".[51] Or, simply put, "In any given inertial frame, the velocity of light *c* is the same whether the light is emitted by a body at rest or by a body in uniform motion".[52]

The first postulate is essentially an extension of the classical principle of relativity restricted strictly to Galilean (inertial) co-ordinate systems in uniform motion relative to one another and with respect to which Galilei–Newton laws are identical. The Galilean principle of relativity, as we might call it, formed a fundamental postulate of the classical *CSD*. We have already noticed how the classical *CSD* came to be undermined by developments in wave optics and electrodynamics in the last decades of the 19th century, thereby giving rise to problems and theories which entailed revolutionary developments embodying methodological variance (= *CSD*-variance), among other things. Einstein's STR, with its two fundamental postulates cited above, constituted the first decisive step in this direction, i.e., in the direction of *CSD*-variance. What the second postulate, on the other hand, might be said to presuppose is this: No physical interaction, or no signal, can propagate *in vacuo* at a velocity $v>c$. Given Einstein's own formulation of these two fundamental postulates of STR, I am inclined to think, with Brent Mundy (1986: 30), that it is historically accurate and philosophically plausible to take the *constancy* and not the *limiting* character of the speed of light as basic. The most appropriate place for the limiting character of the speed of light might be then among the fundamental general *presuppositions* of STR. Such an interpretation obviously entails an important distinction between the basic postulates of a physical theory and its basic presuppositions – a distinction which seems to have suffered from terrible distortions in the recent debates about the completeness of quantum mechanics as a physical theory. The importance of the distinction, however, derives from the following consideration. Every physical theory must make general presuppositions about the interactive order of nature but no physical theory can be expected to serve as a direct source of such presuppositions. The message for the debates just alluded to is quite clear, but I shall here refrain from going into this whole issue.

It is, then, no strange coincidence that the experiments should have led to negative results of far-reaching significance demonstrating – when interpreted in the light of a theory as powerful in its resolving power as STR – that the postulated, or predicted, ether did not exist, and that its very assumption was *ad hoc*. Prompted as it was by the *strategy* of the familiarizing mechanical model-building, its prediction was in no way warranted by the framework (in this case, not the classical *CSD* but Maxwell's theory itself) within which it was actually invoked to deal with the newly unfolding complexities of the optical and electromagnetic phenomena. What Einstein did was to pay attention to these complexities with a view to exploring afresh the methodological scene of problem-formulation and theory-construction in the field of wave optics and electrodynamics.

One might, therefore, ask: What did then the Michelson–Morley experiment really test? One thing which, I believe, it did accomplish *in effect* was to test not the explanatory power but the resolving power of the physical assumptions that had been invoked to render the then current theories of light, electricity and magnetism coherent with the demands of the classical *CSD*. Its negative results showed these assumptions to be *ad hoc* and empty of any T_{RP}, and more so when the STR appeared on the scene of physical problem-formulation and theory-construction. The 'hypothesis' of the ether drift might be, therefore, looked at as a pseudo-developmental structure of objective knowledge in the sense of my (1983). That this is how one should look at these results explains why they did not over-throw or refute the basic framework of the wave theory of light or of Maxwell's equations for the electromagnetic field. For those, if not for Einstein,[53] who gradually realized the failure of the assumptions made *ad hoc* and invoked solely to fit the newly unfolded complexities into the Newtonian scheme (the classical *CSD*), it paved the ground to look for a theory powerful enough to deepen the resolving power of Maxwell's equations for the electromagnetic field. That is, it prepared the stage for a new theory T_2 such that in relation to its rival T_1

$$T_{RP}(T_2) > T_{RP}(T_1)$$
$$\&$$
$$T_{EP}(T_2) > T_{EP}(T_1)$$

Only a theory as rich as STR could accomplish this task, thereby shifting the scene of problem-formulation from the field of the background theories to its own field.

I believe that the chief methodological significance of the Michelson-Morley experiment lay in this: through it, though quite unintendedly, classical physics realized for the first time a state of rule-entanglement making it possible for physicists to lay down afresh the subject-specific rules of problem-formulation and theory-construction in various fields such as optics, electrodynamics and gravitation. This is exactly what the subsequent developments unfolded, resulting in a large-scale *CSD*-variance and *ES*-variance.[54] But this should not be taken to imply that this experiment was a necessary part of the methodological situation of rule-entanglement. All that an experiment can do in such a situation is to monitor the complexities of the natural phenomena in question. But that is precisely what the Michelson–Morley experiment, interestingly enough, did not aim at.

As a rule, all theoretically significant *whether*-questions are prompted by considerations relating to newer theoretical identifications a newly formulated theory makes. This is so for the simple reason that problem-determination in all its developmental stages is a function of the resolving power of a scientific theory. Now, can we say that the question "Whether ether exists as a medium through which light waves must vibrate?" was prompted by the new theoretical identifications due to the wave theory, as against the corpuscular theory of light, and Maxwell's equations for electromagnetic field utilizing Faraday's idea of the *field*? It is customary to trace the theoretical/sentence-basis of this question in these theories themselves. But this is, I believe, as I have already argued above, a wrong approach. The newer theoretical identifications, due to newer field theories, were subsequently confirmed while the negative discovery that ether did not exist had already been made. This shows amply that there was nothing in the resolving power of these newly developing theories themselves which could have warranted experiments in search of the ether-drift. It must have been, therefore, something extraneous to their resolving power which brought in lot of experimentation and the whole debate concerning it. This extraneous factor had essentially to do with the Newtonian tradition to fit everything into a single framework. I have here in mind particularly Newtonian methodological rules of problem-determination and theory-construction as steps in the Newtonian search-cum-discovery procedures, novel complexities of natural phenomena not-withstanding. The *whether*-question of the ether-drift can thus be seen as one that was prompted by the classical physical strategy of familiarizing mechanical models, thereby distorting the resolving power of Maxwell's equations for the electromagnetic field and of the wave theory of light.

We may conclude by going back to the physical problem-situation arising as a result of the joint theoretical background of Copernican revolution, Galilean experiment on terrestrial motion and Keplerian three laws of planetary motion:

> What kinds of forces must be invoked to explain the patterns of motion, terrestrial and celestial, implied by the theoretical identifications peculiar to this theoretical background?

Newton responded to this in terms of the newly introduced concept of universal gravitation, giving rise, however implicitly, to the *whether*-question:

> What is gravitation? Is there a law of universal gravitation as a universal force which acts-at-a-distance such that every body instantaneously attracts the other bodies, however distant these may be, according to the following equation:

$$F = \frac{Gm_1m_2}{r^2}$$

where G is a constant (known as the gravitational constant)? Thus, the very introduction, into the corpus of classical physical theory, of the concept of universal gravitation is prompted by the antecedently known physical problem-situation demanding theoretical identification of the type of universal force which could be invoked to explain the behavior of macroscopic physical bodies in motion, terrestrial and celestial both.

Like Newton's theory of universal gravitation, every physical theory *assumes* or makes certain theoretical identifications of the values of the postulated variables with a view to explaining the antecedently known laws or phenomena of nature. Every physical theory thus possesses a certain resolving power owing mainly to its basic assumptions making relevant kinds of subject-specific theoretical identifications. We have already noticed how, given a theory T_1 in relation to another T_2, $T_{EP}(T_1) < T_{EP}(T_2)$ is demonstrated by the ability of T_2 to solve antecedently known problems arising out of, but not solved by, T_1 – i.e., its ability to entail as its consequences explananda in excess of those entailed by T_1. That $T_{RP}(T_2) > T_{RP}(T_1)$ is demonstrated, on the other hand, by the ability of T_2 to shift the scene of problem-formulation from the field of T_1 to its own field – i.e. by its ability to make novel predictions or raise novel *whether*-questions. Since science aims at the growth of objective knowledge, what we should be interested in is the excess $T_{EP} \cdot T_{RP}(T_2)$ over $T_{EP} \cdot T_{RP}(T_1)$ and hence in appraisals of the type: $T_{RP}(T_2) > T_{RP}(T_1)$ and $T_{EP}(T_2) > T_{EP}(T_1)$. We can say that $T_{RP}(T_2) > T_{RP}(T_1)$ is what is measured by the novel *whether*-questions that T_2 raises and hence by the novel predictions that it makes.[55]

It is just as important for a new theory to be able to explain a previously known but unexplained fact as to predict aspects of natural phenomena hitherto only vaguely suspected or completely unknown. Newton's gravitational theory gave excellent explanation of the regular paths taken by the various planets with one notable exception. Predictions from the law of universal gravitation for the path to be taken by the planet closest to the sun, Mercury, did not agree with subsequent observations. Einstein's GTR was, on the other hand, able to predict or explain exactly why Mercury's path did not agree with Newton's original predictions. To many this result must have been less spectacular than the prediction of the bending of rays of light, but it served just as well to confirm $T_{EP}(GTR) > T_{EP}(NGT)$, with the latter serving to confirm $T_{RP}(GTR) > T_{RP}(NGT)$.

It is in a sense correct that from Newtonian mechanics, and Newton's law of universal gravitation, we can deduce, though not in a straightforward manner, all of Keplerian laws, Galileo's law of free fall, and the law of the tides, just as from the GTR we can deduce not only Newtonian dynamics but also such other things as the hitherto unknown law of the motion of light rays in gravitational fields. It would, however, be more accurate to say that what is derivable from Newtonian dynamics is, among other things, an *improved* set of laws of planetary motion, showing how a planet would move in an elliptical orbit provided there were no sister planets to pull them out of such orbits. Thus, it is in this way that Newtonian dynamics could explain not only the fundamentally elliptic paths, but also the deviations from them, if any (with the exception of the problem of motion of the perihelion of Mercury's orbit known to be advancing by a small but regular amount and unexplainable in the Newtonian system). Similarly, Newtonian dynamics reveals the character of Galileo's law of free fall, which postulates uniform acceleration, as an approximation, or as a special case, since on Newton's theory acceleration of a falling material body increases as it advances to the surface of the Earth. And again we find Einstein's GTR introducing relativistic corrections showing Newtonian dynamics only approximately correct, while improving on it, e.g. in explaining the path of Mercury.

9.2. THE NATURE OF NOVEL PREDICTION: TWO CONCEPTS OF THE PREDICTIVE POWER OF A SCIENTIFIC THEORY

In the sciences and the philosophies of the sciences alike the concept of prediction is, I believe, often treated in a manner that smacks of a hitherto unsuspected ambiguity in our very approach to understanding/analyzing the structure and function of a scientific theory. For example, one of the standard uses of the term 'prediction' links the corresponding concept with explanation as follows. The selected parameters of the behavior of a particular thing under specific conditions at a particular point in space-time continuum, say in a relevant domain of objects or events of subject-specific interest to the scientist, are said to be predictable, either deterministically or probabilistically as the case may be, in so far as we are already equipped/familiar with a set of laws that *explains its* general pattern of behavior. To *explain* an event E is, in this context, in principle or structurally – but not pragmatically or epistemologically – the same as deriving it as a logical/probable consequence from a theoretically and *empirically adequate* premises (say a set of laws and relevant initial conditions) and thereby to *expect/predict* it as a matter of lawful regularity in the general course of natural events. Predictability in this sense then depends on how familiar one is with the particular set of phenomena, one is interested in, in terms of more or less established empirical laws/theories and sets of relevant initial conditions. For example, given the explanatory framework of Newtonian dynamics together with the relevant sets of initial conditions, one can speak of the selected bodies in motion as being predictable as regards their relative momenta and position co-ordinates at any future time whatever.

On a closer look, however, prediction in this sense turns out to be a somewhat trivial notion. So trivial indeed that one is reminded here of David Hume's famous scepticism leading him to consider all our *actual* empirical knowledge as a matter of custom and habit. Prediction in the sense in which it is quite often said to be formally equivalent to explanation is an aspect of science so trivial that it might equally be taken care of by what Hume called custom or habit. Its triviality lies in the fact that it makes no difference to the explanatory framework from which it must be derived; it does not add anything to it. To predict that there will be a sunrise tomorrow is not to give any information over and above those regularities/laws which we are already familiar with. To predict, on the other hand, that there will be no sunrise from tomorrow will be, if backed by appropriate *new* assumptions, giving new information, truly or falsely, relative to what we already possess as our background knowledge. (It is only *trivially* that one can 'predict' that Jones will be dead one day, since any such statement, however vague, is already covered by the empirical law that says that all men are mortal). The long known correlation between the tides and the motion of the Moon provides, I think, a better example. Although already known in antiquity, exact prediction of tides, using the *familiar* correlations, was not possible until Newton introduced the *explanatory framework* of his theory of universal gravitation.

What we have considered so far is the use of the term 'prediction' according to which to 'predict'$_1$ something is equivalent to 'expecting' what is being predicted. But as it turns out that prediction in this sense is determined by how *familiar* one is with the correlations, regularities and laws of nature already formulated or conceived, to predict$_1$ something in a specific domain of interest is to predict just that which is already *expected* under a particular set of *known* or *familiar* laws/regularities and the relevant sets of initial conditions. Considered in this sense then prediction$_1$ in science has the following three features (a–c): (a) it presupposes explanation; (b) it makes no difference to the informative content in the structure of the explanatory system on which it depends; and (c) it is indistinguishable from *expectation* (a kind of Humean habit or custom) of events, given all that it presupposes. On considerations such as these, it seems necessary to recognize how trivial the concept of $prediction_1$ really is, both empirically and theoretically. But this will also depend on the possibility of $prediction_2$ in science in a non-trivially important sense that is anticipated by our discussion above.

Is there then some other non-trivially important sense in which we might speak of a scientific theory's predictive $power_2$ as distinct from its predictive $power_1$? Is there a sense in which its power to *predict* does not correspond to our power of *expecting*, on its basis, the *familiar* pattern of natural phenomena to repeat itself, in the epistemic sense of the term 'familiar'?[56] I think that it is possible to answer this question in the affirmative. But the proposal concerning the concept of $prediction_2$ that I am here going to put forward will be different from the one that comes to surface in Popper's (1972a: 117) conception of the *novel* predictions of a scientific theory, its explanatory power being for him, as for others, identical with its power to $predict_1$ the future course of the familiar natural phenomena. For Popper, as for his followers like Watkins (1984, 1987), novel predictions of a theory remain restricted to what is equivalently referred to as its empirical content or the set of its *singular* predictive implications. On this view, moreover, it makes sense to speak of the *verification* of its novel predictions. However, we might look at a theory's power to $predict_1$ the future course of natural phenomena in its domain of application as a function of its explanatory power (= T_{EP}) and then ask: What is its power to $predict_2$ the *unfamiliar* a function of? The answer that I shall argue for, in what follows, is this: Its predictive $power_2$ is a function of its resolving power (= T_{RP}). In this sense, then, its $prediction(s)_2$ along with the corresponding *whether*-questions, that it itself raises, are always tantamount to resolving the structural-dynamical complexities of natural phenomena at the theoretical level so as to expose or render all the previous theoretical identifications of these phenomena as cases of oversimplifications or just as special cases. But there is more to it than just this. The consideration of the fact that $T_{RP}(STR \cdot GTR) > T_{RP}(NGT)$ illustrates precisely, I think, the distinction that I am proposing here between the two concepts of the predictive power of a scientific theory.[57] The GTR $predicts_1$ quite trivially that, under certain assumptions, Newtonian dynamics (= NGT) will hold true; but it $predicts_2$ non-trivially, among other things, that under conditions more complex than conceived hitherto Newtonian dynamics will break down and so on. (For more on prediction in this latter sense see § 9.4. below).

How sharp is then the proposed distinction between the two concepts of the predictive power of a scientific theory? I have already suggested above that these might be better understood in terms of the more basic distinction between the explanatory and resolving powers of a scientific theory. Let us now develop this distinction in more detail. In the individual sciences, such as the natural sciences, if the scientist wants to describe or $predict_1$ the future course of the behavior of a relatively separate system he can do so most simply and elegantly, provided he has a sufficiently powerful theoretical system (like Newtonian's gravitation theory and his three laws of motion) together with a set of *relevant initial conditions* specifying the initial values of the state-variables of the system that are of subject-specific interest to him. To take a concrete example: In which aspect of Newton's mechanics is its predictive $power_1$ to be found? It is to be found, I think, in that particular aspect of it which enabled Newton himself and enables us even today to calculate the future positions and velocities of a planet in our solar system on the basis of (a) the three laws of motion and (b) the relevant initial conditions that specify the initial position and velocity of the planet in question. Thus, one is not surprised at the common identification of the hierarchical structure (or the logical structure) of a scientific $prediction_1$ in this sense with that of (deductive-nomological) scientific explanation. For the latter too employs, when it is most simple and elegant, a theoretical system including a set of laws on the one hand and a set of initial conditions on the other. But what is generally not realized is this. The predictive $power_1$ of a theory in this sense is a function of that theory's explanatory power, provided 'explanation' in this context is taken in the sense of either strictly causal/deductive nomological explanation, say in the sense of Carl G. Hempel (1965), or probabilistic explanation employing statistical laws instead of the strictly causal ones.

I think that we must, sooner or later, face a very interesting but awkward philosophical question, even if we may broadly agree that one of the things which science aims at is scientific explanation/$prediction_1$ in the sense just considered. The question which we must ask is this: How does the individual scientist, whose scientific interest is guided by this aim, get at the relevant initial conditions themselves, that specify the initial state of the system of subject-specific interest to him? Of course, getting at them presupposes his ability to concentrate on the *system* as a relatively separate system and on certain of its *state-variables* as the relevant state-variables. But how is he able to do all that? How does he, above all, derive or get at the relevant initial values of the state-variables already chosen as the relevant variables of the system? Can the scientist, one must ask here, get at these in absolute independence from the *network(s)* of theories and problems in the background, that may seem to be of little or no immediate relevance to the specific task at hand? I am of the opinion that he cannot do so. But any correct answer to the preceding questions will have to take into account the following complication, which has remained, I believe, unnoticed until now. The precise character or logical form of the statements formulating the relevant initial conditions in the relevant but variable contexts might not be, and need not be, as rigid as is generally thought. It might be, on the contrary, quite variable not only from problem-situation to problem-situation but from one individual science to another. Thus in the context of a Newtonian explanation of planetary motion, the logical form of such statements might invariably be taken to be one of singular observation statements, as is of course usually done by the deductive-nomological-explanation-theorist as well as the practising physicist. But I believe that this need not be the case where either (1) on closer scrutiny the specific set of statements giving the initial conditions are shown to presuppose considerable theoretical identification of the parameters that figure in those statements or (2) it is either hard or impossible to make observations on the *totality* of the system under investigation because it cannot be separated and subjected to observation as a whole. I think that both types of situation obtain in the science of our time. The first is typical of the domain of the elementary particle physics, and the second of cosmology. Elementary particles (quarks, leptons and field quanta such as photons), that can reveal their effective dimensions only as a function of their *interactions* (such as the measurement interactions), do not lend themselves, I believe, to the explanation-theorist's requirement for initial conditions, namely that they be stated by means of the singular observation statements.

How restricted in scope and logical form must then a statement of initial conditions, in a given problem-situation, be before it is recognized as specifying the relevant initial conditions in the given context? This question, which raises a serious problem of the *nature* of initial conditions that are usually taken for granted as quite unproblematic, itself arises because of a largely unexamined assumption of the *separability of physical systems*. I think that it is this very assumption which underlies our usual understanding of the whole *notion* of initial conditions, taken in its role in the prediction$_1$ of the natural course of physical phenomena. And, interestingly enough, there are areas in physics and cosmology of today where neither this assumption itself, pure and simple, nor its simple denial would carry relevant sense. For example, it might be argued that it makes little or no sense to demand *separable* initial conditions specifying the initial values of the selected state-variables of the *same* universe (e.g., the theoretically identified early universe, or any one of the similarly identified later epochs) whose origin the cosmologist tries to understand or explain by reproducing it in his cosmological models. Of course, he does so, making systematic use of the resources of his (or the astronomer's) *present* observations of the very large structures that are interpreted in the light of (the resolving power of) our extant physical theories such as the GTR and high energy particle physics. The usual separability assumption so necessary to the physicist's usual attitude to initial conditions, specifying the initial values of the relevant state-variables of a relatively separated physical system at any arbitrarily selected time, seems here clearly undermined and rendered irrelevant. Again, the physical assumption of separability seems seriously undermined by the problems relating to the measurement interactions and the probabilistic pattern of the predictions of the outcome of an experimental set-up in the domain of quantum mechanics itself.

It was Einstein[58] who introduced *separability* as a physical assumption/principle in his arguments questioning the completeness of quantum mechanics as a physical theory. How are we here to understand the Einstein separability, without getting involved in the controversial interpretations of it in the context of the debates about quantum mechanics? Einstein himself argued that,

> . . . it is characteristic of . . . physical things that they are conceived of as being arranged in a space-time continuum. Further, it appears to be essential for this arrangement of the things introduced in physics that, at a specific time, these things claim an existence independent of one another, insofar as these things 'lie in different parts of space'. Without such an assumption of the mutually independent existence (the 'being-thus') of spacially distant things, an assumption which originates in everyday thought, *physical thought in the sense familiar to us would not be possible. Nor does one see how physical laws could be formulated and tested without such a clean separation*. Field theory has carried out this principle to the extreme, in that it localizes within infinitely small (four-dimensional) space-elements the elementary things existing independently of one another that it takes as basic, as well as the elementary laws it postulates for them.[59] (Italics not mine.)

I am inclined, like Don Howard (1985), to take the Einstein separability as asserting "that spatially separated systems are characterized by separate real states of affairs".[60] But the most crucial question is, I think, this. If the separability principle is to play a basic role in the formulation and testing of physical laws, as Einstein also says it does, how do we decide in a given problem-situation as to whether it has been actually satisfied or not? It seems to me that this repeats our question concerning initial conditions raised earlier: "How are initial conditions rendered ascertainable at all?" This should not be surprising, since any discussion on initial conditions is, I think, bound to raise deeper questions concerning the separability of physical systems in the sense of Einstein. To put it briefly, before the physicist is able to ascertain the initial conditions of a physical system of interest to him, whether in the domain of the very small or in that of the very large, he must either *assume* or actually ascertain, say by means of appropriate procedures of observation etc., the separability of that system from all those larger settings in which it might be otherwise interactively entangled.

If we look at the Einstein separability from the point of view of the general interaction theory in the sense of my (1983),[61] it can then be clearly interpreted as a principle of the *possibility* of physical interactions in nature, including the measurement-interactions in quantum mechanics. It will not be, therefore, surprising if it should assume increasing importance as a postulate of the general interaction theory itself.[62] Notice in this context the recent attempt by Don Howard (1985)[63] to bring considerations of a general interaction-theoretic character to bear on the philosophical and methodological issues of separability and locality, which he distinguishes one from the other, in the context of quantum mechanics. He argues as follows: "the separability principle operates on a more basic level as, in effect, a principle of individuation for physical systems, a principle whereby we determine whether in a given situation we have only one system or two. If two systems are not separable, then there can be no interaction between them, because they are not really *two* systems at all".[64]

Could we not accordingly, one might ask, raise questions concerning the nature of initial conditions of broadly two chief kinds, with one of them (a) expressible as the singular statements of space-time region specific observable correlations of interest to the physicist, and the other (b) as the universal/general statements, and for that very reason hitherto unrecognized ones, describing a proper sub-set of the causally relevant structural/dynamical properties of a whole class of entities (e.g. the early universe as a whole)? Could we not, in other words, conceive of a problem-situation such that the statements which serve as, or determine, the relevant initial conditions are as general in nature as would answer, instead of themselves presupposing some general physical assumption of individuation of physical systems such as the Einstein separability, questions as novel and theoretically significant as the following?

> Assuming the framework of the general theory of relativity on the one hand and that of the high energy particle physics on the other, which initial conditions in the (theoretically identified) early universe are possible? Which kind of initial conditions could be singled out as most natural under this framework, and which as most unlikely?[65]

I think that it is necessary to recognize the separability of otherwise interactively entangled physical systems, as we might call them, not only in the philosophically loaded absolute sense of "spatially separated systems characterized by separate real states of affairs", that one might attribute to Einstein, but in the sense entailed by the principle of the resolving power of physical theory itself, and, therefore, as a relativized but fundamental presupposition of the whole of physics. This relativization should allow us, I think, to lift the *arbitrary* restrictions, whatever, on the logical form of statements which physics, as science generally, might employ as initial conditions in its attempt to predict/postdict the same kind of correlations that it must explain.

The kind of relativization I have here in mind assumes great significance in the context of a problem-situation as complex as the astrophysical one of explaining the theoretically identified facts of stellar evolution or of the formation of black holes. Or, consider the problems of constructing cosmological models of the origin, structure and dynamics of the universe in its theoretically identified different epochs ranging from the singularity epoch to the present epoch.[66] If the cosmologist is to construct plausible cosmological models of our universe in its singularity epoch and various post-singularity epochs extending to the present such that events (like symmetry-breaking) in the *early universe* are taken to have *determined* the nature and evolution of the large-scale hierarchical structures being observed today, it is necessary, I think, to employ *variable* and *general assumptions* concerning the very initial conditions under which alone could the evolution of the various physical laws of force (from the grand unified epoch to the present epoch) and the expansion of the spacetime (of the early universe) be rendered scientifically conceivable and intelligible. What else could then serve as the initial conditions in the case of a theoretically identified early universe evolving into later epochs, if not the very physical laws that governed the symmetry-breakings in it giving rise to the earliest hierarchical structures and interactions between them?

But considerations such as these raise deeper questions of general interaction-theoretic character in the present context, viz., these: What has triggered the evolution of the physical force laws of interaction as we know them at present? How closely inter-related are the two problems of the nature of the initial conditions in the early universe, on the assumption of the most plausible cosmological models that are available now, and the problem of the unification of gravity with all other kinds of interactions, strong, weak and electromagnetic? The processes of symmetry-breaking, causing asymmetries in the order of the early universe, provide the most plausible answer that we can conceive of at present. From the big-bang cosmological point of view, such processes are sought to be understood, in terms of the high energy physics, as processes that occurred immediately after the big bang, in stages, each with the duration of a tiny fraction of a second, say from the grand unified theory epoch to the one second old universe where the electromagnetic and weak forces separated from each other.[67] It is clear from this how, in the construction of cosmological models, a general interaction theory, conceived as a presupposition of physics as a whole, must play a role which is as important as that of physical theory itself.

This gives an indication as to at what point exactly does the resolving power of physical theory enter in the construction of the cosmological models of the universe. Over and above this, cosmology provides, in its turn, the most crucial testing-ground to test a physical theory. Notice in this context the latest optimism among the particle physicists and cosmologists to the effect that cosmology is *now* the main area in which the fundamental theories of particle physics can be (even if indirectly) tested.[68] But one question which we must not forget to ask here is this: In which aspect exactly can they be tested? Without asking this question, it is not possible, I think, to realize that it is not the explanatory power but the resolving power of a physical theory which is of crucial relevance here. (For more on this see § 9.3. below.) I believe that what is true of cosmology in general is equally true of observational cosmology. The resolving power of physical theory can be seen at work here too. Consider, e.g., the discovery which was made by Edwin Hubble in the year 1929 (see § 9.3. below): All distant galaxies are receding from our own galaxy. It would not have been possible to make this discovery if men like Hubble had not resorted to theoretical identification of the observed redshift in the spectra of these galaxies as a Doppler effect. A more accurate picture of the recession of galaxies as a large-scale feature of the universe is given by Hubble's own law relating the velocity v of recession of a galaxy to its distance r as follows:

$$v = Hr$$

where H is known as Hubble's constant of proportionality. The galactic recession thus identified is also characterized by what is known as the property of isotropy: in whatever direction a galaxy lies, Hubble's equation is satisfied by the same value of H.[69]

Now, before one invokes galactic recession as a full-fledged large-scale feature of our universe (the modern epoch) in order to build a cosmological model on its basis, it requires further theoretical investigation. In the present-day standard big-bang cosmology, it is theoretically identified with a general *expansion* of the universe. This is done by invoking the twin properties of *isotropy* and *homogeneity* of the universe together. If by isotropy is meant that there is no preferential direction for the receding galaxies, homogeneity means that the universe has no centre, that there are no preferential reference systems, or that the universe is uniformly filled with galactic systems and that the same large-scale measurements made in different galaxies should give the same typical result. Recession of galaxies interpreted as general expansion of the universe is then taken to be one such result.

I think that it is a distinguishing feature of the big-bang cosmology that, in its task of identifying and locating the singularity epoch and explaining the nature of the present epoch of the universe, it invokes the cosmological principle comprising these twin properties of isotropy and homogeneity. Without doing so it would not be possible, I think, to interpret the recession of galaxies as a general expansion of the universe. But how does it do all this? Any adequate cosmological model, designed to perform a task as difficult as this, must after all proceed from sound assumptions attributing certain definite properties to the universe as it is being observed now. Their constraining effect on the theoretical interpretation of the observable features of our universe on the one hand and on the cosmological model-building itself on the other cannot be, therefore, denied. But, surely, not all such assumptions, not in any case the most important ones, can come directly from observational cosmology itself, the universe as a whole being inaccessible to our inherently limited means of theorizing and observation.

This leaves us with the puzzle: Why single then out just isotropy and homogeneity as the relevant constraining conditions on a cosmological model in an attempt to go back in time from the present epoch to the singularity epoch of the early universe – i.e. to the origin of time itself? This question could assume much deeper significance if cosmology were also to seek models predicting the pattern of the fluctuations in the structural hierarchical systems in our universe at some remote time in its future evolution using as initial conditions what we take to be the large-scale features (e.g. isotropy and homogeneity) of the universe in its present epoch. In either type of situation, we must ask: Can we afford to admit them, as such features, quite independently of the relevant background physical theories such as the special relativistic and general relativistic invariances that are known to physics of our time? These theories tell us a lot about symmetries in nature and the invariance of physical laws. Isotropy and homogeneity as invoked by the standard big-bang cosmological model have to do with symmetries of the overall distribution of galaxies in our universe. It is here then that the resolving power of the background physical theories can be seen at work playing quite a prominent role, the simplest example being Einstein's GTR when considered in its crucial role in cosmology. Among all the known fields of force gravity is one which matters most at the galactic scale, galaxies being electrically uncharged. As a consequence, then, without the background of a theory as powerful as the GTR, cosmology cannot study the large-scale features of the hierarchical structures of the universe, such as the distribution and motion of the galaxies, the density of radiation and dust through the intergalactic space, and perhaps a host of other parameters thereof.

We might quite legitimately then look at the kind of *variance* in the subject-specific initial conditions that I have been here arguing for as yet another interesting case of methodological variance which, I believe, the individual sciences must undergo from one problem-situation to another. No doubt, in doing so we would be repudiating a whole tradition, since initial conditions have always been identified by the physicist and philosopher of science alike with a sub-set of singular observation statements. Moreover, the physicist is supposed to take them as *given*, as *theoretically neutral* – i.e. as something that he cannot *vary* (in logical form) if only because he cannot choose them.

Not-withstanding, then, my own view I have just outlined, there is the view,[70] explicitly held by some physicists, to the effect that "the present state of the world is not 'explained' by physics".[71] In other words, as regards the question of the nature of *initial conditions* and also how the (individual) scientist gets at them, they are thought to lie outside the scope of an *explaining* science, their crucial role in scientific explanation/prediction$_1$ not-withstanding. What this view implies then is this: The initial conditions must always be taken as *given* and never as something pre-determined by the scientific theory itself. That is, the scientist's specification of the relevant sets of initial conditions, while carrying out the task of scientific explanation/prediction$_1$, must be viewed as the kind of step which is invariably possible to take quite independent of any background resources of theory that might be at his disposal.

In the light of our discussion above, it is now possible to reject this view as erroneous on a number of grounds some of which I have just outlined. It is the kind of view that smacks of an excessive observationalism in physics on the one hand and an ignorance, on the part of the philosopher of science, of the nature of the methodological constraints under which the individual sciences must always operate on the other. I have already pointed out above how initial conditions might *vary*, not just in content but in logical form, from one type of problem-situation to another across the individual sciences. The significance of this lies in the fact that it *relativizes* the concept of initial conditions to a whole network of theories and problems in the background – to our background systems of objective knowledge, so to say. Quite the opposite is the case under the dominant tradition which takes it as if it were capable of operating as an *absolute* concept. Let me now argue further to develop this relativization as follows.

Before a scientist, who is concerned with scientific explanation/prediction$_1$, specifies the relevant initial conditions in a given problem-situation, he must already be able to identify a set of state-variables as *the* set relevant for his purpose. Moreover, the specification of the present values of the relevant set of state-variables of a given system will also involve elaborate procedures of measurement, which in their turn will presuppose considerable background resources in terms of appropriate theoretical identifications. It follows from this that any precise specification of the initial conditions in a given problem-situation presupposes procedures of theoretical identification of relevant parameters at the deeper levels of scientific theory – in particular at the level of its resolving power. In other words, the scientist's ability to specify the relevant initial conditions in a subject-specific problem-situation methodologically presupposes not only his ability to make precise measurements of the present values of the state-variables of the given system of his choice – e.g. the relative positions and velocities of the planets in our solar system at a given time, or the observed redshift in the spectra of remote galaxies – but his ability to theoretically identify those and only those parameters of the system without which neither the problem-formulation nor theory-construction would be possible at all. If the Newtonian mechanics was able for the first time to describe the motion of planets with amazing accuracy, employing precisely stated initial conditions, it was, I believe, no less due to the fact that the entire planetary system had already been caught in the network of theory-building, and well-identified as a subject of objective knowledge in terms of, say, Kepler's three laws of planetary motion taken against the background of the Copernican revolution.

As is clear from this example, it is not possible for a science either to identify or to specify the initial conditions of systems of subject-specific interest to the scientist quite independent of the background resources of theory – i.e., in a theoretically neutral and absolutistic manner. We must, therefore, reject the following general belief which is quite deep-rooted and fashionable among physicists and philosophers of science alike: The specification of initial conditions does not strictly form a proper part of the problem-solving/problem-posing enterprise of science. It is, on the contrary, something that must go on *alongside* the explaining procedures of the scientist but without being subject to the processes of the growth of scientific knowledge.

It is appropriate at this point to return to the theme of the two concepts of the predictive power of a scientific theory with a view to sharpening further the distinction I have drawn between them. Thus, given our arguments above, we could say *that while the initial conditions play a crucial role in the structure of scientific explanation/prediction*$_1$ *conceived nomologically and deductively, no comparable role is played by them in those novel scientific predictions*$_2$ *which I hold to be a function, among other functions, of the resolving power of a scientific theory*. On this view then it is the resolving power of a scientific theory itself which contributes to the progressively increasing degrees of precision in the procedures of measurement and of specification, at any given time, of the values of the state-variables (of systems of subject-specific interest to the scientist) for the role of initial conditions. That is, it is the resolving power of our background scientific theories – themselves quite problematic though – which enables the scientist to separate and to specify the relevant initial conditions in a given problem-situation.

To conclude, if there is anything that science aims at it is not just the theories with ever greater explanatory and predictive $power_1$ but the theories with ever greater resolving power and predictive $power_2$. Greater the resolving power of a theory T_2 as compared with a rival theory T_1, more T_2's novel $predictions_2$ as a function of its newer theoretical identifications in otherwise familiar subject-domains will excel those of T_1, thereby leading to newer problems and better reformulations of older problems. Notice, e.g., the GTR's novel identification of a Riemannian space-time with the gravitational field, thereby leading to all the consequences that must follow from the geometrization of the latter. Consider in this context the GTR's $prediction_2$ of *gravitons* as the quanta of gravitational radiation or gravitational waves that are *supposed* to travel at the same speed as light waves. Consider again the theoretical identification of the observed *redshift* in the spectra of distant galaxies first as a *Doppler effect* and then, in the light of that, as *galactic recession*, all this of course on the basis of a whole complex of background theoretical assumptions including the following one: Atoms behave in the same way everywhere in the universe. Theoretical identifications, or separations as we might call them, as clean or as powerful as these are bound in their turn to result in novel $predictions_2$ and, therefore, in the *discovery* of newer empirical laws together with newer problem-areas – i.e. in the newer subjects of objective knowledge with more precisely stated but theoretically non-neutral initial conditions.

9.3. THE METHODOLOGICAL ROLE OF PHYSICAL THEORY IN RELATIVISTIC COSMOLOGY

It was first in my (1976) that it seemed to me quite correct to think that a physical theory must play some fundamental role in cosmology.[72] The latest developments in this field should leave no one in doubt about a very significant interplay between theories such as the GTR and the high energy particle physics as regards the problem of investigating the theoretically separated different epochs in the evolution of our universe. It is now a consequence of our consideration of the two concepts of the predictive power of a scientific theory that physical theory must be accorded a unique role in cosmology. The question which I would like, therefore, to pursue in a little more detail is this: What kind of role is it precisely which we might, therefore, assign to physical theory? How are we to define this role? An important clue is provided by modern relativistic cosmology, particularly when considered in the present state of its problem-situation:

> In summary, at present we are still extremely uncertain as to the basic parameters of the universe. All we can say is that the data are compatible with an extremely simple model of a universe in which expansion is the only motion and in which everything was determined by the initial explosive event. But we are still a long way from being able to ascertain if a theory based on GTR gives a proper description.
>
> The technology for observing faint objects is making rapid progress . . . with the new instruments now becoming available it should be possible to extend the Hubble diagram (magnitude-redshift) into interesting redshift range where different cosmological models lead to quite different predictions.[73]

One of the fundamental assumptions in relativistic cosmology is the assumption of a non-static expanding universe, for which both Newton's theory of gravitation and Einstein's GTR provide ample, and well-corroborated, theoretical basis. This, together with our discussion above, suggests that the unique role of physical theory in cosmology might lie in its resolving power, since in no case can it reside in its explanatory power. That is, it must lie in the field of the physical implications of the novel theoretical identifications that a physical theory makes through its very basic assumptions and postulates.

At any given time, then, a precise, though tentative, formulation and solution of the cosmological problem of the origin, structure and order of the universe must be premised on the well-developed framework of a physical theory. For example, the invariances, the symmetry principles and the fundamental physical constants, in so far as these figure in our physical theories, must play quite an important methodological role in shaping the very theoretical assumptions on which one might base a particular cosmological model. More precisely put, greater the resolving power of a physical theory more likely is it to *constrain* in methodologically significant ways these theoretical assumptions and, therefore, the cosmological model itself. Since T_{RP} (GTR) > T_{RP} (NGT), at present the GTR would seem to be the most suitable choice for this kind of role. The unique order of its resolving power, in this context, can also be gauged from the simple fact that the GTR is essentially a theory with large-scale consequences (other than those that relate to the motion of the perihelion of Mercury, bending of light rays by a gravitational field, and the gravitational redshift)[74] that are testable only in terms of observations of distant events, the strong gravitational fields relevant for relativistic correction being difficult to reproduce on the laboratory scale.[75] It only speaks of its unique resolving power if the physicists and astrophysicists should find themselves assuming the theory more and more while interpreting the distant events involving the extragalactic objects.[76] The same will of course apply to a grand unified theory as powerful and comprehensive as was envisaged by Einstein himself and others.[77] Interestingly enough, the need for such a theory as a unification of all the four kinds of physical interactions arises out of a recognition of the inadequacies inherent in the GTR as a theory of the pure gravitational field. More precisely, it has to do with the problem: "Whether it is legitimate to treat non-gravitational fields in a different way from gravitational ones and whether or not they can affect the geometry as the latter do?"[78]

The other more important constraints have, of course, to do with more direct empirical testability of a proposed cosmological model, or of its basic assumptions. These must, therefore, come from observational cosmology, as it has come to be known ever since Edwin Hubble's discovery, in 1929, of the relationship between redshift and magnitude of nebulae (Hubble's law of the velocity-distance relationship).[79] It would seem, in this context, that one of the most fundamental problems in modern cosmology is the correct theoretical interpretation/identification of the redshifts as these are observed in the spectra of distant nebulae. The assumption here is that only a correct interpretation would provide a sound observational basis for testing and appraising rival dynamical models of the universe.[80]

This unique methodological relationship between physical theory and cosmology is then quite significant. It suggests that the methodological appraisals of physical theories T_1, T_2, . . . need not be restricted to

$$T_{EP} \cdot T_{RP}(T_1) < T_{EP} \cdot T_{RP}(T_2)$$

in the given/intended domains D_1, D_2, . . . On the contrary, it should be possible to appraise these theories in terms of their consequences for cosmology. For cosmology seems to provide an extended ground to test and appraise rival physical theories as to their resolving power. Thus, for example, if we know beforehand, either on the basis of astronomical observations or on the basis of theoretical considerations alone, which possible solutions of the cosmological equations to rule out,[81] it will be possible to appraise methodologically any two or more rival physical theories T_1, T_2, . . . as follows.

$$T_{RP}(T_2) > T_{RP}(T_1)$$

where T_2 is the theory that does, and T_1 the theory that does not, rule out the prohibited solutions/models, *or* where both T_1 and T_2 rule out such solutions with T_2, unlike T_1, serving as a better theoretical framework for an adequate alternative solution. In fact, the latter is what is true of Einstein's GTR when compared with Newtonian gravitation theory. That is, in this case, the following methodological appraisal can be made:

$$T_{RP}(\text{GTR}) > T_{RP}(\text{NGT})$$

For both these theories rule out the assumption of a static universe so necessary to the static cosmological models of the universe, owing mainly to an overlap in their resolving powers. But it is the GTR which definitely provides a much better framework for constructing suitable non-static (dynamical) cosmological models, Einstein's own attempt[82] in 1917 to *complicate* his original gravity equation with the cosmological constant (denoted by the Greek letter λ) so as to produce a static model not-withstanding.[83] The reason why this is so has to do with the fact that the fundamental assumptions of NGT, in comparison to those of GTR, make it unsuitable as a theoretical basis to formulate and solve the cosmological problem. The manner in which its fundamental *assumptions* oversimplify gravitation and restrict its resolving power would have remained unappraised and unrectified in absence of a rival theory of as great a resolving power as GTR.[84] These assumptions relate to (1) the conception of gravitation as *instantaneous interaction* between distant bodies (suspended) in 'empty' space; (2) the conception of space as Euclidian in nature; and (3) the laws of Newtonian mechanics that are valid for bodies moving at ordinary velocities but quite inadequate for describing the recently identified rapidly receding galaxies. The assumption (1) implying infinite velocity in gravitational interaction is contradicted by Einstein's STR according to which no physical interaction/signal, other than that of light itself, can propagate at a velocity $v = c$ (where c represents the limiting velocity of light). What speaks of an overlap in their resolving powers, however, is the fact that both NGT *and* GTR imply the idea of a non-static universe so essential to dynamical solutions of the cosmological equations. Since GTR conceives gravity as a field characteristic of matter giving rise to space-time curvature, thereby shifting quite successfully the scene of physical problem-formulation from the field of NGT to its own field, the following appraisal is in order:

$$T_{RP}(GTR) > T_{RP}(NGT)$$

It is not surprising how in this way GTR gives rise to newer theoretical problems and possibilities inconceivable under NGT – i.e., to newer subjects of objective knowledge.

Excess T_{RP}(GTR) over T_{RP}(NGT) is also demonstrated by the relativistic cosmology which Einstein himself initiated by drawing the cosmological implications of his theory and constructing a cosmological model of the universe as a better alternative to earlier models of it. The earlier 17th and the 19th century static models arose as answers to the following questions respectively:

(1) Whether universe has the structure of an infinitude of stars scattered through infinite Euclidian space?
(2) Whether universe has the structure of a finite 'island-universe' in an infinitude of 'empty' space?

Neither of the two models, implicit in these questions, could be accepted as valid. Indeed, GTR rules out any answer to them in the affirmative.[85] The question which Einstein instead considered, given GTR, was this:

> Whether the structure of the universe is such that it contains a finite number of stars distributed equally through unbounded space?

The very formulation of this question would have been different and perhaps a prohibited one if Hubble's law of velocity-distance relation had been known as early as 1917. However, the key to Einstein's answer was provided by one of his own assumptions in GTR, viz., the presence of matter in 'space' itself produces the curvature of space, thereby giving rise to non-Euclidian space. This enabled Einstein to conceive of the universe as finite in volume though yet unbounded.[86]

We can conclude by saying that what physics aims at are theories with ever-increasing orders of explanatory and resolving powers. The resolving power of a physical theory itself follows a definite pattern of growth. That is, it increases by a process of shifting the scene of problem-determination more and more

(1) from the field of the background/rival theory to its own field; and
(2) to the field of its *extended application* in cosmology that seeks the construction of the formalizing (dynamical) models of the universe.

Here one might quite pertinently ask: How much of the theoretical identification of the problem of the early universe – a problem which is inconceivable on assuming pre-relativistic theories of the universe – do we really owe to the GTR? The answer is this: Without the T_{RP}(GTR), it would have been impossible to identify the early universe with the cosmological parameter of *expansion* (increasing gravitational entropy of the universe in its various epochs) as an aspect of the cosmological problem of the structure and dynamics of the universe. It would have been, moreover, impossible to predict the existence of the cosmic microwave background radiation as an observable *relic* radiation of the initial explosive event at the beginning of the universe, whose blackbody spectrum and isotropy provide an important probe into the nature of the universe then and its expansion since then.[87]

9.4. THE RESOLVING POWER OF A SCIENTIFIC THEORY AS A BASIS OF ITS EPISTEMIC APPRAISAL

Is there then always something in the structure of a good theory, new and old alike, which cannot be destroyed, even by the experimental discrepancies turning up in the future? Perhaps, this is an important philosophical question arising out of the fact that there are theories, such as Newton's and Einstein's, of which, in the opinion of some of the great physicists themselves, it is true to say that there is something in them that *cannot* be destroyed – that there is something in them that will always be admired as a matter of great elegance, beauty and simplicity. For example, this is exactly how P. A. M. Dirac (1980)[88] speaks of the *indestructible*, the essential beauty of Einstein's STR and GTR. Thus of the GTR he says:

The foundations of the theory are, I believe, stronger than what one could get simply from the support of experimental evidence. The real foundations come from the great beauty of the theory. They come from the circumstance that Einstein has introduced new ideas of space which are extremely exciting, very elegant, and these ideas will survive no matter what the future brings before us. These ideas are based on the possibility that one can describe a physical force like gravitation just in terms of the properties of space. They lead one to look upon a physical field in general as just a disturbance of space, like a curvature.[89]

To this Dirac adds further:

It is the essential beauty of the theory which I feel is the real reason in believing in it. This must dominate the whole future development of physics. It is something which cannot be destroyed, even if there are experimental discrepancies turning up in the future. These experimental discrepancies must be looked upon merely as inadequacies in our present theory.[90]

Isn't this something puzzling, especially when it is the physicist of Dirac's eminence saying it? It is indeed puzzling in so far as Dirac's own (1980) account of the performance of Einstein's STR and GTR, strictly from the point of view of their experimental corroboration, is the sort of account which a historian, or a text-book writer, feels obliged to offer. In other words, I find Dirac's appraisal of STR and GTR in terms of experimental test/corroboration non-trivially discrepant with his final assessment of them as the *kind* of theories which must be admired for their (inherent) *indestructible* beauty or elegance. What is only hinted at in the latter is hardly reflected in the former. Dirac's account is, I believe, an excellent example showing how faulty a scientist's appraisal of a given theory can be, with many of its aspects remaining essentially unappraised while being made a subject of great admiration.

While one must, then, with Dirac, admire in the STR and GTR what he does seem to admire most, I find unacceptable the manner of his appraisal of them. This could be generalized, since Dirac's case is a typical example of the general situation in the sciences. How might one describe the salient features of this situation? I have gone into it in some detail elsewhere.[91] Here it should suffice to draw attention to the important point that what is only vaguely hinted at in the general admiration of a good theory, as powerful as Newton's or Einstein's, is what is completely lost sight of in the specific appraisals of it when compared with its rivals. Thus, it is customary among scientists to appraise a theory T_2 as compared with T_1 along the following lines:

(a): T_2 is able to *explain* the antecedently known (*observable*) effect E_1 which T_1 is incapable of explaining. That is, unlike T_1, T_2 'predicts' E_1; and hence E_1 is what must be expected given T_2.
(b): T_2, unlike T_1, also *predicts specific* effects E_2, E_3, E_4, etc. which are not antecedently known
(c): T_2 is acceptable and preferable to T_1, since a number of experiments done give positive results confirming, to a lesser or a greater degree of accuracy, the predicted values of the space-time-region-specific effects E_2, E_3, E_4, etc.

The philosopher of science who finds it pointless logically to distinguish explanation from prediction[92] makes the whole thing look still more simpler:

(d): T_2 is acceptable and preferable to T_1, since $T_{EP}(T_2) > T_{EP}(T_1)$.

It is essentially along these lines that Dirac appraises STR and GTR. Thus, in so far as GTR is concerned, the three kinds of effects that it is known to predict are listed and their experimental confirmation considered.[93] These are respectively described as: (1) the antecedently known but unexplained motion – the perihelion of the orbit – of the planet Mercury; (2) the novel prediction that "light passing close by the sun should be deflected";[94] and (3) the novel prediction "That there should be a redshift in the spectral lines of light that is emitted in a strong gravitational field".[95] To this list, Dirac adds two more which have essentially to do with more recent work on the further *applications* of GTR, although these are not presented as further applications. This aspect of GTR is described by Dirac in terms of (4) the delaying effect of the sun on the waves such as radar waves passing close by it;[96] and (5) the (observable) effect of a star on the radio waves emitted by a pulsar when, as parts of a binary system, the latter moves around the former.[97]

All this, together with the further experimental possibilities in the future, speaks very well of a whole wealth of consequences of GTR at the level of experimental testing of it. What is, however, striking in this account, as in all accounts generally, is this. The predictions of GTR, as of any other theory, are presented as predictions of *specific*, *localized effects* in a particular space-time region experimentally favorable or suitable to it. With the possible exception of (3) above, this is true of all the effects (1), (2), (4) and (5) as described by Dirac. What is, however, lost sight of in this account is this. The GTR, the very fundamental set of assumptions that it makes – e.g., that gravitational fields arise from local distortions of space-time caused by massive bodies and that the trajectories of orbiting bodies are the shortest paths they can take in warped space-time – gives rise to the following theoretical problems, among others:[98]

P_1: Do light rays bend when passing in the vicinity of a strong *gravitational field* by an amount which can be experimentally measured and which may vary depending on the strength of the field in a given case?

P_2: Whether the spectral lines of light emitted in a strong gravitational field are redshifted?

P_3: Whether the periodicity of radio waves emitted in a strong gravitational field is subject to time-delay effect?

P_4: Whether accelerating masses radiate energy in the form of gravitational waves?

P_5: Whether an extremely strong gravitational field has any measurable effect on light waves, either of trapping the photons or of delaying their onward journey?

Thus, what is noteworthy is this. All these problems (P_1–P_5), formulated by *whether*-questions, are T_{RP}(GTR)-determined discovery oriented problems. They relate to effects that must arise *wherever* there are strong gravitational fields. They are not themselves about any space-time-region-specific effects. Each of these, however, gives rise to more specific problems. Thus, it is something like (P_5) which gives rise to problems concerning extremes of gravity of collapsed stars with an escape velocity $v>c$ that are termed 'black holes': Do black holes exist, as regions of space-time out of which nothing, not even a ray of light emitted radially outwards, can escape?[99] With the discovery in 1974 of the binary pulsar PSR 1913+16, the T_{RP}(GTR)-determined P_3 and P_4 give rise to space-time-region-specific problems P_6 and P_7 respectively:

P_6: Whether the periodicity of radio waves emitted by a pulsar is affected/changed as a result of its motion around its radio-silent companion star, when both are parts of a binary system? Whether these radio waves suffer from a measurable time-delay when they graze the silent companion on their way to the Earth?

P_7: Whether the orbiting pulsar and its radio-silent companion emit gravitational waves (gravitational radiation)?[100]

Exploration of gravitational fields stronger than those available in our solar system has in recent years made it possible to test the T_{RP}(GTR), giving rise to novel predictions and problems just considered. Not only this, once such gravitational fields are identified, equations of GTR are found quite indispensable for determining their relevant physical parameters, e.g., the size and orientation of the orbit and the masses of the pulsar and its companion.[101]

It can be shown how favorable solutions of space-time-region-specific problems such as P_7 lead to a confirmation of $T_{RP}(GTR) > T_{RP}(NGT)$. The recent finding[102] that the orbit of PSR 1913+16 around its companion star is slowly shrinking is seen to confirm GTR's prediction that an accelerating mass should radiate energy in the form of gravitational waves. If this is so, then what it confirms is that $T_{RP}(GTR) > T_{RP}(NGT)$. Indeed, the same applies to GTR's prediction relating to the advance of Mercury's perihelion in its orbit around the sun. Being a natural consequence of the space-time curvature in the vicinity of a gravitational field as strong as that of the sun, the confirmation of this prediction confirms that $T_{RP}(GTR) > T_{RP}(NGT)$.

Parallelism between the STR and the GTR cannot be missed. For of the STR also one can say that out of its own basic assumptions there arise interesting theoretical problems, e.g., whether particles moving with a velocity $v>c$ exist or are physically possible? For it is only another way of asking: whether c, the velocity of light, being the limiting velocity for all physical interactions/signals according to STR's limit principle, is also the limiting velocity for all material particles in motion? And, if so, why? Now when I say this, I am already assuming a certain interpretation of STR's principle of the velocity of light as a limiting velocity. That is, I understand this principle as making a fundamental theoretical identification of the maximal velocity with which signals can propagate in space. So interpreted, what STR prohibits are not just velocities greater than c but only signals of a velocity $v>c$.[103] Otherwise, given all the literature on tachyons – suspected particles moving faster than light – one is likely to doubt, with Graham Nerlich, whether the limit principle is a basic postulate of STR at all, or even a consequence thereof.[104]

Now, what is the methodological significance of all this? It is, I believe, essentially this. A good, elegant and mathematically beautiful theory in the sense already hinted at by Dirac (1980) and others is not one which makes specific predictions about particular regions of space-time – something that it must do by way of predicting the outcome of *novel* experiments designed with a view to testing its more basic claims and their consequences. On the contrary, it is one which makes newer theoretical identifications, suggesting possible existence of hitherto unknown parameters, relations, interactions and invariances; and all this without restricting these or the problems arising therefrom to any given space-time region. In other words, it is that aspect of a theory which gives rise to a wealth of theoretical problems hitherto either inconceivable or only inaccurately *determined* under the systems of background theories. It is then the most important aspect of a theory which I have designated as its *resolving power*. It is this very aspect which is testable by novel experimental arrangements whose scope is, however, quite limited. For these experiments depend on the state of technological advance and are significant in a limited way, being space-time-region-specific in character. They are, in other words, an on-going enterprise. Nevertheless, experimental results serve as a basis enabling us to look at $T_{RP}(T_2) > T_{RP}(T_1)$ either as confirmed or disconfirmed. And it is this very aspect of a theory again which continues to 'live' in the newer and more comprehensive theories, if only as a special case. It is, therefore, quite reasonable to look at it, with P. A. M. Dirac, as that aspect of a theory which cannot be destroyed.

But the trouble with the term 'indestructible' as applied in the case of a really good scientific theory is this. Quite unintendedly it suggests that there is something absolute about its very structure, its elegance and beauty. And this is not so. For if there were any such thing, there would be no need at all to compare the rival theories, to look for experimental arrangements to test a given theory or to look for better theories. The fact remains, however, that what cannot be destroyed in a good theory, like the STR and the GTR, is its *simplicity* and *completeness*. That is, even if experimental results showing great discrepancies between the performance of the GTR and what it says turn up in the future, the GTR will survive as a methodologically *simple* and *complete* theory, if not as an *adequate* theory. The possibility that such results might turn up or that the GTR might be superseded by a more comprehensive theory – such as a grand unified theory – in which it 'lives' on as a special case cannot rob GTR of its character as a physical theory that is simple and complete in itself. On the contrary, such a possibility carries deeper significance, which is clearly brought out by the postulates of epistemic structuralism considered before. Epistemic structuralism implies that a good scientific theory is one which serves as a developmental structure of objective knowledge. That is, any theory T_2 and its rival T_1 are good scientific theories only if each gives rise to more problems than it can or is intended to cope with, and in a manner such that: $T_{RP}(T_2) > T_{RP}(T_1)$.

It is perhaps this which Dirac has in mind when he admires the GTR for what is taken to be indestructible in it as a physical theory. My own attempt to identify the 'indestructible' aspects of a scientific theory, however, proceeds in a different way. If science is to undergo *methodological variance* and if objective knowledge is to grow at all according to an interactive pattern, it is necessary to recognize deeper principles of invariance in the background against which this is possible. Thus I look at science, its theories and problems, in terms of what I call theoretical universals. These relate to the principles of invariance against which both theories, say $T_1, T_2, \ldots$ and theoretical problems, say $P_1, P_2, \ldots$ in a given domain, can be subjected to epistemic appraisal essentially in terms of principles of methodological variance along the following lines:

(1) Is $T_{RP}(T_2) > T_{RP}(T_1)$?
(2) Is $T_{EP}(T_2) > T_{EP}(T_1)$?
(3) Whether the problem P_1 is $T_{RP}(T_1)$-determined?
(4) Whether P_1 is just raised *ad hoc*, in the sense of not being $T_{RP}(T_1)$-determined?
(5) Whether P_1 could be reformulated as P_2 in terms of $T_{RP}(T_2)$?
(6) What further problems are entailed or determined by $T_{RP}(T_2)$?

It is interesting, for example, to consider in this context the problem determined by T_{RP}(NGT · NCT) which might be *reformulated* as follows:

> Do not Bodies act upon Light at a distance, and by their action bend its Rays; and is not this action (*caeteris paribus*) strongest at the least distance?
> (*And whether the amount of bending in the case of starlight grazing a distant star of the same size as our sun during a total eclipse is* 0″.87?)[105]

The essentially Newtonian physical problem formulated by the specific *whether*-question in italics (which Newton never raised) is precisely what Einstein must have believed in, in the first instance. What is, however, most remarkable as a most glaring example of T_{RP}-determined methodological variance is the contribution of the non-Newtonian concept of the *curvature of space* to an alternative relativistic formulation of the *whole* problem:

> Do light rays bend in the vicinity of strong gravitational fields because of the curvature of space itself? And is the amount of deflection in the case of starlight grazing the sun during a total eclipse 1″.74, double the Newton value?

Quite independently of the 1919 solar eclipse expedition and the confirmations it yielded, we can consider this as an excellent case of growth of knowledge by methodological variance, since it is the T_{RP}(GTR) which is here at work in shifting the scene of problem-determination to its own field. The methodological scenario here would have been, I believe, hardly different had Einstein left us without knowing the fate of GTR, or had there been no successful 1919 solar eclipse expedition. Interestingly enough, this is also true of methodological variance only vaguely hinted at in the Newtonian formulation of the problem. Thus, methodological variance here remains unaffected by the historical fact that Newton himself *believed* in a problem, directly determined by the T_{RP}(NGT · NCT), if only as a problem of deviation from an assumed physical law, which he did not know how to solve. The one problem which lay beyond the limits of the T_{RP}(NGT · NCT) was then this: Whether light can behave non-mechanically without following the trajectories of particles with mass? It is this, among other things, which set serious limits to methodological variance in Newtonian physics. However, my interpretation here of the Newtonian formulation of the problem of the bending of the rays of light as a step in methodological variance shows how much of the history of science cannot be made sense of in terms of the received frameworks for the methodology of science, particularly where the resolving power of a scientific theory is already at work as a basis of its epistemic appraisal.[106]

Methodological variance in the sense of these examples can be considered as *non-ad hoc*, since it involves significant processes of scientific change at different levels of problem-determination by the resolving power of a scientific theory. Important correlations may, therefore, exist between methodological variance which is *non-ad hoc* and *CSD*-variance (Pandit, 1983) including variance in the search-cum-discovery procedures in the individual sciences.

Does methodological variance in a science have, then, to do with the ability of a (newer) theoretical framework (= *CSD*) to suggest *physically possible* methods of testing specific hypotheses, arising from it, or of answering discovery-oriented *whether*-questions which other theories may raise? The answer is in the affirmative, if we recall the example above of problems arising in Newton's own theoretical system but being subsequently subjected to the T_{RP}(GTR)-determined reformulation. Given T_{RP}(NGT · NCT) < T_{RP}(GTR), it is interesting how Einstein, unlike Newton, was himself able to propose an experimental arrangement – of which 1919 solar eclipse expedition was a result – to test the T_{RP}(GTR). As Newton's example shows, this may not always be the case with every theoretical framework in science. What must have, among other things, prevented the Newtonian framework from suggesting a similar test is, I believe, its assumption that space is Euclidian in character.

What conclusion can we draw then from the principle of epistemic appraisal of a scientific theory on the basis of its resolving power? I think that we can say this: more the ability of a theory to shift the scene of problem-determination from the field of its rivals to its own field, more powerful it will be as compared with them. What is important here is that there are no space-time-region-specific restrictions on the types of problems of very general nature that it can raise. Most important among these problems are those that are formulated by means of the discovery-oriented *whether*-questions with universal theoretical statements as their theoretical basis. Deeply rooted in the search-cum-discovery procedures of a science, the best method of answering them is to formulate their corresponding space-time-region-specific *whether*-questions with a view to setting up experimental arrangements and looking for results which may corroborate/discorroborate the theoretical basis of the original *whether*-questions.

What makes, then, Einstein's GTR more powerful than Newton's gravitation theory? On the basis of the account of the nature of epistemic appraisal in science just presented, it is its ability to give rise to *whether*-questions such as the following: Do light rays bend in strong gravitational fields? If we like we may say that the GTR *predicts*, among other things, a whole theoretical basis of this question which is not addressed to any specific space-time-region in the universe. But what actually enabled the physicists to corroborate the theoretical basis in question was a space-time-region-specific *whether*-question essentially addressed to a physically possible experimental arrangement – the 1919 eclipse expedition – and to its possible outcomes. This particular case is capable of significant generalization, along the lines suggested above, that a really powerful theory is one which is able to determine, formulate and reformulate, *whether*-questions of an unrestricted general scope in its field. Not only this, the theoretical basis of these general *whether*-questions must be testable or capable of corroboration in terms of specific answers to space-time-region-specific *whether*-questions addressed to physically possible experimental arrangements.

More generally, then, a theory T_2 can in this sense be said to be more powerful than another T_1 if and only if $T_{RP}(T_2) > T_{RP}(T_1)$, i.e., T_2 entails a set of universal statements or raises general *whether*-questions with such statements as their theoretical basis, which T_1 is just incapable of doing. The only way to answer such general questions, and to test their general theoretical basis, is to derive, on their basis, *specific whether*-questions in terms of physically possible experimental arrangements in a favorable space-time region. If the outcome of the experiment is favorable, and the answer to the specific *whether*-question in the affirmative, it only suggests a confirmation of $T_{RP}(T_2) > T_{RP}(T_1)$ and not $T_{EP}(T_2) > T_{EP}(T_1)$. If, however, the outcome is unfavorable, with an answer to the specific *whether*-question in the negative, what one can legitimately do is to *extend* the testing ground further by formulating further sets of specific *whether*-questions in terms of newer experimental arrangements preferably in other space-time regions; and all this before one might give up and announce the whole theoretical basis of the general *whether*-question as false or as refuted. Such an extension is necessary as part of scientific method, if only to leave open the future possibilities of re-interpreting the previous experiments and of their outcomes. And all the more so if we keep in view the important fact that all experimental arrangements in science are bound to be space-time-region-specific and hence limited in scope as well as in significance; while the number of specific *whether*-questions derivable from a scientific theory, and of the corresponding singular statements as their theoretical bases, is infinite, and never finite.

The account of epistemic appraisal in science that I have here presented clearly leads to the methodology of theory-problem interactive systems, in the sense of my (1983), which was the subject of discussion in Chapter 8 above. It is thus the kind of account which is bound to disturb not only the inductivist but also those of a strongly deductivist persuation with a commitment to what is popularly[107] known as the hypothetico-deductive model of scientific method. The popularity of the latter among scientists and philosophers of science of a non-inductivist persuation can no longer prevent one from realizing how the most interesting and most powerful theories in science are misrepresented by this model. In any case, what my account shows and explains, among other things, is why it is erroneous to be committed to a deductivist approach, or a falsificationist methodology, pure and simple.

In the hypothetico-deductive model, the appraisal of a theory "relies solely upon deductive consequences (predictions)",[108] where predictions derivable from the theory are generally taken to be (a) *finite* in number; (b) subject to *verification*; and (c) formulated by spatio-temporally *singular statements*. The best statement of this model is available in Popper (1968, 1972a), and is also implied by his three requirements[109] for the growth of knowledge from one theory to another:

(1) *The requirement of simplicity*: The new theory should proceed from some *simple*, *new*, and *powerful*, *unifying idea* about some connection or relation (such as gravitational attraction) between hitherto unconnected things (such as planets and apples) or facts (such as inertial and gravitational mass) or new 'theoretical entities' (such as field and particles).
(2) *The requirement of independent testability*: That is to say, apart from explaining all the *explicanda* which the new theory was designed to explain, it must have new and testable consequences (preferably consequences of a new kind); it must lead to the prediction of phenomena which have not so far been observed.
(3) *The requirement of empirical success*: That the theory should pass some new, and severe, tests.

Concerning this third requirement Popper goes on to say:[110]

Earlier I suggested that science would stagnate, and lose its empirical character, if we should fail to obtain refutations. We can now see that for very similar reasons science would stagnate, and lose its empirical character, if we should fail to obtain *verifications of new predictions*; that is, if we should only manage to produce theories that satisfy our first two requirements but not the third. (Emphasis is mine.)

The model of theory-appraisal[111] that these requirements imply turns out to be an oversimple one which may be stated as follows:

T: The given theory

C_1, C_2, C_3: Spatio-temporally singular statements formulating initial conditions.

P_1, P_2, P_3: Spatio-temporally singular statements formulating the predictions derivable from (T, C_1, C_2, C_3).

O_1, O_2, O_3: Spatio-temporally singular statements formulating the observational outcomes/results of newly set up experimental arrangements aimed at testing the theory.

Among other things, what is noteworthy in this model is this: (a) it presupposes the traditional statement-view of the epistemic structure as a type-identical structure of theoretical statements all falling on a single scale of graded generality/universality; (b) it entails that the growth of knowledge follows a non-interactive pattern of the newer theory *replacing* the older one, a pattern of *conjectures and refutations*, the fact that type-distinct elements/structures are necessary for patterns of growth by interaction not-withstanding; (c) in it theoretical problems are not represented at any level as type-distinct epistemic structures in science; and (d) in it the new predictions of a new theory are essentially construed as being susceptible to *verification* and, hence, as being space-time-region-specific in character (whether intendedly or unintendedly).

Now even Lakatos (1971a),[112] who seeks to improve upon Popper's falsificationist methodology of theory-appraisal, insists on the 'novel facts' or novel predictions as what a progressive as against a degenerating scientific research programme must produce: "A research programme is said to be *progressing* as long as its theoretical growth anticipates its empirical growth, that is, as long as it keeps predicting novel facts with some success ('*progressive problemshift*') . . ."[113] A Lakatosian research programme is thus to be admired in so far as it predicts novel facts. What Lakatos understands by 'novel facts' may be gauged from the type of examples he cites in support of his methodology of scientific research programmes. Thus, of Einstein's programme as an example he says: "This programme made the stunning prediction that if one measures the distance between two stars in the night and if one measures the distance between them during the day (when they are visible during an eclipse of the sun), the two measurements will be different. Nobody had thought to make such an observation before Einstein's programme. Thus in a *progressive* research programme theory leads to the discovery of hitherto unknown novel facts".[114] One wonders whether this is all that matters about a powerful theory; whether being powerful or progressive means just being able to predict novel facts in the suggested sense of *observable effects* in a particular space-time region.

A fresh look at what Popper and Lakatos assume concerning the growth of knowledge from theory to theory or from one research programme to another is called for, particularly from the point of view of the *principle* of the epistemic appraisal of a theory on the basis of its resolving power. It is gratifying, in this context, to find Lakatos (1978) stressing in a footnote a point about scientific theory, although only obliquely, which is much nearer what this principle implies. Lakatos says: "It should be stressed again that we never *know* that a theory entails novel facts. All we may know is that it entails novel corroborated hypotheses. Corroborated hypotheses are the fallible (epistemological-methodological) counterparts of (ontological) facts; corroborated hypotheses may not be 'truly corroborated'".[115]

CHAPTER 10

EPILOGUE

> . . . 'rational criticism' must be applied, not just to the scientist's *words*, but also to his *works* – not just to his theoretical propositions, but also to his empirical praxis – and . . . the canons of rational criticism accordingly involve, not just propositional 'truth' and inferential 'validity', but scientific adequacies and inadequacies of other kinds also.
>
> Stephen Toulmin (1976)

I

At the very outset, we asked first, "What kinds of structures might be assumed to serve as proper objects of our subjectively organized K-states of the minimal identity 'S knows that p'?"; and then, "How we might distinguish K-states in general from states of human interaction?". Assuming an interaction-theoretic point of view, together with a realistic ontological framework as wide-ranging as

the $world_1$ of physical objects/processes;
the $world_2$ of minds/mental processes;
the $world_3$ of linguistically formulated, but language-neutral, abstract problems and theories (= propositional structures)

I have identified these *objects* with abstract epistemic structures best represented by theory-problem *interactive* systems in science: $I(T_1, P_1)$, $I(P_2, T_2)$, $I(T_3, P_3)$, ... My argument here develops against the background-tension (in TE) between (1) their explicit identification with the $world_1$/$world_2$ objects and processes and (2) their implicit identification with propositions about *them*. (1) is traceable, I have argued, to the basic assumptions underlying Plato's famous doctrine of 'Forms' as objects of knowledge. But (2) is a consequence of the traditional standard analysis of all such K-states in terms of those of *justified true belief*. The value of distinguishing such K-states from those of objective knowledge on the one hand and K-states in general from states of human interaction on the other is, then, seen to lie in the resolution they offer of the tension in favor of the answer I have proposed in Chapter 1.

The latter distinction, which challenges the common belief that all knowledge is a form of interaction instructed by bodies of empirical evidence from perceptual experience, is developed in Chapter 2 in the following aspects:

(1) The problem of the nature of human knowledge is essentially different from, and not identical with, the problem of the nature of man's relationship with the world including the social world.
(2) The latter problem relates to interactions which are complex and multi-variable in nature. They are far from being simple or one-sided.
(3) What distinguishes human interaction from all other forms of non-human interaction is precisely that which we call human knowledge, where *knowledge* is understood as an orientating variable of (creative) human interaction.

As a consequence of this distinction, all the TE-theories (Quine's epistemology naturalized being most explicit in the series) with few exceptions (for example, Immanuel Kant's) are exposed in the following respect: they aim at reducing knowledge to one or the other type of state of interaction. But they have not succeeded in this, simply because the types of interaction found relevant invariably presuppose *knowledge* as their orientating variable. Moreover, the distinction serves as a framework for objective evaluation of alternatives to TE-theories. It ensures a constant interaction, at a higher level of organization, between epistemology and the individual sciences like psychology, biology and physics. For the whole problem of the nature of man's relationship with the world with which these sciences are, I think, essentially concerned is itself the most important source of external criticism and evaluation of epistemology which is concerned with knowledge as an orientating variable of (creative) human interaction. In this way our discussion results in fundamental distinctions between *organization*, *information* and *knowledge* on the one hand and *instructed* and *creative interaction* on the other, but all within the framework of a general interaction theory (Pandit, 1983). These develop, later in Chapter 5, into an interesting theme, namely: the fundamental aspects of both language and science lie essentially beyond the kind of deeply conventional, game-theoretic approach to understanding them to which individual men as well as schools of philosophy have attached such a great importance.

The discussion in Chapter 3 pursues the fundamental questions concerning (i) a possible *asymmetry* between science and language and (ii) a possible parallelism between them where the methodological problems of theory-choice and those relating to choice between rival translation-claims about two or more languages are concerned. Here it becomes important to find out the most interesting, and non-trivial, sense in which our recognition of indeterminacy in science and across the languages does not threaten rationality at either of these two levels of appraisal and choice, nor fall back on this or that variety of behaviorism just as happens in Quine's case. As a first step to this end, Quinean indeterminacy of translation is examined in those aspects which restrict it to what he calls standing sentences (= the more theoretical portions of a language) but which premise it on strong behavioral assumptions. But it turns out to be just *criterial indeterminacy* in so far as it is arrived at and formulated by Quine to answer a question that has remained implicit in his own assumptions, viz.: what strictly behaviorist *criteria* must be invoked such that their fulfillment/non-fulfillment will guarantee determinacy/indeterminacy of translation, as the case may? But his assumptions, and therefore the question itself, are rendered impermissible by the recognition not only of a fundamental asymmetry between science and language (with far-reaching consequences against the Kuhn-Feyerabend doctrine of incommensurability) but of language translation/language learning itself as a function of multi-variable interaction.

On the contrary, indeterminacy of a non-trivial sort may reign supreme in all translation and range over language as a *whole* instead of just the so-called non-occasion sentences in Quine's sense. But in order to appreciate just this fact, a prior commitment to Quinean variations on the old theme of behaviorism is not necessary at all. Taken then in this essentially unrestricted non-Quinean sense, it means that there is always more than one way to translate into one's own language *whatever* the native *says* or *asks* in his/her language. But indeterminacy in this sense is best construed as a function of a more fundamental indeterminacy of *problems* and *propositions* themselves, conceived non-mentalistically. Quine's failure to recognize this is traceable to his scepticism concerning *propositions* on the one hand and his strong ideological commitment to behaviorism on the other, both of them largely coming in his way and preventing him from asking the right kind of questions. Thus if we assume that translation is indeterminate in a more holistic sense just alluded to, it is quite reasonable to ask: what must this indeterminacy be a function of? This shows how it is Quine's own assumptions which send him in search of translational determinacy here and indeterminacy there but always in a manner which is highly parasitic on his highly restrictive and reductionistic behavioral criteria.

In order to complete our earlier discussion on the objects of subjectively organized K-states, Chapter 4, the last in Part One, examines the view held by many philosophers that the problem of personal identity and that of the nature of persons arise out of the *existence* and *nature* of self-knowledge, where its existence is assumed as being beyond all doubt. Against this view, I argue as follows: while 'self-knowledge' in the strict sense of TE is impossible, as also any TE-type enterprise concerning it, there is a sense in which persons (= hierarchically organized growing problem-solving knowing subjects) exist and are, moreover, themselves knowable strictly from the point of view of an objectivistic epistemology instead of a subjectivistic one. Thus the concept of a knowing subject – which must remain, I believe, central yet an undefinable primitive within TE – is shown to be the kind of concept which is definable within such a framework. As a consequence, it becomes possible to have a better understanding of the sceptical views of David Hume and Ludwig Wittgenstein.

Turning now to Part Two, here I make it my chief aim to elucidate the novel idea of science as an enterprise comprising *NFCPS* systems,[1] in its methodological and epistemological aspects. Thus, where other philosophers[2] have usually raised questions of its aim and method in their most dogmatically assumed intimate relationship, I raise first, in order of *new* priorities, the question of its *structural identity* in the context of its historical development and then those concerning method, strategy and scientific change.[3] Among these the most important are the following. If methodological conventionalism of Popper has unacceptable anarchistic consequences for science, as becomes clear in Feyerabend's scepticism concerning not only method but science as a whole,[4] and if science is neither a game among other types of games nor just a system of beliefs/linguistic practices among other systems of belief/practice, what kind of epistemological-ontological framework for science must we opt for and what kind of methodology of science, *the latter embedded in the former*?[5] What is the nature of the principles of variance and invariance underlying the theory-problem interactive systems (= *NFCPS* systems) in the individual sciences making it possible for the scientist to subject the states of scientific change and growth of scientific knowledge to critical rational appraisal? What properties make a scientific theory simpler, and therefore better, than its rival in its field? And what is the nature of its epistemic appraisal on the basis of both its range of explanatory power and its resolving power?

As an attempt to find proper answers to these questions, and to those that are related to them, Chapter 5 re-appraises essentialism as a subject-neutral methodological framework for scientific and philosophical inquiry at various levels of subject-specific abstractions and analysis. This assumes much significance not only because methodological essentialism is presupposed by my own approach to different kinds of problems discussed here and in my (1983) but because of the contemporary approaches to the philosophy of science and the philosophy of language, as divergent as those of Karl R. Popper and Imre Lakatos on the one hand and later Wittgenstein, T. S. Kuhn, Paul K. Feyerabend and J. Hintikka on the other, being what they are. Owing to their common commitment to methodological conventionalism, in one or the other version, they all share a persistent assault on methodological essentialism. But which of the two frameworks is it, then, not just reasonable but most fruitful and promising to adopt and follow? An answer based on the current philosophical fashions would undoubtedly go in favor of methodological conventionalism. Without prejudging the issue, however, various methodological conventionalistic positions are, after a detailed examination, subjected to the following criticism:

(1) In every case of its application, methodological conventionalism leads to epistemological anarchism in the sense of Feyerabend's slogan: 'Anything goes'.
(2) Again, unlike methodological essentialism, it is invariably conducive to the sceptical strategy of so changing, in the subject-specific contexts, the very subject of predication/appraisal as to obtain the kind of consequences that are desired from the very beginning. Thus, on the one hand, it leads to the consequence of chaotic incommensurability *within* a given domain of inquiry and, on the other hand, to the consequence of chaotic commensurability – Wittgensteinian family resemblances – *between* that domain itself and every other essentially *distinct* domain whatever.
(3) Methodological essentialism, suppressed at a given level of subject-specific inquiry – obviously by resort to the methodological

conventionalistic strategies – raises its head at a subsequent/higher level where the only way to have it suppressed once again is to face an *infinite regress*.

This results in the proposal of the following twin principles for deciding between alternative methodological frameworks such as essentialism and conventionalism:

(a) Among other things, a proposed framework should be commensurability-preserving as regards the objects – whether abstract or concrete – which fall within *the* domain of one's inquiry. That is, it should be commensurability-preserving at various levels of methodological variance *within* a given science, from one theory-problem interactive system to another such system.

(b) And *it* should be incommensurability-preserving as regards the different subject-specific domains of inquiry themselves – i.e., it should preserve incommensurability between science as a whole and non-science.

Both (a) and (b) are based on the following presupposition: In so far as the domain of objects one wants to inquire into is uniquely distinct from other domains it is incommensurable with them; but the objects within the domain – such as theory-problem interactive systems *in* individual sciences – are commensurable with one another. As a consequence, methodological essentialism survives as the best possible framework for the sciences until better alternatives other than methodological conventionalism are found.

But, assuming this framework, how is empirical science – not as a static corpus or structural hierarchy of theories/hypotheses but as a growing system of the developmental structures of knowledge – possible? An answer to this question is sought, in Chapter 6, by identifying two kinds of methodological assumptions/rules as follows: (a) those which are subject-specific to the individual sciences and necessary to their *search-cum-discovery procedures*[6] (= SD-procedures) of problem-determination, problem-formulation and theory-construction, as we might call them; and (b) those which are highly abstract and subject-neutral with respect to them but which are at work at a more fundamental level making it possible for the scientist not only to *choose* but also to *vary* his assumptions of category (a) in a rational, though essentially unpredictable manner. Under the first category (a), we must distinguish, of course in terms of historical examples from various sciences, two chief types: (1) the structural-dynamical assumptions of complexity and (2) the ideal-type assumptions of simplicity. Under the second category (b), on the other hand, I argue for a theory of theoretical universals as the methodologically invariant properties of all science. This facilitates a critical review of both Kuhn's and Popper's characterizations of scientific revolutions and a search for an alternative to them. Both of them are found inadequate. Thus the Kuhnian characterization not only obliterates the fundamental asymmetry between scientific and non-scientific revolutions but is founded on the kind of oversimplifying strategy which entails a radical change, in the context of scientific appraisal, in the very subject of such appraisal. On the other hand, the Kuhnian scepticism concerning an objectivistic methodology of science is traced back to the mistaken idea of science as philosophically neutral and of philosophy itself as an epiphenomenon of science.[7] The Popperian characterization, based as it is on his conception of science as an aim-oriented inquiry in search of 'truth', is similarly shown to entail an oversimplified view of the structural identity of science with an indifference to its objective character as a complex problem-posing enterprise. Alternatively, scientific revolutions are then defined as those complex states of scientific change which are a function of working our way backwards from the crises-situations of a science entangled in its antecedently given SD-procedures.

Intended, as it is by Popper (1968), as a methodological convention to demarcate empirical science from non-science, what kind of rationality – other than the one its possible consequences including the consequence that it forms the very apex of a whole hierarchy of (Popperian) methodological rules may hint at – is possibly correlated with his falsifiability criterion? In the light of the discussion, in Chapter 7, of theoretical universals as the basic invariant properties of objective scientific knowledge, together with an examination of two of Popper's arguments for the criterion, it is salvaged not as the apex of a whole hierarchy of Popperian rules of the methodological-decision making game but as a methodological requirement embedded in scientific invariance. In this way, it is incorporated into a more comprehensive kind of framework applicable not only to theoretical statements but also to theoretical problems in the individual sciences.

Under methodological conventionalism it is not possible to ask essentialistic questions concerning science at any level of analysis whatever. Moreover, it inevitably ends in methodological anarchism either of a Kuhnian or a Feyerabendian variety. Therefore, it must be replaced by methodological essentialism, as is already argued in Chapter 5 above. This makes it possible to raise, in Chapter 8, an essentialistic question concerning the nature of a falsifiable theory with a view to arguing for the methodology of theory-problem interactive systems as a better alternative to the received frameworks. Thus in the 'game' of science played according to the Popperian methodological conventions there is no move more basic than the theory T being proposed and the following question concerning it: Is T falsifiable? But, once this basic move is made, what one cannot ask are the kinds of question which probe deeper than this Popperian question is supposed to do. For example: What is there in T's structure, if it is falsifiable in Popper's sense, which makes it falsifiable? What is it, besides its logical form, which enables it to satisfy Popper's requirement? Or, what is T's falsifiability a function of? All that is open to Popper, when one asks such questions, is an answer along the following lines: Well, as a matter of *convention*, this is how T must 'behave' before us as a scientific theory. But this is only another way of rejecting these and similar questions as trivial, if only for the reason that falsifiability requirement is considered to be the very apex of the hierarchy of Popperian rules of the 'game' of science.

But the fundamental epistemological and methodological aspects of science lie beyond the purview of methodological conventionalism. Deeper questions concerning it can not only be raised but answered within the framework of methodological essentialism. This becomes clear in our discussion of the simplicity of a scientific theory as a theoretical universal, and of epistemic structuralism as the very epistemological-ontological framework of the methodology of science, making a better understanding of the demarcation between science and non-science, the nature of background knowledge and the unit of epistemic appraisal possible. Theoretical universals assume much significance in this context. They are the principles of invariance which constrain theories and problems in the individual sciences in methodologically significant ways. A further elaboration of the methodology of theory-problem interactive systems as a better alternative to the received frameworks results from this when these are subjected to a fresh examination in terms of the following principle: The methodological models of epistemic appraisal in the individual sciences must be properly embedded in sound models of their epistemic structures and states of scientific change.

All this changes radically the order of priorities philosophers and methodologists have usually accorded to various types of questions concerning science, with questions of its aim and method always posed together before every other question.[8] The new priorities, already in sight, demand putting those questions first which can be considered in the historical context of the development of the individual sciences themselves. To this end, Chapter 9 undertakes a detailed analysis of the questions of *its* structural identity, its SD-procedures of problem-determination, problem-formulation and theory-construction, and those relating to the search for the methodology of rational theory-choice by raising deeper questions as follows: What makes a scientific theory a logically/methodologically complete theory? Which is its more fundamental aspect, its explanatory power or its resolving power? What is the nature of *ad hoc* hypothesis and how do they arise? What do the favorable outcomes of novel experiments prompted by specific *whether*-questions, which are themselves consequences of more general discovery-oriented *whether*-questions, confirm: a theory's excess T_{EP} or its excess T_{RP} over that of its rival? What, in retrospect, is the methodological significance of the famous Michelson–Morley experiment? With the case-studies of Newton's and Einstein's theories assuming much significance in this context, the most fundamental and decisive aspect of a theory, taken in relation to its rival, turns out to be its resolving power, with its explanatory power methodologically parasitic on it. My argument here develops further in terms of a deeper analysis of (i) the types of problems which are determined by the resolving power of a theory (= T_{RP}-determined) and which give rise to discovery-oriented specific *whether*-questions and novel experiments; (ii) the nature of novel prediction in terms of a distinction between the two concepts of the predictive power of a scientific theory; and (iii) the unique role which the resolving power of physical theory plays in cosmology. With this we come at last face to face with the question: Whether there is always something in the structure of a good scientific theory that cannot be destroyed even in the face of adverse experimental discrepancies turning up in the future? With a view to elucidating further the nature of a theory's epistemic appraisal on the basis of its resolving power, I answer this question in the affirmative vindicating not only the view that a logically/methodologically complete scientific theory must possess both an explanatory and a resolving power but the view that it is its latter aspect which must be considered as most fundamental. Thus science cannot aim at some one and only one true theory. On the contrary, it aims at the growth of objective knowledge *reproblematized* instead of *deproblematized*, with a structure always *codetermined* by both theories and problems. Therefore, it can aim at theories with ever-increasing explanatory and resolving powers.[9]

II

In this way, it has been my main aim in this book not just to defend but to develop my own view of the rationality of science, taken in all its relevant synchronic and diachronic aspects, originally set forth in my (1983). It remains to be seen how this view might be appraised when compared with more recent approaches[10] that are admirable for the concern they have shown with questions of rationality and rational reconstruction of science. The well-known Sneed-Stegmüller type set-theoretic structuralism on the one hand and the problem-solving approach of Larry Laudan (1977) and the neo-Popperian approaches of John Watkins (1984) and I. Niiniluoto (1984) on the other are, I think, noteworthy in this context. Let us, then, briefly consider whether there are problems of rationality and rational reconstruction to which they offer non-trivially important, even better, solutions.

As a first step in this direction, it is necessary, I think, to return back to Popper (1968, 1972a, 1972b, 1976) and then to ask: How objectivistic is Popperian objectivism? This hints at possible criticism that his may not be after all objectivism proper when Popper (1968) argues that.

> If we characterize empirical science merely by the formal or logical structure of its statements, we shall not be able to exclude from it that prevalent form of metaphysics which results from elevating an obsolete scientific theory into an incontrovertible truth.[11]

This already seems quite admirably, however unintendedly though, to anticipate the problem of the structural identity of science in its present sense. What I find remarkable about this statement is its underlying unmistakable scepticism about the validity of the dominant statement-view of the structure of science. But if we consider Popper's proposal to solve the problem he has here himself raised, there is no ambiguity as to in which direction he would like to take us. His proposal is as follows:

> . . . empirical science should be characterized by its methods: by our manner of dealing with scientific systems: by what we do with them and what we do to them. Thus I shall try

> to establish the rules, or if you will the norms, by which the scientist is guided when he is engaged in research or in discovery. . . .[12]

We have already examined, in the foregoing chapters, various aspects of the proposals Popper has made in this context. There is no doubt that no objectivistic approach can afford either to ignore or to pay a lip service to the problem of the structural identity of science, especially when the unsatisfactory character of the one-sided statement-view of this identity has been recognized. But what do we, then, mean by objectivism proper in the present context? And whom can we call an objectivist? I think that it has nothing do in the present context either with the evolutionary biological paradigm dominant in Popper's (1972a, 1972b, 1976) evolutionary epistemology or with an epistemology naturalized in the sense of Quine. I do not believe that it is either necessary or fruitful to subject the systems of objective knowledge in all their diachronic aspects to such kind of paradigm. No doubt, evolutionary epistemology places *knowledge* in the biological context of Darwinian theory of evolution by natural selection, assuming the relevance of biology to problems of knowledge. For example, Popper has suggested how extension of the Darwinian paradigm to epistemology leads to a better understanding than is possible otherwise. Thus scientific knowledge can be viewed as possessing a problem-solving structure such that its movement may be described as following an evolutionary *strategy* of trial-and-error-elimination (= conjectures and refutations) as Popper calls it. But evolutionary epistemology fails at one crucial point which should not be over-looked. It attends more to questions of strategy than those of their canons.[13] In nature, it may be a successful strategy for it to trigger variations among species/populations which it then selects for survival. In the field of scientific knowledge, a similar strategy would not be sufficient to trigger growth of knowledge or scientific change. Popper himself recognizes the autonomy of the $world_3$ of objective scientific knowledge and yet fails to recognize the need for a biologically neutral account of how this $world_3$ itself changes while in interaction with the $world_1$ and $world_2$. On the contrary, evolutionary epistemology of Popper has made it easier and simpler for thinkers like Quine to propose a radicalization now in fashion, viz. naturalized epistemology as a causal inquiry into the mechanism of knowledge *as* a system of belief and interaction. This is the idea of theory of knowledge as a proper branch of the empirical sciences like biology and psychology. If we take this seriously, we cannot resist the temptation to ask: Why not then a sociological paradigm instead of Quine's naturalistic causal paradigm? But all these are variations on the same reductionist theme leaving the theme itself unaltered.[14] In saying this, I do not wish to deny interesting relationships that may have always existed and developed over time between *human interests* and human knowledge. But the most important point I wish to make is this:

> Our search for knowledge transcends our pursuit of specific types of human interests which *it* is bound to serve over time. But latent in this search is a permanent possibility of human disaster unexplainable in terms of those specific interests which we may rationally pursue. The primordiality of theory of knowledge is established by the simple fact that even in order for any system of human knowledge to properly inhere in a system of human interests, we have to depend on our theory of human interests that we may cherish as a piece of knowledge and readily modify over time. Thus every case of rationally pursued human interest will remind us of our abiding interest in human knowledge over-riding all our specific types of human interests.

An objectivist will not, therefore, ask: how does our knowledge (or belief) arise? Or, as Quine asks, how it is caused? The locus of causal order relevant to theory of knowledge is sought by an objectivist in the very patterns of its growth instead of its origin. Here it is most important to attack the problem of rationality in two main stages, one relating to the structural identity of science taken in the richer context of its historical development and the other to its growth as constrained by its structural invariances. An objectivist, therefore, asks: What must be the *structure* of objective scientific knowledge such that it is the kind of structure which is at the same time *causally* relevant to the dynamics of its structural growth? It is in this context that my (1983) identification of the *epistemic structures* in science with its developmental structures assumes much significance. This is especially true of the requirement that an adequate model of the growth of knowledge must be an interactive model.[15] It underlines the general interaction-theoretic framework for resolving questions of causal dynamics of the growth of knowledge dispensing with any biological or game-theoretic paradigm. Within this framework, I have been arguing for a different view of knowledge to which it is important

(1) to fix the structural identity of scientific knowledge such that its essential *symmetry* is recovered. The symmetrical growth of knowledge from problems to theories and from theories to problems demands that its systems have the character of problem-posing problem-solving systems. Symmetry-driven science is possible because there exist systems of knowledge which can be properly called theory-problem interactive systems in this sense;
(2) to distinguish, within science, its strategies from canons for permissible strategies. The search for the best possible methodology of rational theory-choice is a search for such canons;[16]
(3) to follow the principle of embedded methodology as a guide in search for the best possible methodology of rational scientific appraisal;[17]
(4) and to view science as a project always in the making and as a successful historical accomplishment. For only then is it possible to appraise philosophical models of rationality in respect of both their rational-reconstructive and normative components.

It must, therefore, remain a puzzling feature of the Popperian objectivism that it should look for the distinguishing features of science not in the systems of objective knowledge themselves but elsewhere: "in our *methods* of dealing with *scientific systems*, in what we do with them and in what we do to them", as Popper says. I think that the very possibility of indefinitely many alternative *methods* of dealing with such systems cannot be ruled out at any given time, especially from the point of view of Popper's own methodological conventionalism. Is then one's decision to choose any one particular method out of so many possibilities always an arbitrary one? It is bound to be so, since on this view there is nothing in science itself which might constrain the actual choice that one makes. This situation does not change even if we might say with Popper (1968) that such decisions "will of course depend in their turn upon the *aim* which we choose from among a number of possible aims".[18] Moreover, what should we expect on this view, when the scientist has to choose one from among the rival theories, T_1 and T_2, in their field? There may be no good reason to prefer T_2 to T_1. On the contrary, each of them might be good and acceptable on one set of methodological conventions (given that there is a definite and unique target-state which science aims at) but unacceptable on a different set of them. But how are we to explain this serious aberration in the Popperian objectivism? The answer is to be found in his strong bent towards a game-theoretic approach to science resulting in the kind of conflation of *method* and *strategy* which has taken an explicit form in some more recent approaches to its rationality. In my view, Larry Laudan's (1977) problem-solving account of the aim of science and theory-acceptance and his (1987) normative naturalistic account of methodology as an enterprise concerning the ends-means co-variances can be taken as a good example. But I shall return to this later.

Let us consider now the neo-Popperian approach to rationality of science as developed by John Watkins (1984). Popper has already suggested, as we have just seen, that the choice of a methodology, or *strategy*, of theory-acceptance will depend in its turn on *the aim* which we choose from among a number of possible aims for science. Either this choice is always arbitrary, made simply on *a priori* considerations, or based on criteria on which there is general agreement among scientists and the methodologists of science. Popper does not, as far as I know, address himself to this problem explicitly. John Watkins (1984) calls the aim he himself proposes the optimum aim for science because he follows a principle of selection based on his five adequacy requirements.[19] Thus any proposed aim for science must:

(1) be coherent;
(2) be feasible;
(3) serve as a guide in choices between rival theories or hypotheses;
(4) be impartial;
(5) involve the idea of truth.

This may go a long way in defeating rationality-scepticism, arising from Popper's methodological conventionalism discussed above, which says that there is never any good reason why we should prefer T_2 to T_1. Thus Watkins identifies scientific rationality with the rationality of an optimum aim for science,[20] as we may call it, provided that the non-arbitrariness of such an aim is demonstrable by means of these criteria. This enables him to make a bold departure from Popper's approach by developing a methodology of rational theory-acceptance based on corroboration estimates but without the Popperian verisimilitude-estimates playing any positive role in it.[21] The main reason for this being the following: "the best theory in its field is the one that is best corroborated rather than the one that is most corroborable",[22] where the best corroborated theory is also the one that best satisfies the optimum aim for science.[23] In a nutshell, a sound methodology of theory-appraisal and theory-acceptance, in order to be non-arbitrary and non-dogmatic, must be governed by, or grounded in, an optimum aim for science. In order to guarantee scientific rationality, against all forms of scepticism, in this sense, the optimum aim must itself be non-arbitrary and non-dogmatic. It must, therefore, satisfy appropriate criteria of adequacy such as proposed by Watkins.[24]

But all this is possible, I think, only if these criteria are themselves, in their turn, not only non-arbitrary in some non-trivially important, non-criterial, sense but deeply constrained by our *common* understanding of the method (= SD-procedures) and the structural identity of science taken in the richer context of its historical development from its infancy to its maturity. The kind of circularity this generates in this type of reasoning raises the same old question again: which of them enjoys a priority over the other, an optimum aim for science or the methodology of rational theory-acceptance?[25] I believe that the only way to have the required non-arbitrariness generated at the relevant levels without circularity is by resort to *theoretical universals* in the sense discussed in the foregoing Chapters.

On the other hand, a Watkins type methodology of corroboration-estimates inherits from the falsificationist methodology of Popper its character as a strategy of T_{EP}-evaluation. Both of them take explanation as the main task of a scientific theory. Viewed, then, as an explanation-theorist's proposal about the *type* of theories that science should aim at, Watkins proposes as an optimum aim for science those explanatory theories that are ever-increasingly (i) deep (ii) unified (iii) predictively powerful (iv) and exact.[26] The methodology of corroboration-estimates without the Popperian verisimilitude-estimates *accordingly* requires that if T_2, as against its rival T_1, has all these properties in a greater measure, then it "should be accepted as the best theory in its field *provided* that no positive reason has been found for supposing" T_2 to be false.[27] The last condition derives from the aim proposed by Watkins as the optimum aim for science, which includes those explanatory theories which are *possibly true* so that they are consistent with available evidence. What are the chances for such a methodology to succeed where other methodologies have either failed or had no success? Since it is governed by, and subservient to, the optimum aim for science, here much will depend upon how non-arbitrary this aim itself is. For example, one might ask, is it not a lip-service to scientific realism to let the Popperian regulative idea of truth *in*, while dispensing with the Popperian verisimilitude-estimates, by simply building this highly elusive idea into the optimum aim itself? Is it not possible to dispense with this regulative idea of truth and yet develop a realistic account of *how* scientific progress is possible and *what* makes it rational?[28] Of course, this may not be feasible so long as we are not inclined to ask questions concerning science in order of *new* priorities[29] I have proposed above. It is not possible here to deal with problems of realism in such detail. Such a task must await another occasion.[30] Here it may suffice to make a few concluding observations about the concept of truth in relation to the changing conceptions of rationality.

According to an old philosophical tradition, truth has to do with the representational character of our knowledge so that the truth of a judgment/assertion, with latter as its very *locus*, lies in its agreement or correspondence with its objects. Here it is quite unnecessary, to rationality of knowledge, to postulate 'truth' as that which we aim at or as something towards which we are moving. Thus the one and only one method of judging a particular knowledge-state (= K-state) is to find out whether it is true in this or some closely analogous sense. This is how we might look at the classical approaches to truth *in* the structure of knowledge itself as these are found in Plato, Aristotle, Descartes and Kant.

The concept of truth has, however, received a clear setback in the contemporary approaches to questions of rationality of our knowledge. It has been, I think, displaced from the center to an ever-widening periphery. Of course, this is not how a contemporary philosopher of truth would like to put it. But I think that Popper's (1972a) view illustrates the point.[31] There can be no criterion, says Popper, to tell that a theory/statement is true even when it, *in fact*, is true, although all our knowledge, including science, aims at truth, and the acceptance of T_2 as the best theory in its field involves the belief that it is closer to truth than its rival T_1.

It is quite clear from this that the concept loses its traditional role as the very mark of rationality of knowledge. What new role does it now assume? It assumes the kind of role, as we are repeatedly being told by Popper and other philosophers, which is more akin to 'good', 'right' and so forth, and which elevates it as the highest, therefore elusive, goal of all knowledge/science. I think that in so far as this may be a simple consequence of the changing conceptions of rationality, we must face it boldly. But the more important question to be asked is this: Which concept can now be taken as the successor-concept,[32] if our extant conceptions of rationality are to be subjected to further investigation? The falsificationist methodology of Popper is itself supposed to provide one answer. But if we remember what Duhem said concerning the disconfirmation of a physical theory, this is easier supposed than done. Moreover, such a methodology, when purged of its inherent shortcomings, must give way to something like neo-Popperian strategy of corroboration-estimates without the Popperian verisimilitude-estimates in the sense of Watkins. For the former is relevant to problems of rationality only so long as it provides strategies of the type that can yield corroboration-estimates as a *basis* for the rational acceptance of theories.

However, development of neo-Popperian versions of the verisimilitude-theoretic methodology of theory-appraisal, notably by Ilkka Niiniluoto (1984, 1987) and Graham Oddie (1986), against Pavel Tichý's (1974) and D. Miller's (1974a, 1974b) objections[33] against Popper's original comparative criterion of verisimilitude[34] deserve a serious attention. Although these objections are no longer considered fatal to Popper's original idea,[35] at first Niiniluoto develops a formal-semantical account of truthlikeness relative to a language L. Thus a given statement P_2 is more truthlike (and has, therefore, greater verisimilitude) than another statement, say P_1, if P_2 is less *distant* than P_1 from a central maximal truth (= the ideal limit) represented by a statement S in the language in which P_2 and P_1 are formulated.[36] This might be said to mimic features of Tarski's classic explication of 'truth' for formalized languages in so far as this involves the definition of a distance function *d* (or a truthlikeness metric M) relative to a certain assumed language-framework L only. Its methodological relevance, if any, to problems of theory-appraisal is not, therefore, obvious. At best, it might suffice as explication of certain aspects of the original idea hidden in Popper's basic assumption that science aims at truth. However, Niiniluoto also develops an epistemic theory of estimated truthlikeness, in addition to real/semantic truthlikeness, for scientific hypotheses, T_1, T_2, . . . with the concept of evidence *e* (and, therefore, a whole confirmation theory itself) playing an indispensable role in it. In final epistemological analysis, then, these estimates turn out to be as follows:

> T_2 *seems* to be more truthlike than T_1 on evidence *e* if $V(T_1/e) < V(T_2/e)$.[37]

V-estimates (= verisimilitude-estimates of a scientific theory) conceived in this fashion might be said to explicate aspects of the original idea underlying Popper's requirement that, at any time, growth of scientific knowledge must lend itself to an objective assessment in terms of a candidate-theory's degree of approximation to truth, or its degree of verisimilitude.[38] Here one might ask: how, then, are we to distinguish between the two methodologies, developed by Watkins and Niiniluoto, *within* the framework of Popperian realism? On the other hand, the basic problem of scientific rationality remains essentially untouched under Niiniluoto's approach considered in its semantic as well as epistemic version. In its epistemic version, V-estimates would have to be based on empirical evidence as much as corroboration-estimates usually are. Even if a methodology of V-estimates, without Watkins-type corroboration-estimates, were possible as a serious alternative, it would leave open the possibility of these estimates behaving asymmetrically with respect to one and the same theory's performance in its field when considered over a certain period of time and relative to the variable bodies of relevant empirical evidence. Given a proper confirmation theory fed with indefinitely varying bodies of evidence, V-estimates of rival theories might be revisable and even reversible.[39] But rational reconstructibility of scientific progress from past to the present presupposes that rational appraisal of a scientific theory in terms of relevant parameters is *symmetric* with respect not only to varying bodies of relevant evidence but its performance during its scientific career. Thus what makes a methodology of scientific appraisal of theories itself rational and non-arbitrary has little, or nothing, to do with the role variable bodies of evidence may play in the revisable V-estimates of them.

How successful is, then, Niiniluoto's attempt to rescue Popper's realistic theory of the growth of scientific knowledge from its serious difficulties? I think that it does no more than reproduce, albeit in a more sophisticated fashion, the basic pattern of Popperian realism, notably by assuming, first, that science aims at truth (= one and only one true theory) and, then, that the growth of scientific knowledge is a function of the *distance* every theory, in a *finite* series of theories, covers in approximating this ideal. Of course, a significant deviation from Popper's approach, if any, appears here in the semantic relativization of real truthlikeness of scientific statements, as also of the central maximal truth they all aim at, to a certain well-defined language-framework L.

However admirable this may be as a feature shared by other types of linguistic conception of scientific progress, it is subject to the criticism that it must lead to a language-sensitive proliferation effect in the following sense:

> The degree of truthlikeness of a statement P in L is no longer the same (= preserved) when P is considered in another different language-framework L′ which may just be an expanded version of L.

But this consequence is not surprising at all. Quite contrary to Popper's claim, recently various claims have been put forward to the effect that Tarski's semantic conception of truth supports not scientific realism in Popper's sense but relativism as built into the linguistic conception of science.[40] Thus a relativist will hold that of one and the same statement/sentence P it could be true to say that P is true in L but false in L′. But this is precisely what one cannot assume at the very outset if one wants to further improve upon Popperian approach. The problem arising from Popper's approach is, I think, essentially as follows: What sense does it make to suggest that science aims at truth (= one and only one true theory about the world or about nature)? If it should make any sense at all, it implies that science is governed by the ultimate goal of achieving complete knowledge of nature. *When* it does achieve such a state of knowledge completely *deproblematized*,[41] it would simply cease to grow any more. For then it should be already in a state of perfection quite incommensurable with its earlier states of imperfect knowledge. The summit would have been then reached, to use a metaphor of Popper.[42] But any theory of the growth of scientific knowledge with incommensurable K-states among its consequences must be rejected not only for the reason that it has inductivistic (therefore *finitistic*) implications for scientific method but also for the reason that from the very outset it must make strong *a priori* truth-theoretic assumptions concerning science.

We have been concerned with the view, held by Popper and his followers, that science aims at truth and that the progress which a science can make from an earlier to a later theory is a matter of their comparative degrees of *truthlikeness*. This is the only way, it goes on to add, for 'truth' to play its central role in scientific realism as an account of *how* scientific progress is possible or *what* makes it rational. But, as we have seen, this is not the last word on scientific realism, nor on the problem of truth and truthlikeness. And who can question the possibility that this problem itself may be some other problem in disguise and, therefore, not so central to scientific realism as it is generally thought to be. On the contrary, I think that it is possible to view science alternatively by shifting the priorities among the fundamental questions concerning it.[43] Since the truth-theoretic approach of Popper and his followers presupposes a dogmatic attitude to the epistemic-structural identity of science, the kind of new priorities I have in mind for developing a more dynamic realistic view of scientific progress may be indicated by the following ordering of these questions:

(a) What is (the epistemic-structural identity of) science?
(b) Does science progress and knowledge grow according to some universal rational pattern?
(c) What constrains our search for the best possible methodology of rational theory-choice in science?
(d) Does science have an aim which, together with the scientist's search for the best possible theory in its field, may be said to constrain this?

No doubt, this breaks with an old tradition of putting the question 'What does science aim at?' before every other question and, then, constraining everything else by saying that science aims at truth; and that the more truthlike its theories are the more progressive they will be. On the alternative view I myself have proposed, the structural identity of science is codetermined by its type-distinct epistemic structures as abstract as *theories* and *problems*. This implies clearly that 'truth' may be altogether irrelevant to scientific realism. Any form of scepticism concerning the former does not, therefore, necessarily imply scepticism concerning the latter. On the contrary, realism of scientific theories may lie elsewhere. It lies, I suggest, in their ability to give rise to those problems which can serve as a best guide to experimentation in newer fields of subject-specific scientific interest. Growth of knowledge follows, accordingly, an *interactive* pattern instead of either an inductive or deductive pattern. The resulting theory-problem interactive systems must be subjected to epistemic appraisal in terms of parameters that are highly relevant but have nothing to do with truthlikeness. Through his famous metaphor likening the scientific community to 'sailors who have to rebuild their ship on the open sea, without ever being able to dismantle it in dry-dock',[44] Otto Neurath may have meant to hint at the possibility of developing scientific realism within the statement-view of the structural identity of science but *without* truthlikeness playing any role in it.[45] If we are to make sense of Neurath's picture of science, we must recognize, with him, that "The possibility of science becomes apparent in science itself. . .".[46] Even in this picture, a subtle dynamics builds up between the metaphor of the ship on the open sea and the metaphor of the dry-dock, with the latter hinting at *the* world presupposed by science though unrecognized in its important role as the only possible frame of reference.[47] My interpretation here assumes much significance if viewed in light of the more recent shift in metaphor Hilary Putnam (1981a) has proposed by suggesting a relativistic variation by, first, likening science not to a single boat but to a *fleet* of boats and, then, putting not just science but the whole culture in each boat.[48]

'Truth' may be, as Putnam (1981a, 1983)[49] and Nelson Goodman (1978)[50] have suggested, akin to 'good', 'right' and so forth. But neither our recognizing this nor our questioning the role of truthlikeness in *scientific realism* necessarily leads to scepticism concerning *it* if developed as a more dynamic framework for the growing sciences. Thus it should be possible to remove all those deep-rooted confusions on which is based the kind of scepticism recently advocated by Goodman (1978). In his view, we cannot test a scientific theory by comparing it with a world undescribed. On the contrary: " . . . All we learn about the world is contained in right versions of it; and while the underlying world, bereft of these, need not be denied to those who love it, it is perhaps on the whole a world well lost".[51] But it is here necessary, I suggest, to distinguish between the *world(s)* which a theory may refer to and *the* world which, if it does, it must *presuppose*.[52] The doctrine, proposed by Goodman, that *the* world, which science deals with, is better taken as a world well lost is a result of a confusion between these two kinds of world. Since, one might argue, science does not have directly to deal with *the* world that it must presuppose as its very frame of reference, there is no question of our taking it as a world well lost. Since, on the other hand, the *worlds* (= ontologies) its rival theories refer to are the very same *worlds* which they themselves discover against the background of *the* world as their frame of reference, there is again no question of our assuming a sceptical stance towards scientific realism in the proposed sense. In order for science to be possible at all, and for its rival theories to be able to refer to those very *worlds* which they themselves create, it is, I believe, *necessary* that they become, in variable degrees, *anchored* to *the* world as their common frame of reference. Without *the* world thus presupposed it would not be possible for them to carve out, against it, as it were, their respective ontologically determinate *worlds*. Viewed in light of our arguments in Chapter 8 (§ 8.2.), this is the same world of unknown ontological complexity which a proper theory of T-simplicity needs if a non-trivial simplicity-ordering for scientific theories, symmetric with respect to their performances over time, is to be possible at all. Since my explication of T-simplicity in § 8.2. is itself, in a sense, a simple consequence of the theory of the structure and growth of scientific knowledge I myself have proposed, it might be interesting to appraise the methodology of theory-problem interactive systems in the following respect. Unlike other methodologies, it allows an irreducible role to T-simplicity as a parameter in theory-appraisal such that of any two or more rival theories in their field it becomes possible to decide which one of them is a better candidate for rational acceptance on the basis of a comparison of their explanatory powers (= T_{EP}) on the one hand and resolving powers (= T_{RP}) on the other.

III

Asymmetry (or the asymmetrically ordered pairs, triplets and so forth) instead of *hierarchy* (or hierarchically ordered/organized structures) is, I believe, central to both the structure and growth of scientific knowledge.[53] This fundamental idea, though still commonly unrecognized, can be illustrated by the following simple consideration. If science aims at ever better theory-problem interactive systems

$$I(T_1, P_1), I(P_2, T_2), I(T_3, P_3), \ldots$$

not only empirically more and more adequate but methodologically simpler and complete, we can *always* assert, on the basis of the relevant parameters of appraisal, that one of them is a better candidate for rational acceptance in its field than its rival, without the converse being possible at any time. In so far as we can go on with this kind of operation of their asymmetrical ordering in their field, we can also gauge the deeper symmetry in the growth of knowledge as a *function* of the underlying asymmetries in its structural organization. This is the reason why I attach great importance to the following question:

> What kind of structural organization must the objective K-states possess if it is to play a crucial role in the very causal dynamics of the growth of scientific knowledge? Or, what kind of factors can be admitted as determining or co-determining the structural identity of these K-states such that the epistemic structures of science are its developmental structures at the same time?

I am aware that this is not at all the kind of question which the philosophers of growth of knowledge usually like to ask. The reasons why this is so look varied. But they are all variations on the preconceived ideas of growth of knowledge cast in one mould of reductionism, instrumentalism and conventionalism or other. This is also true, I believe, of the different philosophies of growth of knowledge, whether instrumentalist or realist, that are currently in fashion. I shall consider first the view put forth by Larry Laudan (1977, 1981, 1987).

According to the view held by Laudan (1977), "*the aim of science is to maximize the scope of solved empirical problems, while minimizing the scope of anomalous and conceptual problems*".[54] (Italics are his.) Given this 'mini-max strategy', as he calls it, "*progress can occur if and only if the succession of scientific theories in any domain shows an increasing degree of problem-solving effectiveness*".[55] (Italics are his.) In other words, it is Laudan's (1977, 1981) view that

(1) science aims at solving as many problems as possible over a period of time.[56]
(2) Therefore, at a given time, the best theory in any field is one which *solves* a greater number of problems than its rivals do individually.

Although, on this view, science no longer aims at truth (= one and only one *true* theory), it has a definite aim. Thus the steps (1) and (2) are connected with each other in the same fashion as in the realistic approaches of Popper, Watkins and I. Niiniluoto. In all of them, methodology of science is invariably derived from its aim which is usually determined in an *a priori* fashion independently of its historical development. In Laudan's case, therefore, methodology of rational theory-appraisal does not look for theories which are true or truthlike or corroborated but for those with high problem-solving effectiveness.[57] The growth of knowledge must be accordingly measured in terms of their comparative problem-solving effectiveness. But how can a scientist determine this in the case of a historically given theory T? Laudan requires him to assess the number and importance of the empirical problems which T solves and, then, to deduct therefrom the number and importance of the anomalies and conceptual problems which T generates.[58] Moreover, "what matters for purposes of theory evaluation are only those problems which have been solved, not necessarily by the theory in question, but by *some* known theory".[59] But there is a complication which Laudan himself introduces into this conception. This is his unsuccessful attempt at the very outset to deny that scientific problem-solving is the same thing as explaining facts just because of certain supposed asymmetries between those *epistemic attitudes* which scientists might adopt with respect to them over time.[60] "The only kind of facts which can possibly count as problems are *known* facts", says Laudan.[61] But doesn't a scientist sift and determine facts just as he determines problems? Both facts and problems can have a common history. Moreover, the fact that *problems* as a subject of scientific discovery can *develop* over time is itself an important fact about science but overlooked or ignored by Laudan when he observes that "problems recognized as such at one time can, for perfectly rational reasons, *cease* to be problems at later times".[62]

How rational or useful is, then, Laudan's methodology of comparative problem-solving effectiveness? It is not essentially different from the traditional conception of science as an explanation-seeking problem-solving enterprise with maximization of explanatory powers of theories as its highest aim. But unlike the explanation-theorist's conception, it allows no role to truth or truthlikeness in theory-evaluation. Without an appropriate successor concept to replace truthlikeness, it turns out to be yet another variation on the old doctrine of instrumentalism. In particular, it is quite one-sided in so far as it pays no attention to that problem-posing power of a scientific theory which I call its resolving power. This results in Laudan's choice of 'the solved problem', empirical or conceptual, as 'the basic unit of scientific progress'.[63] Problems as a subject of scientific discovery and, therefore, as a function of the resolving power of scientific theory do not figure here as units of appraisal at all.

In a recent paper,[64] Laudan (1977) has sharpened his attack on the rational reconstructionist models of rational appraisal in science.[65] Having once argued that "*the historical episode itself remains untouched and unexplained* – except to the extent of its faithfulness to the *a priori* reconstruction",[66] Laudan assumes a *finitistic framework* of *goal*-oriented rationality for the sciences already built into his problem-solving model. Thus earlier he had argued that "the norms of rational evaluation change through time . . .".[67] And, therefore,

any theory of norms, if it is to be applied historically, must take into account the fact that previous scientists had norms of their own (often different from ours) which cannot be ignored in explaining their cognitive stances with respect to the theories of the day.[68]

According to Laudan (1987), this affects not only the *aim* of the methodological enterprise but its very nature. Methodology must be naturalized so that it becomes an inquiry into the permissible kinds of means-ends co-variances, where data concerning them are drawn from the historical records of past science.[69] In order to discover "the most effective strategies for investigating the natural world", what it needs are just "data concerning which strategies of inquiry tend to promote which cognitive ends".[70] It is not possible here to pay that attention to Laudan's proposal of normative naturalism, as he calls it, which it truly deserves. I shall instead restrict my criticism to a brief comment as follows.

Can methodology be naturalized at all? Is the search for the best possible methodology just a search for the most effective strategies? If yes, then methods employed by scientists in the past and being employed by them now are nothing but strategies analogous to individualistic schemes of practical syllogism. Scientific rationality is not essentially different from other forms of rationality in our cultural life. This view is not only mistaken but based on a confusion of method with strategy on one hand and of scientific rationality with a goal-oriented rationality on the other. If this kind of confusion has become a commonplace, it is because of the neglect of the principle of embedded methodology which says:

A sound methodology of rational appraisal of theory-problem interactive systems in science to facilitate rational choice of the best possible candidate from among them must be embedded in sound models of the structural growth of scientific knowledge itself.[71]

Epistemology remains, therefore, in the center of focus as a basis for any methodology of rational theory-appraisal in science. Of course, Newton's strategies need not be exactly the same as Einstein's. But at the more fundamental level of *method*, there will always remain an essential core of basic assumptions/rules shared by them, as by other great physicists of the past. Laudan's criticism of the rational reconstructionist models of scientific rationality commits the same error of confusing method with strategy.[72]

In conclusion, let us consider briefly Suppes-Sneed-Stegmüller's set-theoretic structuralism (structuralism for short).[73] I shall here refrain from going into the details of its development within the rational reconstructionist tradition of axiomatizing a physical theory before raising methodological and conceptual problems concerning it.[74] As Stegmüller (1979) has himself admitted, it has changed beyond recognition in response to various kinds of criticism ever since the publication of his *The Structure and Dynamics of Theories* (1976, German edition 1973). I shall instead restrict myself to a brief critical comment on some of its salient features to be found in Stegmüller's own account (1979).[75] As possible points of broad agreement, I think that its rejection of the statement-view addresses not the epistemic structural identity of science (in my sense or in an analogous sense) but the view "that scientific progress could be defined as convergence towards truth".[76] But as I have argued above, it is possible to develop a realistic account of scientific progress in a way which is consistent with this rejection.[77] On the other hand, as to the very task of explicating scientific progress, we must agree with Stegmüller when he says that (i) it is *no trivial task*; (ii) it is primarily a *logical task* and not something to be accomplished by psychological, sociological, and historical tools; and (iii) the explication would have to be a *precise* one.[78]

Explicitly instrumentalistic in orientation, structuralism starts from the mathematical cores, together with empirical interpretations, of historically given physical theories such as statistical mechanics and thermodynamics. Instead of treating them as structures of statements, or classes of sentences, whether true or false, it subjects them to informal set-theoretic axiomatization in order to exhibit the 'logical' relations of reduction between them. Thus it is interested in rationally reconstructing those cases of intertheoretic *reduction*, whether strict or approximative, in which *it* constitutes a *one-many* relationship of progress from the theory reduced to the reducing theory.[79] But, one might ask, how is it possible at all to conceptualize scientific progress in terms of intertheoretic reduction in this fashion, from rigid body mechanics to particle mechanics, thermodynamics to statistical mechanics, geometric optics to electromagnetic theory, and classical particle mechanics to relativistic mechanics (provided that velocities are smaller as compared to the velocity of light)? Have these theories not developed over time? How to reconstruct this development even if it may consist in just those hypotheses, laws and applications which are developed within the framework of the given theory? Structuralism does not address itself directly to problems of scientific rationality. Thus it is interested neither in the scientist's search for the best possible theory in its field using the kind of *SD*-procedures that he does use nor in the philosopher-methodologist's search for the best possible methodology of rational theory-choice.[80] Yet it recognizes the challenge posed to it by problems arising from the important fact that scientific theories develop over time.[81] If structuralism has changed beyond recognition from its earlier versions to later ones, it is in order to meet this particular challenge as well as that arising from Stegmüller's earlier (1976) "excessive emphasis on Kuhn-reconstruction" and lack of a clear recognition of the ambiguity of the 'non-statement view' and so forth.[82]

Where is it exactly that statement-view of science, of which there may be many versions, fails but structuralism promises success in generating solutions of well-identified problems and puzzles? The problem of the structural identity of science as I understand it is a problem recognized by neither of them as clearly as is necessary. But the theories of science and scientific progress they develop have interesting consequences for a correct and critical understanding of this problem. What about other problems, then, such as the problem of theoretical terms and their controversial Ramsey-eliminability, the problem of identity and equivalence of theories, the problem of intertheoretic reduction, the problem of incommensurability and so forth? Structuralism questions the adequacy or correctness of solutions of these and related problems proposed by philosophers within the statement-view of the type represented by Carnap's formal language approach to rational reconstruction of physical theories as formalized systems with one single big application (= Carnap's partially interpreted formal deductive systems). But it does not, I think, question the presuppositions behind these problems which receive sharp formulation within the statement-view. On the contrary, sharing the same presuppositions, it regards these problems as of fundamental importance to special as well as general philosophy of science but beyond adequate solution within the essentially ambiguous statement-view. Thus, in search of better solutions to the same old problems, it changes the strategy of Carnap-type formalistic approach in favor of the new strategy of Suppes-Sneed-Stegmüller-type sytematic axiomatic reconstruction of physical theories within informal set theory. Earlier, this was thought necessary because Carnap-type approach to axiomatization was itself thought to be a special case of systematically developed statement-view. But now the situation looks more complicated as a result of those changes which structuralism has undergone thanks to the criticisms of Paul Feyerabend.[83] "The new approach suggests", says Stegmüller, "differentiating strictly between *theories* or *theory-nets* on the one hand and *empirical claims*, associated with theories and theory nets, on the other hand".[84] But, then, he adds: "The method of theory nets which replaces the former procedure of expanded cores . . . demonstrates that structuralist approach and the statement-view are, contrary to the *prima facie* impressions, not very remote from each other. . . . What a physical theory 'has to say' at an historical time *t* is, basically, just the result of 'collecting together' all these particular claims belonging to the theory-net at *t*".[85] Here it is also relevant to our understanding of what type of change structuralism is undergoing to notice Stegmüller's suggestion that "logical, epistemological, *and* historical aspects must be taken into account in order to answer"[86] the following question being attributed to Kuhn: By what criteria can we *identify* a historically given theory?

Given its declared concern with the mathematical aspects of physical theories on the one hand and the assumption of essential ambiguity of axiomatization on the other, structuralism is interested in their rational reconstruction through axiomatization using only informal logic and informal set theory. Given a physical theory T (= MF · PI), consisting of mathematical formalism MF together with a physical interpretative system PI, T can be reconstructed outside Carnap-type formalization as a mathematical structure together with an open set of intended applications. In Stegmüller's own words: "If the mathematical structure of a theory is represented within a Suppes-type axiomatization by a set-theoretic predicate S, we may, in a first approximation, identify a physical theory with an *ordered pair* <M, I>, whereby M is the extension of S, i.e., the set of possible models of this structure, and I the open set of 'really' intended applications".[87] Thus while it shares its basic aim of laying bare the mathematical structure of a historically given physical theory with Carnap-type formal language approach, it is intended as a better alternative to the very method of accomplishing this first step of rational reconstruction.[88] Using informal logic and set theory, it aims at replacing Carnap-type meta-mathematical method of formal axiomatization.[89] But what about possible similarities/dissimilarities between them as regards the very structuration of scientific knowledge due to the dynamics of theory-problem interaction at different levels of SD-procedures of problem-formulation and theory-construction? In both types of approach, theoretical problems as subjects of scientific discovery do not enjoy any epistemological-ontological recognition as theories (as explanatory nets) do. There is, therefore, no question of rational reconstruction, either in Carnap's or in the structuralist's sense, being extended to dynamics of theory-problem interaction itself. But I think that no non-trivial account of the structural identity of science as a whole, as also of scientific change, can ignore their importance for our understanding of theory-problem interactive systems of the type

$$(1) \quad \langle T(= MF \cdot PI), P(= P_1, P_2, P_3, \ldots) \rangle \begin{array}{l} \nearrow T_{EP} \\ \searrow T_{RP} \end{array}$$

This raises a serious possibility that structuralism itself might be just a variation on the traditional hierarchical construal of a theory as an explanatory net and of progressive scientific change either as evolutionary or reductive in nature. (1) hints at possible directions of scientific change (= knowledge change) a theory-problem interactive system, $I(T, P) = I[T(= MF \cdot PI), P (= P_1, P_2, P_3, \ldots)]$, can undergo. These have to do with correlated theory-change and problem-change in the sense considered earlier.

Explication of progressive scientific change (always identified with progressive theory-change) is, then, a task structuralism inherits from contemporary rational reconstructionist tradition of Carnap and other philosophers of science. The task must be faced on the evolutionary scale of the development of a theory, or its special laws, over time on the onc hand and on the revolutionary scale of one particular theory replacing another theory with their incommensurability coming in the way of their rational appraisal on the other. The latter is construed by structuralism as a task of reconstructing intertheoretic reduction. It is the former which is more difficult, since it involves reconstructing how a particular physical theory, say classical particle mechanics, has developed over time as new special laws and applications have been discovered within its framework. Here Stegmüller (1979) abandons his earlier reconstruction in terms of a core K of a given theory T, $K_T(K)$, its expanded core E, $EK_T(E)$, and its domain of intended applications I, $App_T(I)$.[90] Attention now shifts to possible hierarchical order of physical theories, one presupposing the other. In particular, it becomes important to view them, given their clear axiomatic forms, as hierarchical structures consisting of the fundamental laws, general laws and more specific laws (= empirical claims or hypotheses). To this end structuralism develops a new method of hierarchical reconstruction in the language of abstract theory nets and core nets, instead of the earlier expanded cores, as hierarchical nets of abstract theory elements.[91] Since the term 'theory-element' is intended to refer to both theories and laws, it becomes possible for structuralism for the first time to do away with the method of expanded cores in reconstructing progressive scientific change in terms of theory-evolution over time, from its fundamental laws to special laws, by chains of specialization on one hand and by their assignment to particular intended applications on the other.[92] Thus in the language of hierarchical theory-nets and core-nets, a main theory T, represented by the ordered pair (K, I) consisting of the core K and the set I of intended applications, develops over time as a hierarchical structure consisting of the fundamental laws, general laws and more specific laws, with new conditions added to them all. For example, one can look at a special law, (K′, I′), as a result of (K , I) by theory specialization and K′ as a result of K by core specialization, as Stegmüller puts it.[93] In order to reconstruct Thomas Kuhn's theory of normal science, Stegmüller extends this analysis to theory-evolution in scientific communities by a suitable pragmatization of the language of hierarchical theory-nets and core-nets.[94] But I shall here refrain from going into this. What is of immediate interest to us is whether this kind of structuralist analysis of theory-evolution leads one directly to a systematic and useful analysis of revolutionary scientific change, quite irrespective of what kind of support it derives from its Kuhnian normal science reconstruction. Stegmüller (1979) himself vaguely hints at such a possibility when he says: "The concept of a theory-net induces generalizations in which several *kinds of* theories in *hierarchical order*, vaguely alluded to in the presupposition relation, are explored. Roughly speaking, within such an order the M_{PP}'s of a given theory are the M_P's or M's of an *underlying* theory. In the course of the systematic study of such hierarchical structures important new intertheoretical relations may be discovered".[95] (Italics his.)

Let us suppose for sake of discussion that the structural identity of a science consists in its theories, T_1, T_2, . . . T_n, reconstructed as structural hierarchical systems in the sense of set-theoretic structuralism, instead of classes of sentences formulated in different languages with altogether different vocabularies. In what sense is, then, a scientific revolution progressive, if it is taken to involve a radical Kuhnian paradigm change from T_1 to T_2? Here T_2 is the theory which replaces the given theory T_1 in the course of a scientific revolution. Instead of questioning their Kuhn-incommensurability, structuralism concedes it as unavoidable under the statement-view of theories as classes of sentences.[96] For the very language in which a theory is formulated will, in the course of a scientific revolution, vary radically from the theory dislodged to dislodging theory itself and render them incomparable.[97] On the other hand, however, structuralism permits analysis of "the internal structure of different theories, as expressed in their core-nets and theory-nets", in the *same* informal language.[98] This shows that the question structuralism seeks to answer is not the one I have asked above but the following: What structure must a rationally reconstructed scientific theory possess such that T_1 and T_2 can be ordered in terms of a unilateral reduction relation R between them, given that T_1 and T_2 considered as classes of sentences are Kuhn-incommensurable? It is to this type of question that Stegmüller addresses when he says:

> *Within the structuralist framework it is possible to introduce inter-theoretical relations that apply to 'theories' which are 'incommensurable', in the sense of having very different 'theoretical super-structures'*[99]. (Italics his.)

The reason why it assumes such importance has, then, mainly to do with the structuralist's pre-conceived idea of the growth of knowledge, in situations of revolutionary theory-change, as a function of R, under some appropriate interpretation of it, that binds theories like T_1 and T_2, in their mathematical structural cores, into a structural organization which is again essentially of a hierarchical kind.[100]

But can R, either strong or approximative, capture the physicist's best intuitions concerning growth of knowledge by revolutionary theory-change from T_1 to T_2 in a given field? If R is construed as a strong one-many relation, from reduced theory to the reducing theory, it is said to hold between specific theory-nets N_1 and N_2 such that R strongly reduces the basic theory-element T_1 of N_1 to the basic theory-element T_2 of N_2 and every theory element of N_1 to some theory-element of N_2.[101] To compare them as steps in this growth is for structuralism, as for many others, to reduce T_1 to T_2 in this or an analogous sense. Thus reduction-theoretic approach to understanding the growth of scientific knowledge, under revolutionary theory-change, overcomes the challenge of Kuhn-incommensurability.[102] To different types of intertheoretical reduction there are generally correlated different types of deductive explanation of one theory by another.[103] Revolutionary progress in science is not, therefore, discontinuous. But Kuhn-incommensurability says, in effect, that revolutionary progress is progress not by reduction of one theory into another but by a discontinuous process of one theory being replaced by a radically different one. The two theories are incompatible and incommensurable with each other.

Let us suppose that the reduction operation R on revolutionary physical theories replacing older theories is not only possible but feasible in the sense that they lend themselves to it in a desired manner as demanded by structuralism. In the given case, then, let T_1 reduce to T_2. But can we still talk of T_2 as being better than T_1 in some non-trivial sense determined by the relevant parameters of epistemic appraisal? Or is the ground of comparison between them lost with the loss of asymmetry due to reduction? One might, of course, resort to relevant parameters of epistemic appraisal (such as simplicity, explanatory power, resolving power and so forth) to compare them until T_1 is reduced to T_2. If one is at all to compare Kepler's laws of planetary motion with Newton's theory of gravitation, it can be only as a first rational reconstructive step to approximatively reducing them to latter. Successful reduction of T_1 (Kepler laws) to T_2 (Newton theory) should result in a structural hierarchical ordering of the former with respect to the latter such that there is no ground left to compare them critically. For after reduction, there is no more any point in saying that T_2 is better than T_1, since T_1 must now, on some appropriate interpretation of R, find its best explanation in T_2, the two together forming a whole structural hierarchical *system*. But how can this reduction-theoretic structural hierarchical organization of successive theories in science be correlated with the scientist's search for the best possible theory and the SD-procedures he must employ to this end? This may not be an easy task, since at any time his SD-procedures do not reduce just to a search for a theory which reduces its predecessor in its field to itself. How are we, then, to understand this consequence of the loss of comparability of rival theories by resort to irreducible parameters of epistemic appraisal. The structuralist strategy of accepting the challenge of Kuhn-incommensurability on one hand and generating comparability of rival theories through R on the other must fail for a number of reasons. First, acceptance of Kuhn-incommensurability as a consequence of the statement-view of theories is not only dogmatic but a partially wrong diagnosis. As I have argued in my (1983), it has little, or nothing, to do with the statement-view. It is well-known that Kuhn himself does not subscribe to this view taken in any of its traditional versions. On the contrary, it arises as a consequence of Kuhn's own sceptical strategy which he employs in order to bring about a kind of paradigm-shift, to use his own terminology. I have designated it as the strategy of changing the subject of epistemic appraisal in science from a *scientific theory* to the *scientific community* as a whole which holds the theory. Of course, it can be shown that this itself presupposes a conception of a scientific theory modelled on a language and a language community. Like all paradigm-shifts, a powerful strategy such as this can easily be employed to bring about radical consequences such as Kuhn-incommensurability. Is not structuralism itself, as one might ask, a similar exercise aimed at a paradigm-shift so that the subject of appraisal/predication changes from scientific theories as classes of true-or-false propositional structures to those construed as non-propositional set-theoretical hierarchical structures alluded to above? The problem of Kuhn-incommensurability, it may be true, no more arises now, since rival theories will be related to one another by an intertheoretic reduction relation R. Or, is there no *real* problem of comparability of rival theories left now as an unintended consequence of structuralism?

I have been arguing against the kind of strategy with which structuralism seeks a structural hierarchical organization of rival scientific theories with the reduction operation R as a key principle to explain scientific progress in situations of revolutionary theory-change. The very supposition of its success entails a loss in those intertheoretical relations, based on relevant parameters of epistemic appraisal, which organize rival theories as asymmetrically ordered pairs, triplets and so forth.[104] It cannot, therefore, cope with all those complexities of the growth of scientific knowledge which result from revolutionary theory-change. One can, of course, employ it as a tool in the tradition of structuralism in order to analyze those aspects of intertheoretic relations which bear on the specific empirical problems of explanation of one theory by another. Thus its usefulness in this restricted sense (e.g., for the purpose of treating T_2 as an approximative explanation of T_1) might be even beyond doubt. Neither Kuhnian analysis in terms of paradigm-change nor the structuralist analysis in terms of one-many intertheoretic reduction can, therefore, offer adequate explication of the dynamics of the growth of scientific knowledge through revolutionary theory-change. It can be shown, in support of this claim, that neither of these strategies are ingredient in the scientist's own SD-procedures of finding out the best possible theory in its field at a given time. But such a task must await another occasion.

There may be various types of intertheoretical relations in addition to those of a presuppositional and reductive kind alluded to above. I think that all of them are subject to fundamental irreducibility of a theory's never exhaustible scientific implications. Although at no time we can claim to have completely understood a theory's implications, intertheoretical relations in different directions exist as a subject of scientific investigation and rational reconstruction. But why should this assume importance to a proper conceptualization of the growth of scientific knowledge whether over a period of theory-evolution or in course of a scientific revolution? It is not so easy to answer this question, since there might be not just one single way but alternative ways of analyzing these relations, structuralism itself being an alternative to Carnap-type rational reconstruction. Which one of them does one pursue and to what end? Which one of them does one pick out as relevant to the task of explicating growth of scientific knowledge under different states of scientific change? These seem to me deep and difficult questions which have remained essentially unasked so far. For in the formal axiomatic deductive approach of Carnap, these relations have been assimilated to logical relations of inference between a reducing theory and the reduced theory. But structuralism takes them to be non-inferential relations of inclusion (= reduction) between their mathematical structural cores (and their respective applications) such that a science such as physics tends in the long run to assume a structural organization of an hierarchical kind.

In my view, it is a mistake to so oversimplify these relations as to paint them as binding all successive theories, T_1, T_2, T_3, . . . T_n, in their field into a *backward branching network* either of whole classes of statements or of hierarchically organized set-theoretic mathematical structures. There is, on the contrary, something more important but missing in this picture which makes them possible and dynamically significant.[105] First, they are of a continuing synchronic interest to us from the point of view of empirical testing of a theory's consequences. Thus one might fruitfully resort to tools of formal or informal rational reconstructions with a view to comprehensive testing of these consequences, which is in itself an on-going but a never-ending process. But in so far as they are of any dynamical interest to us, they must themselves be viewed as relations that *grow* over time. This is so because they are mediated by *problems* which they either raise or solve over time. And, interestingly enough, it is this particular aspect which seems to take care of those inexhaustible implications of a theory which its historical development alone can bring to light.

Secondly, both *theories* and *problems* are, in my view, the kind of dynamic structures which I have called the developmental structures of scientific knowledge. Thus if a scientist does not consider which types of problems his theories T_1 and T_2 determine, or solve, he would never know how they are mutually related in the interactive growth of knowledge. Thus there is no simple, straightforward and easily formalizable dynamic connection between them. Each is irreducible in its own right yet connectible to other in a manner that is essentially determined by the complexes of problems which they either solve or themselves raise. This dynamical idea of a theory's irreducibility amidst its connectibility is, I believe, best conceptualized by the methodology of theory-problem interactive systems, $I(T_1, P_1)$, $I(P_2, T_2)$, $I(T_3, P_3)$, . . . which has been a subject of discussion in the foregoing Chapters. In so far as it is based on the important assumption that *theories* and *problems* are the developmental structures of objective scientific knowledge, such systems are not susceptible to axiomatization of any known kind whatever. This is so partly because the structural identity of scientific knowledge, which this methodology presupposes, is codetermined by theories and problems in their various types of interactions. Reproblematization of knowledge in this sense reveals that it has a richer structural identity than is required under the received views. The tendency to oversimplify and deproblematize knowledge has been so deeply rooted in us that we like to imagine that its structural identity is solely determined by theories construed either realistically as classes of true or false sentences or instrumentalistically as non-linguistic, type-identical set-theoretic mathematical structures.[106] Hierarchical reduction of earlier structures to later structures, like earlier machines to later more sophisticated ones, suggests itself, then, as the most rational alternative to other forms of reduction as the key principle of scientific progress. But all of them appear, and disappear, as variations on the same theme of explanation-theoretic reduction of an earlier theory to a later theory.

Thirdly, and lastly, we can quite agree with Stegmüller (1979) that which of the alternative approaches to understanding scientific change and scientific rationality one prefers "depends on the respective aims and problems".[107] Thus we can say, with Stegmüller, that structuralism "seems to be a most promising tool for clarifying various prevalent but yet unexplained notions as well as for solving quite a number of puzzles which have been haunting the philosophy of science for a long time".[108] One of these problems is known as the problem of theoretical terms. By relativizing theoreticity to a physical theory T, instead of a language L in the fashion of Carnap-type double language model for science, structuralism has reduced it to the problem of T-theoretical terms.[109] In this sense, then, there may be no serious problem of theoretical terms or of their eliminability within a realistic view of science. But it may at the same time remain a *puzzle* for structuralism because of its commitment to an instrumentalistic view of science.

NOTES

CHAPTER 1

[1] G. L. Pandit (1983)
[2] It is, I believe, for the same reason why Quine is led to propose his thesis of a naturalized epistemology – for him a branch of biology – as a radical alternative to TE. For my criticism of this thesis see G. L. Pandit (1983).
[3] "As to distinguishing between knowledge claims (of the form 'X claims to know that p') and knowledge states (say, X knows that p)", as Fetzer points out (in personal communication), "generally a knowledge claim will turn out to be a true claim just in case X knows what X claims to know. But . . . it might be true that X claims to know that p even though it is false that X knows that p". In his (1983a:406), he argues as follows: "X may *claim* to know that *p* for any of an enormous variety of reasons (to deceive, to mislead, to entertain, to amuse, to reassure, to soothe, and so forth) *without* knowing that *p*; and conversely." See also James H. Fetzer (1981: 5–7).
[4] See James H. Fetzer (1981: 3–16).
[5] For this distinction see G. L. Pandit (1983)
[6] Cf. Jaako Hintikka (1965: 56–59)
[7] The very fact that Descartes resorts to this strategy entails that he had propositions in mind, these being the only type of candidates to which such a strategy could be legitimately applied.
[8] In the Cartesian scheme, the objective knowledge is constructible in the form of propositions about the essential natures of things in the world. It is significant how Descartes works out the implications of the subjectively known proposition allegedly expressed by 'I exist' so as to confront us with the proposition that the 'self' is essentially a non-extended thinking thing. In an act of subjective knowledge of this proposition, Descartes then asserts 'I am a non-extended thinking thing', and so forth.
[9] These background beliefs might include those that have *problems*, instead of propositions, for their objects.
[10] This tension is discernible, I believe, in the logical empiricist debates on the nature of the empirical basis of knowledge, which are marked by considerable disagreements as to whether in the process of validation propositions can be directly compared with 'facts'. It is Otto Neurath who, unlike Moritz Schlick, advocates the view that propositions can be compared only with other propositions constituting the empirical basis, and not with 'facts' directly, thereby (allegedly) landing himself in a coherence theory of 'truth'. I think that the basic insight in Neurath's view can be still retained without leading to any such awkward consequences, particularly if commitment to the traditional problem of validity/justification is given up. For more of Neurath's approach see Otto Neurath (1983: 52–57) and C. G. Hempel (1980: 193–199).
[11] OE is, as I now think, presupposed by SE ontologically. OE's domain is restricted to objective epistemic structures as abstract as theories and problems in the sciences. Since it is open only to a knowing subject to enter into various kinds of cognitive relations with these structures, we can reasonably assign the domain of these relations (of subjective knowledge) to SE.
[12] Many interesting examples of subjective knowledge in the sense of rational belief in *problems* can be cited from various sciences, particularly physics. A single look at the

whole series of questions which Newton poses in his *Opticks* (4th ed. 1730, Book three) will convince one about the significance of subjective knowledge in the sense suggested here. Consider, for example, his Query 1: "Do not Bodies act upon Light at a distance, and by their action bend its Rays; and is not this action (*caeteris paribus*) strongest at the least distance?". Consider again Einstein's belief in the problem of a unified field theory: What form might such a theory take mathematically if it were to result in the unification of all the known kinds of physical force or interaction? Or, again, consider his belief in the problem of a finite though unbounded universe: Is a universe which is finite in volume but unbounded conceivable? Although at one time developments in observational cosmology persuaded Einstein to give up his belief in the last problem, all these cases, Newton's included, exemplify subjective knowledge in the sense of rational belief in scientific problems (and in whatever their sentence-bases might imply) without being able to solve them.

[13] For a detailed discussion turn to Chapter 4 below.

CHAPTER 2

[1] Consider in this context the systems of criminal law which invariably insist on eye-witnesses, *die Augenzeugen*, ('present' at the actual scene of the crime, both before and during the (commission of the crime) as *the standard evidence* to prove/disprove a crime and then to bring about justice. In such systems the very conditions for proving the crime and identifying the real criminal include the condition of perceptual evidence of the assumed 'omnipresent' eye-witness as a pre-requisite for punishment and justice. To point this out, however, is not to belittle the great merits of such systems. There is, e.g., a lot to learn from them, particularly when there seems to be nothing in these systems corresponding to (the types of) 'self-evidence' concepts that have bedevilled philosophy since antiquity.

[2] The following statement of A. M. Quinton (1955: 28) is noteworthy as a statement in a nutshell of the philosopher's epistemological worries concerning perception: "The problem of perception is to give an account of the relationship of sense-experience to material objects. This relationship has traditionally been seen as logical, a matter of showing how belief about objects can be established or supported by what we know in immediate experience. For, it is held, only our knowledge of experience is direct, immediate, by acquaintance; what we know or claim to know about objects is indirect, derivative, by inference from what we know directly. Consequently if our beliefs about objects are to have any secure foundation, it must consist in what we know directly by acquaintance, about sense-data. From this starting point philosophers have gone on to present varying accounts of the type of inference involved".

[3] In the words of B. A. Farrell (1970: 51–52): "When one puts on scientific spectacles and takes a look at man's capacity to see and hear, one is constrained, in the first instance, to regard a human being, not as a mind or as a person, but as an organism. One thinks of physical energy of different sorts and properties impinging on the organism, and being absorbed and transduced in various ways by different parts of the body – parts which are built to do this job. These parts are the receptors – the eyes in the case of light energy, the ears in the case of energy from atmospheric disturbances".

Again he (1970: 64) says: "When one investigates human perception by the methods of science, one is concerned throughout with the flow and processing of energy in and by a human organism. This processing seems to have a hierarchical character . . . The human organism also comes to tie inputs and its processing of them to certain sorts of output. It does so largely as the result of interaction with the environment. The ways in which the mature human organism conceptualises the world around it is determined (in part at least) by what has gone into store as the result of past interaction. This store is open to modification in the light of 'failures in pay-off' when interacting with the world".

[4] Cf. J. F. Rychlak (1977: 354).

[5] J. F. Rychlak (1977: 354) makes a similar point when he says: "To have a grasp of what is taking place demands a broader frame of reference, within which we see not only that which stands before us as reality but . . . the possibilities and likelihoods not yet "there" for perceptual delineation. Telic animals are aware animals because they are constantly looking forward to such nonliteral aspects of experience. And of course even a potentially telic animal is unaware when he cannot organize the developing and unfolding possibilities of a situation in which he finds himself . . .

The person who looks to the skies and sees sunshine but considers the likelihood of rain based on subtle cues in the clouds rimming the horizon is reflecting an awareness . . ."

[6] In a conversation with Don Howard, I am led to think of a *denumerably complex*, or an infinitely complex, ontology of *the world*, that science such as physics must presuppose if it is to deal with it at all, as of fundamental importance to an adequate non-probabilistic view of theoretical simplicity in science. For a discussion turn to chapter 8, § 8.2. below.

[7] P. G. Bergmann (1970: 74). According to Bergmann (1970: 73): "Just how we utilize the readings of certain macroscopic instruments to arrive at interpretations, saying that we have 'observed' a given property of an atomic system, is an integral part of our theoretical scheme of things". For his attempt to answer the question why science makes the observability assumptions that it makes see his (1970: 75).

[8] Cf. P. G. Bergmann (1970: 72).

[9] It is not my intention here to go into the questionable completeness of quantum mechanics as a physical theory of a non-classical kind. I think that the most challenging problems (of locality, separability and quantum-mechanical measurement) that arise in the wake of its own development are, *in essence*, problems relating to a general interaction theory at the level of the basic presuppositions of physical theory (see Pandit, 1983; and § 9.2. below). Due to the limitations of space here, a detailed discussion of these problems must await another occasion.

[10] I owe the formulation of b_3, as those of several other related points, to a discussion with James H. Fetzer (in personal communication).

[11] To quote Karl Marx (1859: 8–9) in original German: "In der gesellschaftlichen Produktion ihres Lebens gehen die Menschen bestimmte, notwendige, von ihrem Willen unabhängige Verhältnisse ein, Produktionsverhältnisse, die einer bestimmten Entwicklungsstufe ihrer materiellen Produktivkräfte entsprechen. Die Gesamtheit dieser Produktionsverhältnisse bildet die ökonomische Struktur der Gesellschaft, die reale Basis, worauf sich ein juristischer und politischer Überbau erhebt, und welcher bestimmte gesellschaftliche Bewußtseinsformen entsprechen. Die Produktionsweise des materiellen Lebens bedingt den sozialen, politischen und geistigen Lebensprozeß überhaupt. Es ist nicht das Bewußtsein der Menschen, das ihr Sein, sondern umgekehrt ihr gesellschaft-

liches Sein, das ihr Bewußtsein bestimmt". See also Maurice Godelier (1972: ix).

[12] See K. R. Popper (1961: v).

[13] Cf. *ibid.*, v–vi.

[14] Cf. *ibid.*, vi.

[15] Quine (1969: 69–90), I believe, commits the same mistake of confusing knowledge with a relationship of interaction. For his conception of epistemology as a branch of biology or psychology rests on this confusion.

[16] For this concept see G. L. Pandit (1983: 65–67).

[17] Cf. *ibid.*

[18] For discussions relevant to this topic see Garland E. Allen (1978: 187–228); and Jacques Monod (1977: 98–113).

[19] An interesting attempt to introduce the concept of a biological field in developmental biology is to be found in C. H. Waddington (1966: 105–230).

[20] G. L. Pandit (1983: 65–67).

[21] See my discussion in (1983: 61).

[22] See my discussion in (1983: 68).

[23] I think that only a knowing subject as complex as a human being can risk everything in a field of interaction as new as interstellar travel and exploration, even without being sure of his return to the planet Earth. Human interaction in this sense is creative in that it has knowledge for its orientating variable and it clearly falls outside the purview and scope of the formalizing models of game theory.

[24] For further criticism of Popper's construal of the growth of knowledge as a Darwinian process of competition between theories themselves see L. J. Cohen (1980: 175–180).

[25] For a discussion of this concept see G. L. Pandit (1983: 60, 62, 66, 68).

[26] For details regarding game theory and its basic concepts see J. von Neumann and O. Morgenstern (1944), A. J. Jones (1980), Carole Eisen and Martin Eisen (1979). For a discussion of the achievements of game theory – originally projected to establish a general method for solving all rationality problems – positive as well as negative, see John Watkins (1970: 199–206).

[27] For a discussion see G. L. Pandit (1983: 60–62).

[28] See G. L. Pandit (1983: 60–77).

[29] For this concept see G. L. Pandit (1983: 60, 62–63, 79).

CHAPTER 3

[1] See Wolfgang Stegmüller (1976, 1979). For a discussion turn to chapter 10 below.

[2] T. S. Kuhn (1971: 146). For Kuhn's more recent work where he, unlike Quine, distinguishes translation from interpretation see T. S. Kuhn (1983: 673, 681–83).

[3] See G. L. Pandit (1983: 63).

[4] In his (1975a: 90), Quine writes: "I am persuaded that alternative manuals of translation can exist, incompatible with each other, and both of them conforming fully to the dispositions to behaviour on the part of the speakers of the two languages. The two manuals would agree on observation sentences but conflict in some of the standing sentences. Each manual, being a manual of translation, purports to specify the equivalence relation between sentences and their translations, and neither manual is right to the exclusion of the other".

[5] According to Quine (1971: 297–298), "*Occasion* sentences, as against *standing* sentences, are sentences such as 'Gavagai', 'Red', 'It hurts', 'His face is dirty', which command assent or dissent only if queried after an appropriate prompting stimulation. Verdicts to standing sentences *can* be prompted too: stimulation implemented by an interferometer once prompted Michelson and Morley to dissent from the standing sentence 'There is ether drift' . . . But these standing sentences contrast with occasion sentences in that the subject may repeat his old assent or dissent unprompted by current stimulation when we ask him again on later occasions, whereas an occasion sentence commands assent or dissent only as prompted all over again by current stimulation. Standing sentences grade off toward occasion sentences as the interval between possible repromptings diminishes . . ." In his (1979: 156), Quine speaks of *occasion* sentences as follows: "These are the sentences that admit of verdicts and truth values not once for all but from one occasion of utterance to another, depending on what is going on in the neighbourhood. They are sentences like 'It is raining', 'This is read', 'That is his uncle', 'He owes me money', 'There goes a rabbit'. Historical truths are not among them, nor are scientific hypotheses, nor credos, nor slogans. Now we might measure the clarity of an occasion sentence by the readiness of witnesses to agree in their verdicts on it from occasion to occasion. By this standard 'It's raining' and 'This is red' rate high; 'There goes a rabbit' not quite so high; 'That's his uncle' rates lower; 'He owes me money' lower still. There are three possible responses – assent, dissent, and abstension – and we may also distinguish degrees of hesitation. The great value of this standard of clarity lies in its linking of language to nonlinguistic reality". Quine (*ibid.*) adds, then, that the occasion sentences "that pass this clarity test with high marks are what I call observation sentences".

[6] W. V. O. Quine (1975b: 302–303).

[7] *Ibid.*, p. 303; see also W. V. O. Quine (1970: 178–182).

[8] W. V. O. Quine (1960, chapter two). Quine develops his thesis essentially as a thought experiment in radical translation, i.e., "in translation of an initially unknown language on the strength of behavioural data". See W. V. Quine (1979: 167).

[9] Cf. W. V. O. Quine (1975b: 304).

[10] Cf. *ibid.*, p. 304.

[11] Cf. *ibid.*, p. 304.

[12] *Ibid.*, p. 304.

[13] *Ibid.*, p. 304.

[14] In Quine's (1975a: 88) own words: "In what behavioural disposition then does a man's knowledge of the truth conditions of the sentence 'This is red' consist? . . . it is the disposition to assent or dissent when asked in the presence or absence of red. Query and assent, query and dissent – here is the solvent that reduces understanding to verbal disposition. Without this device there would be no hope of handling language down the generations, nor any hope of breaking into newly discovered languages. It is primarily by querying sentences for assent and dissent that we tap the reservoirs of verbal behaviour".

[15] "The stimulus meaning of a sentence for a subject", writes Quine (1971: 296), "sums up his disposition to assent to or dissent from the sentence in response to present stimulation. The stimulation is what activates the disposition, as opposed to what instills it . . ."

[16] Cf. P. K. Feyerabend (1978a: 17–23).

[17] Cf. G. Harman (1975: 14). In Quine's (1975a: 90–91) own words: "Of course, transla-

tion must go on. Indeterminacy means that there is more than one way; we can still proceed to develop one of them, as good as any. And, in a more theoretical mood, we must still consider what counts as evidence for *an* acceptable translation relation, even if the relation is not unique. The evidence will be behavioural still, of course, even though the relation is no simple coinciding of behavioural dispositions, as it was in the case of the equivalence of observation sentences. We have to examine relations of inter-dependence between verbal dispositions: systematic interdependences between dispositions to assent to standing sentences and dispositions to assent in certain circumstances to observation sentences . . ."

In his (1979: 167), Quine restates his thesis of the indeterminacy of *radical* translation as follows: "I argued that the translations would be indeterminate, in the case of sentences at any considerable remove from observation sentences. They would be indeterminate in this sense: two translators might develop independent manuals of translation, both of them compatible with all speech behaviour and all dispositions to speech behaviour, and yet one manual would offer translations that the other translator would reject. My position was that either manual could be useful, but as to which was right and which wrong there was no fact of the matter". Quine (*ibid.*) clarifies his position further in these words: " . . . I speak as a physicalist in saying that there is no fact of the matter. I mean that both manuals are compatible with the fulfilment of just the same elementary physical states by space-time regions".

[18] W. V. O. Quine (1975a: 91).

[19] Consider, e.g., what Quine (1975a: 90) says: "This indeterminacy of translation is unsuspected in mentalistic semantics, because of the facile talk of meaning. Sentences have meanings, and a translation is right if it has the same meaning. Mentalistic semantics requires that one of two conflicting manuals of translation be wrong, though it conforms to every speaker's dispositions. Mentalistic semantics thus sets a false goal, which, even though vague and ill defined, tends to obstruct other lines of thought".

[20] This is, in fact, Quine's (1975a: 91) own assumption. It is intimately related to his more general assumption that mind is a system of dispositions to behaviour, thereby falling in line with the dispositional philosophy of mind of Gilbert Ryle.

[21] See K. R. Popper (1972b: 106–115).

[22] See G. L. Pandit (1983: 18–19, 79–80). See also ch. 8 below.

[23] Cf. W. V. O. Quine (1963b: 41–42)

[24] W. V. O. Quine (1975a: 89).

[25] W. V. O. Quine (1971: 323).

[26] For this distinction see G. L. Pandit (1983: 174; and 1981: 29–30).

[27] Axiomatization of a physical theory, if resorted to, is what reduces to the minimum its methodological content-indeterminacy. How far it may help reduce indeterminacy of physical theory in second intension – indeterminacy of the translation of the foreigner's physical theory as under-determined by translation of his observation sentences – is worth debating. Cf. W. V. O. Quine (1970: 178–183).

[28] It is one thing to explore *ideologically neutral conditions*, necessary and sufficient, under which specific *translation-claims* could be appraised as to their content-determinacy or indeterminacy, and quite another thing to impose an ideologically non-neutral *criterion* – such as the behavioural criterion of Quine – on translation from one language to another. For a closely similar line of thinking see H. Putnam (1981a: 113–114).

[29] Does Quine's scepticism about propositions smack of a strategy of changing the subject of predication in all the relevant contexts, both semantical and methodological? That *propositions* are sought to be eliminated and replaced in all contexts by *sentences* shows that it does. For more on this type of strategy see G. L. Pandit (1983: 1–2, 131–135, 164).

[30] Quine's occasion sentences approximate to, I believe, what might more appropriately be called *interaction-sentences*, i.e., sentences which report/codify the results of interaction between the knowing/perceiving subject and the world around him. Their significance is not only closely similar to that of scientific experiments but transcends the occasion of the stimulation so much emphasized by Quine's view of occasion/observation sentences. For Quine, each such sentence can be taken in its individual capacity and queried under current stimulation as to one's assent to/dissent from it. And this implies that their significance is exhausted on the occasion of their production by one's assent to/dissent from them under current stimulation – by their stimulus meaning. But I think that this is far from the whole truth about these sentences. Like experimental results, observation sentences too serve as sources of feedback, negative or positive, for our inherently indeterminate theory-problem complexes already at work in the background. John Wettersten's (1982: 475) reading of Quine's occasion sentences as those "which are specific to contexts in which humans interact most intimately" is co-incidentally closer to mine.

[31] For my discussion on pragmatic and (non-pragmatic) theoretical simplicity turn to ch. 8, § 8.2. below.

CHAPTER 4

[1] In G. L. Pandit (1983) and in G. L. Pandit, "The Concept of Person" (unpublished) respectively.

[2] For a recent formulation and detailed discussion see S. Shoemaker (1963: 2–9, 211–260). Bernard Williams (1973: 1) puts it thus: "There is a special problem about personal identity for two reasons. The first is self-consciousness – the fact that there seems to be a peculiar sense in which a man is conscious of his own identity".

[3] For Popper's $world_1$, $world_2$, $world_3$ ontology see K. R. Popper (1972b: 106–148).

[4] See S. Shoemaker (1963: 9, 13).

[5] See S. Shoemaker (1963: 221–260).

[6] Consider, e.g., Wittgenstein's (1968: §§ 243–308, 404–410) and the growing literature on the subject that these have given rise to.

[7] See S. Shoemaker (1963: 211–225).

[8] This can be seen more clearly in the light of our discussion in Chapter 2 above.

[9] See S. Shoemaker (1963: 215).

[10] Consider in this context Shoemaker's (1963: 215) thesis to the effect that all P-statements and M-statements "are based on the criteria for the truth" of their "non-personal" (sentential) components. Thus, e.g., the P-statement that "I see that there is a Japanese cedar in the garden", "if known to be true or asserted with justification, must be based on the criteria for the truth" of its component, viz. "There is a Japanese cedar in the garden".

[11] Consider e.g., what Strawson (1964: 103) says of Hume and Wittgenstein: "It was the entity corresponding to this illusory primary concept of the pure consciousness, the ego-substance, for which Hume was seeking . . . when he looked into himself and

complained that he could never discover himself without a perception . . . it was this entity of which Hume vainly sought for the principle of unity, confessing himself perplexed and defeated . . . Finally it is this, perhaps, of which Wittgenstein spoke, when he said of the subject, first that there is no such thing, and then that it is not a part of the world, but its limit".

12 David Hume (1962: 301–302).

13 *Ibid.*, p. 302.

14 Cf. L. Wittgenstein (1968: §§ 293, 304).

15 See P. F. Strawson (1964: 94–95, 103, 133–134).

16 L. Wittgenstein (1968: § 281).

17 L. Wittgenstein (1968: 410).

18 It may be that language-learning follows a pattern which is not neutral with respect to conventionalism and essentialism; and it may be, on the contrary, taken to favor the former rather than the latter. But the very 'fact' that it follows one particular kind of pattern rather than some other could be used to turn tables against conventionalism. For it would favor the essentialist view of language-learning as following one and the same pattern everywhere.

CHAPTER 5

1 I think that it is in *TLP* that the Western essentialistic tradition of philosophical analysis culminates, finding the most sophisticated expression that we know of in the entire history of this tradition. It is here that the doctrine of the one and only one essential structure is advocated and formulated at three different but corresponding ontological levels of (i) the world of languages; (ii) the abstract world of propositions/thoughts/meanings; and (iii) the world of facts:

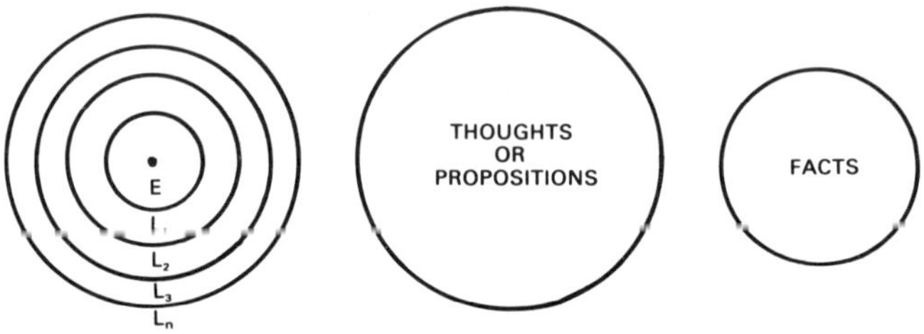

Fig. 5.1. where E is the essential structure common to all languages L_1, L_2, L_3. . . . . L_n.

2 Consider, e.g., the following questions;

(a): *Galilean what-question*:
What is the structure of motion of a terrestrial object which falls to the ground or which is thrown in a certain direction (projectile motion)?

(b): *Keplerian what-question*:
What is the structure of motion of a heavenly body while it moves around the sun?

(c): *Newtonian what-question*:

What is the relation between the forces that heavenly bodies exert upon one another and those that cause terrestrial objects to fall towards the centre of the Earth?

Questions (a), (b) and (c) respectively led to Galilean experiments in terrestrial mechanics, Keplerian three laws of planetary motion and Newton's unifying inverse-square law that between any two particles, small or large, there is a force of gravitation inversely proportional to the square of the distance between them. Consider now the following *why*-questions:

Newtonian why-questions:

(d): Why do terrestrial objects falling to the ground obey Galilean law of free fall?

(e): Why do planets in motion around the sun obey Keplerian laws?

[3] R. Carnap (1965: 179).

[4] *Ibid.*, pp. 133–4.

[5] Cf. *ibid.*, p. 167.

[6] R. Carnap (1974: 256).

[7] Carnap continues (*ibid.*, footnote 4): "In my view greater clarity often results if discussions of whether certain entities are real are replaced by discussions of preference of language forms". This view is defended by him in considerable detail in his essay "Empiricism, Semantics and Ontology" in his (1956: pp. 205–250).

[8] R. Carnap (1956: 206–213).

[9] *Ibid.*, p. 206.

[10] *Ibid.*, p. 206.

[11] *Ibid.*, p. 206.

[12] *Ibid.*, p. 207.

[13] *Ibid.*, p. 207.

[14] Cf. *ibid.*, p. 207.

[15] *Ibid.*, p. 207. See also p. 208.

[16] See K. R. Popper (1961: 26–34).

[17] *Ibid.*, pp. 28–29. A formulation closely similar to this occurs in K. R. Popper (1972a: 20).

[18] See Popper (1961: 29).

[19] *Ibid.*, p. 29.

[20] *Ibid.*, p. 29.

[21] *Ibid.*, p. 119, footnote 2.

[22] Cf. *ibid.*, pp. 29–30, 130–136.

[23] *Ibid.*, p. 136.

[24] *Ibid.*, p. 136.

[25] See K. R. Popper (1972a: 103).

[26] Cf. *ibid.*, pp. 103–105.

[27] *Ibid.*, p. 105.

[28] *Ibid.*, p. 105.

[29] *Ibid.*, p. 107.

[30] Cf. *ibid.*, p. 106.

[31] *Ibid.*, p. 106.

[32] K. R. Popper (1972b: 195, footnote 6).

[33] *Ibid.*, p. 194.
[34] *Ibid.*, p. 195.
[35] *Ibid.*, p. 195.
[36] *Ibid.*, p. 195.
[37] K. R. Popper and J. C. Eccles: (1977: 100).
[38] It is also possible to argue in favor of methodological essentialism from the point of view of language-learning in children as follows: Quite at an early stage in language-learning, it is *necessary* that children are either capable of asking or actually do ask *what*-questions of the following type: What is this? What is that thing called? And so on. What is noteworthy is this, that one invariably answers such questions in the following manner: "This is (called) a chalk", "That is (called) a kite"; and so on. Ideally, such answers, amounting to instructions in language, could be seen as ostensive elucidations centering around paradigms or samples of what a given concept being learned must be applicable to: "*This* and similar things are *called* a chalk". This brings out the complexity inherent in language-learning by means of *what*-questions and answers to such questions. It lies in what might be called a primordial formalization – a process which throws a whole external objective world open to the growing child for interaction at various levels. That language-learning in this sense does take place through a process of ostensive elucidation – a kind of explanation – shows how deep-rooted methodological essentialism is in our language; and how difficult it is to uproot it not as a metaphysical doctrine but as a methodological framework at the adult-levels of interaction with a whole complex world including human beings themselves.

Science, on the other hand, begins with subject-specific problems which have their theoretical bases in the corresponding subject-specific assumptions about a world already "preformed by language", to use a phrase coined by Ernst Cassirer. This implies that science begins with problems whose very formulations presuppose formalizations of a primordial kind, making minimal theoretical identifications available. Thus the following basic assumptions can be made: (1) That our (ordinary natural) language imposes a primordial formalization/order on the world; and (2) that without being brought under the formalization of a primordial kind – the principles of this formalization being built into our language – the world would not be available to the individual sciences for a specialized scientific understanding of it.
[39] See L. Wittgenstein (1968: §§ 3, 18, 23, 65–77, 89–119, 293–304).
[40] Cf. *ibid.*, §§ 60–64.
[41] See L. Wittgenstein (1968: §§ 1–24, 46–88, 108).
[42] K. R. Popper (1968: 52).
[43] *Ibid.*, p. 54.
[44] *Ibid.*, p. 53. See also p. 280.
[45] Consider what Popper (1968: § 30, p. 106) says: "Basic statements are accepted as the result of a decision or agreement; and to that extent they are conventions. The decisions are reached in accordance with a procedure governed by rules." But what about these rules themselves? For Popper, as we have already seen, they too are accepted as conventions.
[46] In his (1974a: 245) I. Lakatos raises the question: "How can one criticize the rules of the scientific game?" And then goes on (*ibid.*, p. 248) speaking of his own amendment of Popper's definition of science "so that it no longer rules out essential gambits of actual science". Lakatos claims (*ibid.*, p. 248) to have done this "primarily by shifting the

problem of appraising theories to the problem of appraising historical series of theories, or, rather, of 'research programmes', and by changing Popper's rules of theory-rejection". "First", writes Lakatos (*ibid.*, p. 248), "*one may 'accept' not only basic but also universal statements as conventions: indeed, this is the most important clue to the continuity of scientific growth*". "My modification", Lakatos continues (*ibid.*, p. 249), "then, presents a very different picture of the game of science from Popper's. The best opening gambit is not a falsifiable (and therefore consistent) hypothesis, but a research programme". For more on Lakatos' improvement on Popper's conception of 'the game of science' and on Popper's own by way of his 'Tarskian turn' see Lakatos (1971a: 99; 1974a: 242–45, 249–56; and 1974b: 311–24). For more on Popper's game-theoretic conception of science and its influence on other thinkers see Gerard Radnitzky (1976: 524–25).

[47] Jaakko Hintikka (1982: 39) observes: "Recently, it has turned out that a most fruitful approach to Wittgensteinian language-games is to take the game aspect seriously, that is, to construct language-games so that they are also games in the strict sense of the mathematical theory of games". For this see, e.g., Jaakko Hintikka (1979: 1–26). Here he propounds the thesis (P. 1) "that the most instructive actual examples of ties or links between a part of language and that aspect of the world which it can convey information about (or otherwise facilitate our dealings with) are to be found in what I have ventured to call *semantical games*". Conceived as language-games of seeking and finding (p. 16), as a link between language and reality, semantical games in Hintikka's sense are essentially Wittgenstenian games played not *in* language but by *means* of language (P. 12). My criticism of the game-theoretic approach to language applies with equal force to more recent attempts in the same direction, such as in Lauri Carlson (1983).

[48] The discipline called 'the theory of games' was first conceived by John von Neumann and Oskar Morgenstern in their (1944). Of this theory C. H. Waddington (1977: 161) writes: "For some time the theory of games became a very fashionable subject. More recently, it has gradually been realized that it can do very little in the way of giving practical recipes of how to act within conflicts or games. It certainly does not provide any infallible rules for winning at chess, bridge or poker, let alone games where moves require physical skill, like golf or tennis. At one period it was extensively used by the American military in planning their strategy in the Korean and Vietnam wars – not perhaps very persuasive testimony to its value. However, it does provide some interesting suggestions about how to analyze situations of conflict, sufficiently to get an insight into the type of thinking that is most likely to lead to a useful result". The theory of games, in any case, deals with situations of conflict, perhaps from the simplest to the most complex, that can be represented, theoretically or mathematically, as games in which every move and every outcome (losing or winning, e.g.) is rule-governed and hence 'rational', its main aim being "to tell you how to play safe in situations of this kind". (Cf. *ibid.*, p. 163). But one must conclude, with Waddington (*ibid.*, p. 171): "Although Games Theory provides some interesting concepts and Tools for Thought, I do not think that by itself it is a very effective tool for actually dealing with complex situations".

[49] There is, however, a third possible sense of 'game' traceable to Carnap's and young Wittgenstein's conception of languages as calculi – a conception which both extend to science. Quite consistent with the logical empiricist programme of analyzing *the structure* of the language of science, Carnap (1937: 5) holds the view that the logical syntax of language is concerned with that part of it "which has the attributes of a calculus". One

must accordingly understand his statement (*ibid.*, p. 7): "The *logic of science* (logical methodology) is nothing else than the syntax of *the language of science*". At another place (*ibid.*, pp. 167–8) he says: "By a language we mean here in general any sort of calculus, that is to say, a system of formation and transformation rules concerning what are called *expressions*, i.e. finite, ordered series of elements of any kind, namely, what are called *symbols*". Similarly, in his *Tractatus Logico-Philosophicus* Wittgenstein based his approach to various problems on the assumption of language as a calculus – as consisting of names and combinations of names. To both Carnap and Wittgenstein, an assumption as simple as this must have appeared as the only clue to the ideally structured languages free from all sorts of 'imperfections' associated with the ordinary natural language.

[50] See K. R. Popper (1968: 54)

[51] *Ibid.*, p. 54.

[52] *Ibid.*, p. 54.

[53] Popper (*ibid.*, p. 50) is afraid that the kind of analysis that it involves "does not elucidate those aspects of the empirical sciences which I, for one, so highly prize". I believe that Popper has in mind such aspects as the susceptibility of empirical statements to criticism and revision.

[54] Cf. *ibid.*, p. 49.

[55] *Ibid.*, p. 50.

[56] See K. R. Popper (1972b: 106–22).

[57] K. R. Popper (1968: 78–81). It is not possible here to consider whether and how far the practising scientist might adopt a cautious and critical approach to constructing *ad hoc* hypotheses as and when occasion arises. It seems to me that any such construction must be ruled out completely.

[58] See K. R. Popper (1968: 53).

[59] See K. R. Popper (1972b: 106–22).

[60] See T. S. Kuhn (1970a).

[61] I now realize how little attention I paid in (Pandit: 1983) to Kuhn's (1970b) own observations such as this (*ibid.*, p. 1): "On almost all the occasions when we turn explicitly to the same problems, Sir Karl's view of science and my own are very nearly identical".

[62] T. S. Kuhn (1970b: 4–5).

[63] "It is important", explains Kuhn (1970b: 5, footnote 1), "to notice that when I describe the scientist as a puzzle solver and Sir Karl describes him as a problem solver the similarity of our terms disguises a fundamental divergence".

[64] *Ibid.*, pp. 4–5.

[65] *Ibid.*, p. 5, footnote 1. For my discussion on 'puzzle-solving' as against 'problem-solving' see Pandit (1983: 13, 65, 87–8, 132).

[66] See T. S. Kuhn (1970a: 150).

[67] See P. K. Feyerabend (1978b: 7–10, 39–40, 73–79, 98–107). Feyerabend, e.g. writes (*ibid.*, p. 106): "Science is one Ideology among many . . ." At another place he says (*ibid.*, p. 79): "*rationalists and scientists cannot rationally (scientifically) argue for the unique position of their favourite ideology*". Again (*ibid.*, p. 103): '*non-scientific ideologies, practices, theories, traditions can become powerful rivals and can reveal major shortcomings of science if only they are given a fair chance to compete*". Of Feyerabend's anarchism I. Lakatos (1978: 107) says "one school of thought on the appraisal problem can be traced back to the Greek tradition of Pyrrhonian scepticism, and is now known as

'cultural relativism'. Scepticism regards scientific theories as just one family of beliefs which rank equal, epistemologically, with the thousands of other families of beliefs. One belief-system is no more 'right' than any other belief-system; although some have more *might* than others. There may be *changes* in belief-systems but no *progress*. This school of thought, temporarily muted by the stunning success of Newtonian science, is today regaining momentum particularly in the anti-scientific circles of the New Left; its most original and colourful version is Feyerabend's '*epistemological anarchism*'. According to Feyerabend, philosophy of science is a perfectly legitimate activity; it may even *influence* science. *Any* belief-system . . . is free to grow and influence any other belief-system; but none has epistemological superiority". Lakatos concludes (*ibid.*, p. 108): "The sceptic thus *denies* the possibility of producing any acceptable solution to the problem of appraising scientific theories". But what Lakatos fails to notice is this: That scepticism in the sense of epistemological anarchism must embrace even philosophy of science whose aim it is to solve the appraisal problem as a problem of scientific rationality.

[68] Cf. *ibid.*, pp. 73–86.

[69] *Ibid.*, p. 106.

[70] I. Lakatos (1971a: 99).

[71] I. Lakatos (1970: 133).

[72] *Ibid.*, p. 135. See also pp. 136–7.

[73] John Worrall (1975: 2/1).

[74] *Ibid.*, p. 2/1.

[75] *Ibid.*, p. 2/1.

[76] *Ibid.*, p. 2/1.

[77] See G. L. Pandit (1983: 25–6, 34–6, 38–9, 41–2).

[78] *Ibid.*

[79] J. Worrall (1975: 2/28). "According to this methodology", writes Worrall (*ibid.*, pp. 2/24–5), "major scientific achievements consist not merely of a set of statements about the world. They consist rather of a set of such statements *together with* some ideas about how to 'fill in', make more precise, draw consequences from these statements, and how to elaborate on and modify them when difficulties arise".

"Lakatos called this set of ideas", Worrall continues, "the 'positive heuristic' of a programme. It guides the production of specific theories within the programme".

[80] See Larry Laudan (1977: 11, 80–81, 99, 109, 120). See also Larry Laudan (1981: 144–155); and B. Baigrie and J. N. Hattiangadi (1981: 85–95).

[81] L. Wittgenstein (1958: 17, 81 respectively).

[82] *Ibid.*, p. 81.

[83] Further confirmation comes from Wittgenstein's (1974b: Part I, § 134) remark: "No one will deny that studying the nature of the rules of games must be useful for the study of grammatical rules, since it is beyond doubt there is some sort of similarity between them".

[84] See L. Wittgenstein (*PI*: § 66).

[85] See L. Wittgenstein (*PI*: § 69).

[86] L. Wittgenstein (*PI*: § 67).

[87] L. Wittgenstein (*PI*: § 494).

[88] Cf. L. Wittgenstein (*PI*: §§ 11, 12).

[89] Some of these confusions are so deep-rooted in the scientist's and the philosopher's approaches to science that they still go unnoticed, leaving them dogmatically fascinated by

the "fact" that "most of our thinking is analogical and metaphorical". [Cf. J. Ziman (1978: 26)]. "Much of science", says J. Ziman (1978: 98), "has arisen through theoretical conjectures, model building and confirmation of predictions". "What is not always recognized", he writes (*ibid.*, p. 23), "is that a model being drawn from another field of knowledge than that to which it is to be applied, carries a certain amount of pre-existing understanding of its own properties. The Rutherford-Bohr picture of the atom as a planetary system of electrons in orbit around a nucleus owes its *strength* not only to the basic principles of classical physics, but also to our *familiarity* with just such systems in astronomy. The *knowledge* conveyed in such a metaphor goes much further than the technique of solving a few differential equations; it contains a large element of experiential and intuitive understanding . . ." (Italics are mine.) After considering Darwin's "model of interspecific competition as the motive force of organic evolution" in a similar manner, Ziman (*ibid.*, pp. 23–24) continues: "These tacit features of a well-conceived model both restrict and enlarge its capabilities. The restriction is advantageous, for it spares us from self-contradictory conjectures. The fact that the model system is actually realized in its original sphere implies that its defining properties are coherent and mutually consistent . . . The defect of many a theory built around a hypothetical mathematical relationship – for example, the non-linear field equations to which Heisenberg gave the later years of his life – is that one simply does not know in advance whether it will 'work' at all . . . By confining his imagination to realizable models, the theorist avoids speculative schemes that fail through sheer inconsistency and automatically keeps within the bounds of consensibility." As to how, in Ziman's view, the imagination is enlarged by such models see his (1978: pp. 24–25).

[90] Cf. M. G. Bulmer (1980: 7).
[91] Cf. *ibid.*, pp. (7–8).
[92] See G. L. Pandit (1983: 151–8).
[93] Cf. *ibid.*, pp. 59, 190.
[94] Cf. *ibid.*, pp. 60, 63–6, 77–8, 137.
[95] See J. Ziman (1978: 25).
[96] Cf. *ibid.*

CHAPTER 6

[1] Albert Einstein to Georg Jaffe, January 19, 1954, quoted in John Stachel (1986a: 377).
[2] The concept of the epistemic structures of science as the developmental structures of objective knowledge is introduced and considered in terms of a general interaction theory in my (1983).
[3] An extreme form of relativism is developed by Paul Feyerabend in his (1978a, 1978b and 1981c). For my discussion turn to Chapter 8 below.
[4] I introduce here the terms 'subject-specific' and 'subject-neutral' in order to distinguish between methodological rules which belong to the special sciences respectively and those which are abstract and neutral with respect to their specific subject-matters. It is the former which vary from time to time, with the latter providing an invariant methodological background for methodological variance in this sense.
[5] I think that another reason why such assumptions might be viewed as assumptions of

complexity is that these can often be seen to entail comparative estimates of complexity of the following kind:

1. The Newtonian assumption: The phenomena of a physical body in motion are more complex than the Aristotelian physical assumptions concerning such motions can capture. The *simplest* cases of motion conceivable would be those of which the law of inertia speaks and these approximate to what might be regarded as self-explanatory cases. The Aristotelian assumptions must, therefore, be discarded as a case of over-simplification.
2. The non-Newtonian assumption: The complexity of the phenomena of electricity (and magnetism) is far greater than can be accounted for/explained within the Newtonian framework of action-at-distance forces between essentially unchangeable material particles. The Newtonian assumptions must accordingly be varied or discarded.
3. The Einsteinian assumptions: The structural-dynamical complexities of the phenomena of physical/electrodynamical bodies or systems in motion are far greater than can be explained/captured within the Newtonian absolutistic conceptions and over-simplifications. These complexities are such as warrant the assumption of structure-laws in place of the Newtonian (non-structure) action-at-distance force laws. One must naturally consider here the fundamental assumptions of Einstein's STR and GTR introducing, as they do, new concepts such as those of mass as a function of velocity (and therefore a variable quantity) on the one hand and gravitational field, space-time curvature, and the equality of the gravitational and inertial mass on the other. I think that the scientific revolutions which are exemplified by Einstein's STR and GTR would be impossible without the kind of structural-dynamical assumptions that entail not just comparative estimates of the complexity of the natural phenomena but rejection of the existing systems of descriptions as cases of over-simplifications.

[6] The concepts of T_{EP} and T_{RP} as the very methodological bases of epistemic appraisal of theories in science are introduced and discussed in G. L. Pandit (1983: 85–110).
[7] For a discussion of what I call the unique resolving power of ideal-typical assumptions see G. L. Pandit (1983: 98–101).
[8] Cf. Nicolas Ruwet (1973: 5).
[9] For Chomskyan mentalism and some of his own recent formulations of such questions see Noam Chomsky (1976: 3–35, 137; 1977: 1–3, 32–37; and 1972).
[10] For Chomsky's discussion of the general linguistic theory that specifies the language universals see his (1976: 3–35). Here (p. 29) he says: "Let us define "universal grammar" (UG) as the system of principles, conditions, and rules that are elements or properties of all human languages not merely by accident but by necessity – of course, I mean biological, not logical "necessity".
[11] Cf. Noam Chomsky (1977: 165–167).
[12] See Jacques Monod (1977: 81).
[13] An excellent discussion formulating this problem and its theoretical solution is to be found in Jacques Monod (1977: 65–78).
[14] For a detailed discussion see Jacques Monod (1977: 101).
[15] *Ibid.*, p. 101.
[16] Cf. *ibid.*, p. 102.
[17] The problem-formulations (a) and (b) are Monod's (1977: 102).

[18] I introduce here the term 'embedding process' in order to refer to the kind of relation between any two or more *type-distinct systems-specific processes* such that (a) any *succeeding* process is inconceivable without the process that *precedes* it; (b) each process is characterized by a certain *invariance* of its own; and (c) the processes satisfying these conditions (a–b) give rise to what might be called *interactive* biological organizations of varying complexity – organizations that are capable of entering into significant multivariable interactions, whose variables range over complex systems more or less similar to/different from these organizations.

[19] Here *genetic replication* refers to the mitotic cycle of a single cell giving rise to "two daughter cells which are genetically identical to each other and to the cell from which they arise"; *cellular differentiation* to the process by which the genetically identical cells diverge giving rise to "morphologically and physiologically distinguishable cell lineages"; and *histoorganogenesis* to those processes by which "differentiated cells aggregate to form tissues of unitary function. These tissues then associate to make up organs". For detailed discussion and formulation see F. J. Gottlieb (1966: 2–3).

[20] The way is thus paved for the rediscovery of the Mendelian concept of the gene as the unvarying bearer of the hereditary traits and to the theoretical identification/elucidation of its structure in bio-chemical terms. For relevant details see Robert Rosen (1972: 215).

[21] The structure of the DNA molecule, in this model, is captured as one of a double helix: "a ladderlike molecule twined around itself in a spiral". For relevant details see James D. Watson (1968: 210–211).

[22] For details see James D. Watson (1968: 22–23, 119).

[23] As James Watson (1968: 119) later on recounted, their experiments "established that a key feature of the infection of a bacterium by a phage was the injection of the viral DNA into the host bacterium. Most important, very little protein entered the bacterium. Their experiment was thus a powerful new proof that DNA is the primary genetic material".

[24] In the words of Garland E. Allen (1978: 225–226): "Although they saw the importance of using the structural and biochemical information about DNA, Watson and Crick were primarily informationists at heart. Their interest was guided from the start by Schrödinger's intriguing questions about genes and the secret of life. The most important question became, "How does the structure of the gene become transformed into the structure of proteins?" This was the key to understanding life, because this was how life differed from most non-living phenomena with which scientists were familiar. Structure was important, but it was subservient in Watson's and Crick's minds to function. Every structural feature had to be consistent with the biological demands placed on the DNA molecule. It was this insistence on relating structural and biochemical knowledge to the requirements of biological function that allowed Watson and Crick to put together the formerly separate structurist and biochemical approaches".

[25] Cf. James D. Watson (1968: 51–55, 77).

[26] Cf. James D. Watson (1968: 69).

[27] This work is acknowledged by James D. Watson (1968: 72, 168–169, 225–226).

[28] See James D. Watson (1968: 51–52, 165). The assumption (2) does not surface in the literature, as far as I know, as a clearly stated assumption. But I think that without an assumption as methodologically constraining as this, the discovery of the double helix would not have been possible.

[29] Consider in this context what Watson and Crick (1953: 737) said while describing their

own finding: "It has not escaped our notice that the specific pairing we have postulated immediately suggests a possible copying mechanism for the genetic material". What they are here referring to is described in the same work (p. 737) as follows: "The novel feature of the structure is the manner in which the two chains are held together by the purine and pyrimidine bases. The planes of the bases are perpendicular to the fibre axis. They are joined together in pairs, a single base from one chain being hydrogen-bonded to a single base from the other chain, so that the two lie side by side with identical Z-co-ordinates. One of the pair must be a purine and the other a pyrimidine for bonding to occur . . ."

[30] For the relevant details of this model see James D. Watson (1968: 196–203).

[31] See James D. Watson (1968: 184–185).

[32] See James D. Watson and F. H. C. Crick (1953: 737).

[33] See James D. Watson (1968: 196).

[34] "Always pairing adenine with thymine and guanine with cytosine meant", argues Watson (1968: 196), "that the base sequences of the two intertwined chains were complementary to each other. Given the base sequence of one chain, that of its partner was automatically determined. Conceptually, it was thus very easy to visualize how a single chain could be the template for the synthesis of a chain with the complementary sequence".

[35] That Chomsky (1977: 36) makes subject-specific ideal-type assumptions should be clear from his statement as follows: "It seems to me that the idealization of grammar as a cognitive structure with certain properties and principles, interacting with other structures, is a reasonable working hypothesis, justified by the success in discovering explanatory principles specific to grammar". "Study of language is", he (p. 36) continues, "part of a more general enterprise: to map out in detail the structure of mind. On the *assumption* just noted, we might continue to regard grammar as a *separable component* of this more general system and proceed to investigate its specific properties". (Italics are added.)

[36] In other words, Euclidean geometry and the special theory of relativity hold only in an empty universe or only in an approximate way. Cf. A. Einstein (1960: 59–61).

[37] Oversimplification – a function of abstraction, idealization, etc. – is, as it were, built into science conceived as a theoretical problem-solving enterprise with an essentially variable network of canons and strategies to anchor it across different problem-situations. That this is so would also seem to follow from David Bohm's cosmological principle of 'The qualitative infinity of nature', which assumes considerable relevance in the present context. It says, in his own words (1957, 133): "Any given set of qualities and properties of matter and categories of laws that are expressed in terms of these qualities and properties is in general applicable only within limited contexts, over limited ranges of conditions and to limited degrees of approximation, these limits being subject to better and better determination with the aid of further scientific research . . . beyond the above limitations on the validity of any given theory, the possibility is always open that there may exist an unlimited variety of additional properties, qualities, entities, systems, levels, etc., to which apply correspondingly new kinds of laws of nature . . . nature may have in it an infinity of different kinds of things".

[38] Consider, for example, a subject-specific assumption of complexity as fundamental as the following: Language learning is a function of multi-variable interaction between a species-specific language responsible cognitive structure/state (LRCS) and other cognitive

structures on the one hand and the environmental variables on the other. The assumption, once made, would send one in search of explanatory hypotheses in terms of the structural dynamical variables and the processes of interaction of an appropriate kind. Chomsky's assumption "that the grammar of a particular language, e.g., LRCS (English), is uniformly represented as a single invariant system in the mind of each English speaker" serves as a very good example of an ideal-type assumption of simplicity, since what it abstracts from is a vast scene of variations and complexities that characterize not only the speech communities themselves but the individuals within each community. Cf. N. Chomsky and E. Walker (1978: 17).

[39] As Chomsky puts it (1977: 164): "Linguistics may be regarded as that part of human psychology that is concerned with the nature, function, and origin of a particular 'mental organ'."

[40] See G. L. Pandit (1981b).

[41] Cf. N. Chomsky and E. Walker (1978: 17).

[42] See E. H. Lenneberg (1967: 371–394); and N. Chomsky (1967: 397–442).

[43] Cf. N. Chomsky and E. Walker (1978: 18).

[44] Cf. N. Chomsky (1977: 1–21).

[45] See P. K. Feyerabend (1978a, 1978b).

[46] See T. S. Kuhn (1970a: 35–42, 92).

[47] Cf. T. S. Kuhn (1970a: 92, 96, 109–111).

[48] See T. S. Kuhn (1970b: 4–5). See also H. Putnam (1981b: 70).

[49] Cf. T. S. Kuhn (1970b: 4–5). See also H. Putnam (1981b: 69). However it seems to me that the pragmatical concepts of a *paradigm* and a *dominant theory* are so interchangeably usable that it hardly helps in defining the one in terms of the other.

[50] Cf. T. S. Kuhn (1970b: 4–5; p. 5, footnote 1).

[51] Consider, e.g., Kuhn's (1970a: 92) definition: Scientific revolutions are "those non-cumulative developmental episodes in which an older paradigm is replaced in whole or in part by an incompatible new one". For discussion see G. L. Pandit (1983: 126–136, 201). Again in his (1974: footnote 18), Kuhn remarks: "In *The Structure of Scientific Revolutions*, particularly Chapter X, I repeatedly insist that members of different scientific communities live in different worlds and that scientific revolutions change the world in which a scientist works. I would now want to say that members of different communities are presented with different data by the same stimuli. Notice, however, that that change does not make phrases like 'a different world' inappropriate. The given world, whether everyday or scientific, is not a world of stimuli".

[52] For details see S. I. Benn and G. W. Mortimore (1976: 177); J. von Neumann and O. Morgenstern (1944); and A. J. Jones (1980).

[53] In more extreme cases, not just of divergent but seemingly conflictual religious ideologies of any two or more people (cultures), consider their effects as opposite in nature as those of perceiving a Schwein in every instance of good luck ('Du hast Schwein gehabt!' is how one is likely to be greeted in such a case, according to a German Sprichwort) and an evil in every instance of the natural kind Schwein. One might well wonder as to how is it after all that the members of one and the same animal species (say *das Schwein* in our example) are perceived by the members of one particular cultural community as most admirable creatures but by those of another as most abominable (abscheulich). It is fairly well-known that the latter, having *learnt* to hate it, would be too

happy to see the object of their abhorrence getting totally out of existence such that it no longer forces itself into their day-to-day perceptions with the same force with which it might leave its traces in their mythology (very much like *der Drachen* as a symbol of Böse in European fairy-tales but as an opposite symbol in Chinese cultural tradition). Not surprisingly, then, there is really no place for instances of such a *natural kind* in their ontology because there is none in their ideology. Quite the reverse is true of the members of the other community. It is then ultimately to one's ideological up-bringing and group-membership that we must trace the fact that the members of one and the same natural kind may be perceived, by one cultural community, as an exemplar of good luck but, by another, as one of abject abhorrence.

[54] See T. S. Kuhn (1970b: 4–5; p. 5, footnote 1). It should here suffice to remark that Kuhn's thesis is tantamount to assimilating scientific (epistemic) appraisal to historical appraisals and we may, therefore, characterize it as involving a kind of methodological reductionism.

[55] I discuss this strategy in considerable detail in my (1983).

[56] T. S. Kuhn (1970b: 6, 10–11, 19). Kuhn's remarks in this context are often contradictory. Consider what he (1970b: 6) says: "I suggest then that Sir Karl has characterized the entire scientific enterprise in terms that apply only to its occasional revolutionary parts". At another place in the same work (p. 6): "If a demarcation criterion exists (we must not, I think, seek a sharp or decisive one), it may lie just in that part of science which Sir Karl ignores". And consider now (*ibid.*, p. 19): "Sir Karl has erred by transferring selected characteristics of everyday research to the occasional revolutionary episodes in which scientific advance is most obvious and by thereafter ignoring the everyday enterprise entirely".

[57] *Ibid.*, p. 6.

[58] This is admitted by Kuhn (1970b: 7).

[59] *Ibid.*, p. 10.

[60] In this context one might wonder what does Kuhn's *normal-revolutionary* distinction for science achieve. The post-revolutionary and pre-revolutionary paradigms are held to be mutually incommensurable. But, as we have noticed, on Kuhn's own account they are, in effect, structurally identical. In his (1974: 460, 463), Kuhn traces a systematic ambiguity in his own use of the term 'paradigm' in his (1970a) in terms of, as he claims, at least two types of usages. I say 'systematic', since according to Kuhn himself one of these confers on 'paradigm' what he now calls a global sense "embracing all the shared commitments of a scientific group; the other isolates a particularly important sort of commitment and is thus a subset of the first". Kuhn now redesignates the 'paradigms$_2$' in the latter sense as exemplars, and locates these as proper *pragmatical* subsets in 'paradigms$_1$' in the former sense but now redesignated as 'disciplinary matrix'. "Constituents of disciplinary matrix include", says Kuhn, "most or all of the objects of group commitment . . ." most importantly 'symbolic generalizations', 'models' (= preferred analogies) and exemplars (= paradigms$_2$). Since an understanding of the operation of at least these three components of the disciplinary matrix is what Kuhn holds necessary to a proper understanding of how "a scientific community functions as a producer and validater of sound knowledge", such an understanding should have for its object, so it seems to me, what Kuhn calls normal science.

[61] Kuhn (1970b: 7) himself recognizes a frequent "coincidence" between "Sir Karl's line

of demarcation" and his own, if "only in their *outcome*" and not in "the *process* of applying them".
[62] For a fuller discussion see G. L. Pandit (1983: 126–136, 201).
[63] See P. K. Feyerabend (1981a: 21–25). Consider in this context the following historical generalization made by Feyerabend (*ibid.*, p. 23): "the 'meaning' of a hypothesis often becomes clear only after the process that led to its elimination has been completed . . ."
[64] R. Carnap (1963: 55).
[65] See R. Carnap (1937: 51, 164, 277, 298–301, 312, 321).
[66] Cf. J. Locke (1961: xxxv).
[67] For Quine's relentless criticism see W. V. O. Quine: (1963b), (1960: ch. 2), and (1975b: 302–319).
[68] See K. R. Popper (1968: 86).
[69] See P. Duhem (1954: 187).
[70] A Duhemian theoretical system is in my view best conceived as always consisting of a newly proposed theory and a set of background theories which bear a presupposition-relation of logical support to the former.
[71] See K. R. Popper (1972a: 229, 233).
[72] Although it is generally agreed that all sciences simplify, I think it would be more correct to say that all sciences oversimplify in essentially unintended ways.
[73] For detailed and critical discussion on this see Chapter 9, section 9.1. below.
[74] See G. L. Pandit (1983: 54–81).
[75] Further discussion of the far-reaching consequences of this principle of scientific rationality for the methodology of the sciences follows in Chapters 7 and 8 below.
[76] It seems to me now that the framework of objectivistic epistemology that I proposed in (Pandit: 1983) is essentially a framework of *theoretical universals* in this sense.

CHAPTER 7

[1] K. R. Popper (1968: 41).
[2] *Ibid.*, p. 40.
[3] This anticipates my deviation, in what follows, from the (Popperian) custom of taking the asymmetry between falsifiability and verifiability to be simply a logical one. Indeed, this deviation will form an essential part of my criticism of Popper.
[4] I am, of course, assuming here as a framework principle, proposed in (Pandit: 1983), that a sound methodology of science must be embedded in a sound objectivistic theory of scientific knowledge. My conception of an objectivistic epistemology, though closely similar to Popper's, does not entail an antagonism between it and the traditional subjectivistic epistemology in the sense of Popper's sceptical attacks on the latter. See K. R. Popper (1972b: 106–150). In his (1972a: 402), Popper says: "My thesis is, simply, that the theory of subjective belief is utterly irrelevant to the philosophical theory of knowledge. Indeed, it is destructive of the latter if the two are mixed up (as they still are, in accordance with tradition)".
[5] K. R. Popper, (1968: 41).
[6] Cf. *ibid.*, p. 41.
[7] Cf. *ibid.*, p. 76.

[8] Cf. *ibid.*, p. 41.
[9] Cf. *ibid.*
[10] Cf. *ibid.*, pp. 86–87, 269.
[11] Cf. *ibid.*, p. 100.
[12] Cf. *ibid.*, p. 100.
[13] Cf. *ibid.*, p. 47. Here Popper says: "*there can be no ultimate statements in science*: There can be no statements in science which cannot be tested, and therefore none which cannot in principle be refuted, by falsifying some of the conclusions which can be deduced from them". At another place he says (p. 104): " . . . any basic statement can again in its turn be subjected to tests, using as a touchstone any of the basic statements which can be deduced from it with the help of some theory, either the one under test, or another. This procedure has no natural end".
[14] Cf. *ibid.*, p. 104.
[15] Cf. *ibid.*, p. 105. For Popper's detailed characterization of basic statements see his (1972a: 385–388).
[16] See P. K. Feyerabend (1965: 212–13, 219, footnote 3).
[17] See K. R. Popper (1968: 86).
[18] Cf. *ibid.*, p. 41.
[19] Cf. *ibid.*, p. 41.
[20] For Duhem's argument see his (1954: 187). I think that Duhem's argument that in science – more precisely, in physical science – it is not an individual hypothesis but a whole logically inter-related complex network of hypotheses which can be confirmed/disconfirmed by the positive corroborative/negative experimental or observational results respectively leaves Popper's falsificationist methodology of scientific appraisal essentially unscathed. For what this argument accomplishes is the change from the individual theory/hypothesis to a complex system of inter-related theories as *the unit of appraisal*: We may not know which particular theory is refuted/confirmed; but we can know which *theoretical system* is. And one could quite reasonably admit this Duhemian shift in the unit of appraisal as valid without pushing it to the ultimate limits of a Quinean holism under which the unit of appraisal suffers from a fundamental indeterminacy. And one could perhaps at the same time devise ways – such as axiomatization as suggested by Popper (1968: 71–2) – of canonically so reformulating a complex theoretical system as to facilitate the detection of the refuted components in it.
[21] See K. R. Popper, (1968: 53). The sense in which science is here supposed to be a game becomes clear from what Popper says (p. 54): "Just as chess might be defined by the rules proper to it, so empirical science may be defined by means of its methodological rules. In establishing these rules we may proceed systematically. First a supreme rule is laid down which serves as a kind of norm for deciding upon the remaining rules, and which is thus a rule of a higher type. It is the rule which says that the other rules of scientific procedure must be designed in such a way that they do not protect any statement in science against falsification". I think that any characterization of science based on such a comparison must be rejected as involving an unnecessary oversimplification that originates in a game-theoretic approach.
[22] *Ibid.*, p. 53.
[23] *Ibid.*, p. 52.
[24] Much of what Paul Feyerabend has written right from his (1965) to his (1978a, 1978b)

seems securely built upon the implications of Popper's philosophical works. It is therefore no surprise if Feyerabend moves, seemingly unpredictably, from his earlier theoretical pluralism, mostly implicit in Popper's (1968), to his latest epistemological anarchism (implicit in Popper's work again).

[25] A 'phenomenon' which T. S. Kuhn in his (1970a) lauds as the most significant aspect of science that one can perceive should one choose to attend closely to the history of the sciences.

[26] For a discussion of psychologism in this sense see my (1971: 86).

[27] Epistemic structuralism is originally introduced and discussed in my (1983).

[28] Popper (1972b: 197) seems to be coming closer to recognizing some of those properties of scientific statements/theories which, in my (1983), I call theoretical universals:

(1) " . . . our theories must be *universal*, that is to say, must make assertions about the world – about all spatio-temporal regions of the world".

(2) " . . . our theories make assertions about structural or relational properties of the world; and that the properties described by an explanatory theory must be, in some sense or other, deeper than those to be explained".

(3) "The 'depth' of a scientific theory seems to be most closely related to its simplicity and so to the wealth of its content".

Again, he comes closer to being a theoretical universalist in my sense in his ontological doctrine of $world_1$, $world_2$ and $world_3$ (Popper: 1972b), theoretical universals being the invariant properties that characterize the complexes of problems and theories in their capacity as the developmental structures of objective knowledge. Even otherwise, in his (1968, 1972a) Popper remains, I believe, an essentialist in disguise, methodological conventionalism advocated therein not-withstanding.

[29] The concept of an interactive model of the growth of knowledge is originally introduced in my (1976: 409–36) and discussed in (Pandit: 1983).

CHAPTER 8

[1] Lakatos' attempt to go beyond Popper, recognizing the unrecognized epistemic complexities in science is interesting, no doubt. But it lands him in what he calls the methodology of scientific research programmes, without any attempt to overcome methodological conventionalism. On the contrary, he approves of it.

[2] This is particularly true of J. Hintikka. See J. Hintikka (1982: 39–59); and also E. Saarinen (ed.) (1979). Popper himself is foremost among the small philosophical minority believing in their independent ontological status.

[3] The reason being his methodological conventionalism in the philosophy of sciences. In a nutshell, this is the view that what we call science is a matter of convention or decision. Conventions are bound to change in the course of time. But need our demarcation criteria for science too change everytime as a result of the changing conventions? As we already noticed, for the paradigm-theorists like Kuhn and for the methodological conventionalists like Popper the answer to this question is in the affirmative.

[4] See G. L. Pandit (1983: 141, 162–163, 178, 182, 201).

[5] Even Hempel (1966: 40–45) recognizes that there are great difficulties in formulating objective criteria of the simplicity of a scientific theory.

[6] See M. Reichenbach and Robert S. Cohen (eds.) (1978: 162).
[7] See, e.g., Harold Jeffreys (1973: 31, footnote). According to him, " . . . the prior probability is intended to express simply the probability at the start of investigation and may have been influenced by many previous investigations".
[8] Harold Jeffreys (1973: 38–39).
[9] Harold Jeffreys (1973: 39).
[10] Cf. Harold Jeffreys (1973: 39); and W. C. Salmon (1967: 122–123).
[11] I. J. Good et al (1984: 123).
[12] I. J. Good et al (1984: 125).
[13] I. J. Good et al (1984: 125).
[14] See Nelson Goodman (1966).
[15] K. R. Popper (1968: 140).
[16] K. R. Popper (1968: 141).
[17] K. R. Popper (1968: 142).
[18] K. R. Popper (1972a: 61).
[19] K. R. Popper (1972a: 61).
[20] See G. L. Pandit (1983: 94).
[21] Elliott Sober (1975: VII–VIII).
[22] *Ibid.*, p. 19.
[23] *Ibid.*, pp. 16–17.
[24] *Ibid.*, p. 18.
[25] *Ibid.*, p. 18.
[26] *Ibid.*; see also pp. 1–5.
[27] *Ibid.*, see pp. 19–23. This development proceeds in terms of his interpretation of questions – the frame of reference for the simplicity-appraisals – as sets of natural predicates, with a view to avoiding any kind of relativism.
[28] *Ibid.*, see p. 18.
[29] *Ibid.*, see pp. 18, 161–175.
[30] *Ibid.*, p. 161.
[31] Very early Hans Reichenbach said: "The special significance of the concept of simplicity deserves a more detailed investigation within the context of the theory of probability". See Maria Reichenbach and Robert S. Cohen (eds.) (1978: 164). See also pp. 162–164; and H. Reichenbach (1938: 373–387, § 42).
[32] That this is the view most dominant among philosophers may also become clear from how Wesley C. Salmon (1967: 126) looks at the problem: "Although no one can say just what simplicity is, everyone seems to agree that it is a very desirable characteristic of scientific hypotheses. A certain type of simplicity lends beauty and elegance to the hypothesis and ease to its application, but such economic and aesthetic considerations are secondary when an issue of truth or falsity is at stake. At this point a different type of simplicity is invoked. We place more confidence in simple than in complex hypotheses for purposes of explaining a given body of fact. We judge the simpler hypothesis more likely to be true. We have learned by experience that this works, and at the same time we have learned by experience to avoid oversimplication".
[33] See B. van Fraasen (1980: 90). He recognizes here what I call T-simplicity as "a criterion in theory choice, or at least a term in theory-appraisal", but he thinks that all attempts "so far to explain precisely what people could possibly mean when they call a

theory simple, or simpler, have failed". He argues then as follows: " . . . some writings on the subject of induction suggest that simple theories are more likely to be true. But it is surely absurd to think that the world is more likely to be simple than complicated (unless one has certain metaphysical or theological views not usually accepted as legislative factors in scientific practice). The point is that the virtue, or patchwork of virtues, indicated by the term is a factor in theory-appraisal, but does not indicate *special* features that make a theory more likely to be true (or empirically adequate)". Similar scepticism is expressed by Abner Shimony (1976: 584) as follows: " . . . it is a commonplace that the prior probability of a simple hypothesis should be greater than that of a less simple one, but many methodologists have despaired of the utility of this rule just because of the difficulties of finding non-arbitrary apriori criteria for simplicity . . .".

[34] Not-withstanding my optimistic view of Elliot Sober's approach to T-simplicity, he no longer seems inclined to believe in the theory of simplicity he himself propounds in his (1975). But why is it so? He now thinks (personal communication) that "it is wrong-headed in its attempt to talk about simplicity without developing a concept of evidential support", whereas in his (1975), Sober takes this concept as an undefined primitive. Unlike Sober, however, I am sceptical of any approach which seeks to explicate T-simplicity by resort to concepts of evidential support, prior or posterior probability, falsifiability – concepts that might be usefully employed elsewhere for other purposes. Notice, in this context, how James H. Fetzer (1981) is also very sceptical about the prospects for any prior/posterior probability approach to the question of T-simplicity. He takes the view (personal communication), with which I am in essential agreement, that "the most of the presumptive desiderata that simplicity has been supposed to reflect – such as prior probability, falsifiability, and the like – are at best derivative and secondary to the underlying desideratum of elegance (explanatory elegance, really) where a theory whose lawlike sentences concern *all and only* the nomically (or the causally) relevant properties that bring about the phenomena of interest achieves (strict) maximal specificity (= simplicity = elegance), relative to that class of phenomena". On this view, then, it is correct to require the simpler of the two rival theories to satisfy what might be called a principle of maximal specificity in a higher degree than the other theory does. In other words, simpler theories are those which achieve explanatory elegance in greater degree because they achieve, as Fetzer puts it, "the broadest scope of application on the basis of the smallest number of" nomically/causally relevant factors. But while all this can be accepted as correct, the meaning of T-simplicity would still remain, I think, unexhausted in all its richness, given my argument for a two-dimensional simplicity, one rooted in the explanatory power and the other in the resolving power of a scientific theory. See also James H. Fetzer (1983b: 25–32).

[35] Elliott Sober (1975: 174).

[36] Elliott Sober (1975: 173).

[37] I argue for a simplicity-theoretic, instead of a verisimilitude-theoretic, realism in my (1988b).

[38] A detailed discussion on this point is developed in my (1988b).

[39] A proposal as to how such a framework is possible is formulated in my (1988b).

[40] It might not be completely idle to think that Einstein (1934) was perhaps the first physicist-philosopher of science to have anticipated as early as 1918 the idea of T-simplicity somewhat analogous to the one intended here. Nor might it be idle to address

questions of T-simplicity in the sense suggested here to the current debates about quantum mechanics which were initiated by Einstein as early as 1935. (See the concluding part of the discussion below).

[41] See G. L. Pandit (1983: 65–79).

[42] For details about these two aspects or functions of a theory see G. L. Pandit (1983).

[43] Einstein, Podolsky, Rosen (1935).

[44] J. S. Bell (1964, 1966). For details see Max Jammer (1974), B. d' Espagnat (1971) and John F. Clauser and Abner Shimony (1978).

[45] See. J. S. Bell (1964, 1966).

[46] See, e.g., Erhard Scheibe (1986). There is a view, held by some, that quantum mechanics supports instrumentalism but undermines realism. This view is, I think, seriously mistaken, if only for the following reasons. First of all, not only is the problem-situation of quantum mechanics not as simple as that, its interpretational problems are yet to be resolved in a rational manner. Secondly, however, realism is wrongly taken in this context, as in other contexts, as a rigid doctrine. The question of its interpretation or reinterpretation, which becomes more meaningful if considered in relation to that of quantum mechanics itself, is never faced boldly. Recent interesting discussions are, however, to be found in James Fetzer (1983), M. Stöckler (1986) and D. Howard (1985).

[47] G. L. Pandit (1983: 18–21, 23–24, 27–28, 30, 34, 38–39).

[48] K. R. Popper (1972a: 390). While clarifying how the severity of the test, in a given case, depends on the novelty of the tested prediction relative to background knowledge, Watkins (1978: 34) points out that "Background knowledge is, roughly, all that 'knowledge', which, whether or not true, is being taken as unproblematic in the testing of a particular theory (or theories)".

[49] Paul K. Feyerabend (1978a: 17) writes: "Science is an essentially anarchistic enterprise: theoretical anarchism is more humanitarian and more likely to encourage progress than its law-and-order alternatives". At another place (1978b: 195) he says: "Once philosophers of science were intelligent and well informed people and philosophy of science a rather interesting subject . . .". And to this he (1978b: 201–202) then adds: "All the distinctions of the discipline (context of discovery/context of justification; logical/psychological; internal/external, and so on) have but one aim: to turn incompetence . . . into expertise . . .".

[50] See Paul K. Feyerabend (1978b: 195–205). In his (1980: 281), Feyerabend says: "Mach, Einstein, Bohr use philosophy as an instrument of research that guides research but is also transformed by it while philosophers pontificate on a process they do not understand and would not know how to change. Small wonder they want to persuade us that Mach's ideas are as vapid as their own". One might agree with all that Feyerabend (1981b: 80–98) has to say concerning the work of Ernst Mach and its influence on Einstein's work. But at the same time, one should not fail to notice considerable influence of Ernst Mach (1976) on Feyerabend's own (1978a, 1978b) approach to science – on his epistemological anarchism. I think that such influence is also discernible in Feyerabend (1981b: 81). Indeed not only Feyerabend's but also Kuhn's approach to science reminds us of Mach's own (1976: XXXV, 189, 223) historical-genetic method of presenting physical theories – a method on which Mach finds himself in fundamental agreement with Pierre Duhem. Notice, for example, what Mach (1976: 223) says: "Schematizing the cognitive stages may perhaps benefit further enquiry when similar situations recur, but there can be no widely effective instructions for enquiry by formula".

[51] See K. R. Popper (1972a: 226–231); (1972b: 29–30). For a critical discussion on Popper's conception of the aim of scientific inquiry see Isaac Levi (1983: 153–174).
[52] See G. L. Pandit (1983: 44–45, 80–81).
[53] See G. L. Pandit (1983: 80–81).
[54] P. K. Feyerabend (1978a: 17).
[55] P. K. Feyerabend (1978a: 23).
[56] *Ibid.*
[57] Cf. *ibid.*
[58] P. K. Feyerabend (1978b: 39–40).
[59] *Ibid.*, p. 188.
[60] P. K. Feyerabend (1978a: 23).
[61] G. L. Pandit (1983: 141, 162–163, 178, 182, 201).
[62] H. Margenau (1982: 187).
[63] *Ibid.*, p. 188.
[64] G. L. Pandit (1983: 91, 102–105, 108, 195–196, 198).
[65] Popper (1972b: 202) speaks of *the problems of depth* and cites Newton's theory as an example of an explanatory theory which is deep in the sense that it unifies Galileo's and Kepler's laws: "But far from being a mere conjunction of these two theories – which play the part of *explicanda* for Newton's – it corrects them while explaining them". A theory with theoretical depth in the sense of this example is thus one which is not *ad hoc*; and which is capable of independent testability (cf. *ibid.*, p. 202). Although Popper (1972b: 202) invokes "the principle of correspondence" (due to Bohr) as a sufficient condition of depth – viz., that a new theory of a higher level of universality should explain some older theory by correcting it and hence by containing it as an approximation – he does not commit himself to a fuller analysis of the concept of theoretical depth. The kind of analysis I have here in mind is exactly what is taken care of by the concept of the resolving power (T_{RP}) of a scientific theory in the sense of my (1983). It must, therefore, go beyond an explanation-theoretic approach.
[66] Consider in this context various historical cases. The 'problem' of ether as a carrier/medium of the electromagnetic field is a good example. This 'problem' can be seen to appear as a result of an attempt, made *ad hoc*, to *complicate* the newer theoretical identifications of the optical and electromagnetic phenomena that were made in the last decades of 19th century. The complications introduced were believed to give results desired from the very beginning – viz., the description of the phenomena in terms of Newtonian mechanical laws. *Ad hoc* hypotheses and the 'problems' arising therefrom can thus be seen to hinder theory-problem interactions, rooted as these are in reduction-strategies operating in the sciences from time to time. Behind every concept that we may employ in the formulation of a scientific problem there must be theoretical assumptions which are made either implicity or explicity. In case, however, a theoretical problem is found to employ a concept without the methodologically embedding problem-solving/problem-posing structure of a theory, it must be subjected to a proper scrutiny. Problems with their bases in assumptions made *ad hoc* will, on the contrary, only prevent the methodologically progressive problem-formulation from taking place. The physical problem-formulation in classical physics employing 'concepts' of absolute space, ether and so forth is a good illustration.
[67] Methodological essentialism in the sense defended in Chapter 5 above can, in its turn,

be seen to imply methodological isomorphism between science and philosophy of science, if not all the way but in the following restricted sense: (a) philosophy (and methodology of science) must likewise formulate its problems with their logical forms restricted to *what*-questions, *whether*-questions and *why*-questions; and (b) methodological isomorphism in this restricted sense extends further to the construction of formalizing models in sciences and in philosophy of science.

68 The explanation-theorist's method of hypothesis – the method of Galileo, Descartes, Pierce and others – involves the assumption of the statement-view of the epistemic structural identity of science. It is not surprising that it leads one either to an inductivist or to a hypothetico-deductive view of science and scientific knowledge. As against such an over-simple view of science and its method, the methodology of theory-problem interactive systems (of scientific knowledge) leads to some very interesting consequences, radically changing our traditional approach to science. This methodology is, I believe, exemplified in Newtonian dynamics and optics, in what these *assume* and *lead to*. It is also interesting that Newton's extra-scientific pursuits in metaphysics, theology and alchemy have hardly any bearing on one's proper understanding of Newtonian science.

69 See also G. L. Pandit (1983: 17–19, 38–39, 40–42, 44).

70 I. Lakatos (1978: 221–222) writes: "*Rationality (and here I follow Popper) is not concerned with propositions in isolation . . . but with the way we modify them and hence, their relation to other propositions*. This is one of Popper's greatest innovations. But in Popper's work . . . the new is slightly marred by the remains of the old. His demarcation criterion is formulated in terms of falsifiability or unfalsifiability of *propositions* rather than in terms of the progressive or degenerating character of problem-shifts (i.e. series of propositions which are the result of progressive or *ad hoc* modifications).

This shifting of the *locus* of scientific appraisal from propositions to problemshifts introduced a historical dimention into scientific appraisal. (The problem of when a hypothesis-*replacement* is '*ad hoc*', i.e. irrational, degenerating, bad, has never been discussed with more attention and detail than by Popper and myself)".

71 The inductivist approach to science and scientific knowledge is an excellent example of the kind of doctrine that can never fulfil this important condition. This is true of all the following three aspects of inductivism: (a) Science aims at the establishment of empirical laws/generalizations that are verifiable as to their degree of probability or likelihood to be true; (b) a statement/theory is scientific if and only if it is verifiable in terms of an essentially meaning-invariant framework of observational language; and (c) the empirical generalizations in science can be arrived at and verified essentially on the basis of such a framework.

72 The thesis that scientific rationality has nothing to do with 'rational belief' can be found in Popper (1975: 84–90), Pandit (1983: 126, 128, 183) and also to some extent in Carnap (1950: 37–51). Lakatos (1978: 220) argues in support of it as follows: "Science progresses through the competition of research programmes, not simply through conjectures and refutations. But a programme is a complex entity, a special case of a problemshift (that is, a *series* of propositions), plus mathematical and observational theories *and* heuristic techniques which provide the tools for forging ahead. A research programme as a whole cannot be either true or false. How can one 'rationally believe' that a *programme* is 'likely to be true'? Grünbaum may retort that the scientist *might* rationally believe or disbelieve the programme's *hard core*. But there is no need to believe (rationally or non-rationally)

in the hard core of the programme on which one is working. Newton disbelieved his own action-at-distance programme in its realist interpretation; Maxwell elaborated kinetic theory and Planck quantum theory with actual disbelief".

CHAPTER 9

[1] See A. Einstein, B. Podolsky and R. Rosen (1935: 777–780), A. Einstein (1948), A. Einstein (1949: 81–87; 681–682) and A. Einstein (1960: 157).

[2] See A. Einstein, B. Podolsky and R. Rosen (1935: 777–780).

[3] John Stachel (1986b: 2).

[4] This illustrates, on the one hand, how the $T_{RP}(T_2)$ corrects the $T_{RP}(T_1)$ or how $T_{RP}(T_2) > T_{RP}(T_1)$, since $T_{RP}(GTR) > T_{RP}(NGT)$. That, on the other hand, the STR or its fundamental assumptions are essentially ideal-typical assumptions should follow from the fact that, as Einstein (1960: 76) puts it: STR "cannot claim an unlimited domain of validity; its results hold only so long as we are able to disregard the influences of gravitational fields on the phenomena (e.g. of light)". Cf. p. 17.

[5] Consider in this context how Einstein (1960: 77) himself describes the scene of physical problem-formulation in the field of GTR: "In the example of the transmission of light just dealt with, we have seen that the general theory of relativity enables us to derive theoretically the influence of a gravitational field on the course of natural processes, the laws of which are already known when a gravitational field is absent. But the most attractive problem, to the solution of which the general theory of relativity supplies the key, concerns the investigation of the laws satisfied by the gravitational field itself".

[6] Lack of reference to Soldner's (1801) paper in the literature on the gravitational bending of light rays is ironical and regretably so in view of the role it plays not just in developing a *problem* first entailed by the T_{RP}(NGT.NCT) but in anticipating the later developments in this field due to the T_{RP}(GTR). For possible historical factors that might have contributed to this see P. Lenard (1921: 593–604) and Stanely L. Jaki (1978: 927–950). All references here to Soldner's paper are to its translated version in Jaki (1978).

[7] Isaac Newton (1952: 339, Query 1). Indeed all of Newton's (1952: 339–376) 'Queries' (1–31) are negatively formulated *whether*-questions that formulate significant physical problems with each having its own propositional/theoretical basis in the Newtonian theoretical assumptions and laws. For example, consider the following:

Query 5.: (p. 339): Do not Bodies and Light act mutually upon one another; that is to say, Bodies upon Light in emitting, reflecting, refracting and inflecting it, and Light upon Bodies for heating them, and putting their parts into a vibrating motion wherein heat consists?

Query 6.: (pp. 339–340): Do not black Bodies conceive heat more easily from Light than those of other Colours do, by reason that the Light falling on them is not reflected outwards, but enters the Bodies, and is often reflected and refracted within them, until it be stifled and lost?

Query 30.: (p. 374): Are not gross Bodies and Light convertible into one another, and may not Bodies receive much of their Activity from the Particles of Light which enter their composition? . . .

Query 31.: (pp. 375–376): Have not the small Particles of Bodies certain Powers, Virtues,

> or Forces, by which they act at a distance, not only upon the Rays of Light for reflecting, refracting, and inflecting them, but also upon one another for producing a great Part of the Phaenomena of Nature? . . .

It is noteworthy how the Query (30) seems to anticipate the T_{RP}(STR) whose fundamental theoretical assumptions make it possible to ask and answer in the affirmative the question: Whether mass is convertible into energy and energy into mass? The famous equation $E = mc^2$ is the result that unifies the law of the conservation of mass and the law of the conservation of energy into a single whole.

[8] "An atom", argues Einstein (1960: 130–131), "absorbs or emits light of a frequency which is dependent on the potential of the gravitational field in which it is situated". The problem of redshifts is accordingly formulated by him (1960: 132) as follows: "Whether the displacement of spectral lines towards the red by the gravitational potential does exist?" An affirmative answer would, thought Einstein (*ibid.*, p. 132), not only confirm GTR but enable us to use our theoretical interpretation of redshifts in calculating the masses of the heavenly bodies. This itself speaks of the unique resolving power of GTR.

[9] See A. Einstein (1960: 129–32).

[10] The electric and the magnetic field was conceived by Faraday, in the words of Ronald W. Clark (1973: 87) : as "a region of space in which certain physical conditions were created and through which forces were transmitted; somewhat like ripples through an invisible jelly, the electromagnetic waves of light being propagated through the field in straight lines at a finite speed, and the pull of the magnet for iron fillings being a property of the field which the magnet had itself created".

[11] Of this experiment J. Ziman (1978: 25) says: " . . . the Michelson-Morley experiment showing that the velocity of light seemed independent of the velocity of observer was much more serious for the model of light as a 'vibration of the luminiferous ether' than it was for the electromagnetic theory in which light is just a wave-like solution of Maxwell's equations, which Lorentz could easily modify to give the correct experimental result".

[12] L. Tondl (1973: 131).

[13] *Ibid.*, p. 134.

[14] *Ibid.*, p. 134.

[15] *Ibid.*, p. 134.

[16] *Ibid.*, p. 135. Here '≡' abbreviates equivalence.

[17] Cf. *ibid.*, p. 135.

[18] Cf. A. Einstein (1960: 63–65). For more discussion as to how this problem arises, see Jayant Narlikar (1977: 152–53).

[19] Cf. A. Einstein (1960: 79, 82, 98) and J. Narlikar (1977: 157).

[20] Cf. A. Einstein (1960: 130–32) and J. Narlikar (1977: 158).

[21] This is not surprising since according to GTR itself, in Einstein's (1960: 127) own words, "half of this deflection is produced by the Newtonian field of attraction of the sun, and the other half by the geometrical modification ("curvature") of space caused by the sun".

[22] See Ronald W. Clark (1973: 224); and A. Einstein (1960: 128).

[23] Cf. *ibid.*, p. 228. "It was not until the afternoon of Thursday, November 6th, 1919", in the words of Ronald W. Clark (1973: 227), "that the Fellows of the Royal and the Royal Astronomical Societies met in Burlington House to hear the official results of the two eclipse expeditions. Dyson read the reports on behalf of himself, Eddington and David-

son". Dyson concluded as follows (cf. *ibid.*, p. 228): "Thus the results of the expeditions to Sobral and Principe leave little doubt that a deflection of light takes place in the neighbourhood of the sun and that it is of the amount demanded by Einstein's generalized theory of relativity as attributable to the sun's gravitational field".

[24] Thus Einstein (1960: 75) wrote: "Although a detailed examination of the question shows that the curvature of light rays required by the general theory of relativity is only exceedingly small for the gravitational fields at our disposal in practice, its estimated magnitude for light rays passing the sun at grazing incidence is nevertheless 1.7 seconds of arc. This ought to manifest itself in the following way. As seen from the earth, certain fixed stars appear to be in the neighbourhood of the sun, and are thus capable of observation during a total eclipse of the sun. At such times, these stars ought to appear to be displaced outwards from the sun by an amount indicated above, as compared with their apparent position in the sky when the sun is situated at another part of the heavens. The examination of the correctness or otherwise of this deduction is a *problem* of the greatest importance, the early solution of which is to be expected of astonomers". (Italic is added.)

[25] *Ibid.*, p. 75.

[26] *Ibid.*, p. 76.

[27] *Ibid.*, p. 76.

[28] Cf. *ibid.*, pp. 125–6. Red being the colour with the longest wavelength in the visible part of the spectrum of light, "redshift" refers to the systematic increase in the wavelength of light coming from distant nebulae. Cf. J. Narlikar (1977: 74–5.).

[29] A. Einstein (1960: 123–4).

[30] See G. L. Pandit (1983: 161–2, 164).

[31] Cf. Gerald Holton (1969: 137).

[32] See, e.g., Abraham Pais (1982: 21, 111–137); Jeremy Bernstein (1973: 48–52); Ronald W. Clark (1973: 83–112); R. S. Shankland (1963: 47–57); and Gerald Holton (1969: 133–197).

[33] This is quite surprising since Einstein's (1960: 1–57) own account of the development of STR is remarkable for the clarity with which it captures the essential aspects of the methodological situation.

[34] It is of considerable methodological interest to note in this context how in his *Two New Sciences* Galileo formulates a theoretical problem concerning light: Is the speed of light instantaneous? Or does its propagation require time? This problem was solved only in the middle of 19th century.

[35] For a discussion of this requirement see G. L. Pandit (1983: 102, 108, 152, 154–6, 158–9, 161).

[36] This is corroborated by Einstein's (1960: 1–57) own account of the theoretical developments culminating in his 1905 paper on STR. Gerald Holton (1969: 195) also puts forth his view that "the role of the Michelson experiment in the genesis of Einstein's theory appears to have been so small and indirect that one may speculate that it would have made no difference to Einstein's work if the experiment had never been made at all".

[37] Maxwell authored a whole article 'Ether' in *Encyclopedia Britannica*, 9th edition, vol. 8, 1878.

[38] In the words of Gerald Holton (1969: 136): "Michelson's ingenious solution was to let two light beams from the same source simultaneously run a round-trip race along two paths which had effectively the same length in the laboratory but were laid out at

90 degrees, thereby causing the two light beams to be differently affected by their relative motion with respect to the ether. But on bringing the two beams together to compare by their interference pattern these relative effects, Michelson's apparatus (an 'interferometer') surprisingly gave what is usually called a negative or null result. More accurately, it gave within experimental error the result that would have been expected on the basis of a quite different hypothesis – namely, that the ether is not stagnant but somehow does get dragged along with the earth and so has no measurable motion or drift with respect to the earth". For further details regarding this experiment see W. K. H. Fanofsky and M. Phillips (1955); Abraham Pais (1982: 111–137); Ronald W. Clark (1973: 83–94).

[39] The last time this experiment was repeated by Michelson was in 1929. One such repetition was, interestingly enough, prompted by Lorentz-Fitzgerald contraction hypothesis, that aimed at explaining the null result so as to rescue the ether hypothesis. In the words of Einstein (1960: 58): "Lorentz and Fitzgerald rescued the theory from this difficulty by assuming that the motion of the body relative to the aether produces a contraction of the body in the direction of motion, the amount of contraction being just sufficient to compensate for the difference in time mentioned above". The question arises here whether the Lorentz-Fitzgerald aether-theoretic contraction hypothesis was itself, as an auxiliary hypothesis, introduced *ad hoc*? As Adolf Grünbaum (1973: 387–388) points out, it cannot be regarded as logically *ad hoc* in the context of the aether theory. "An auxiliary hypothesis", argues Grünbaum (p. 387), "introduced in response to new evidence embarrassing to an established theory, is *ad hoc* in the logical sense within the framework of the theory modified by it, if it does not, in principle, lend itself to any independent test whatever. Since the theoretical terms of the postulates of sophisticated empirical theories are given only a partial or incomplete physical interpretation by the rules of correspondence at any given time, these terms (and the concepts they represent) are "open" in the sense of admitting the adjunction of further rules of correspondence to the given theoretical postulates. Hence the class of all observational consequences of a physical theory is not a clearly circumscribed class, unless we specify and delimit the rules of correspondence which, in conjunction with the theoretical postulates, constitute the given established theory". For Grünbaum's distinction between logical *ad hoc*-ness of a hypothesis, relativized by him to a specific delimited set of rules of correspondence, and psychological *ad hoc*-ness of it see Adolf Grünbaum (1973: 387–397).

[40] In the words of Ronald W. Clark (1973: 90), the results showed that "light which travelled back and forth across the ether stream did the journey in exactly the same time as light travelling the same distance up and down the ether stream".

[41] Attempts to rescue the ether hypothesis by invoking *ad hoc* an ether-drag hypothesis was equally unacceptable if only for its awkward implication suggesting *distortions* of the paths of the beams of light coming from stars and other sources – something not only undetectable but calling for unnecessary complications in the astronomical data. Cf. Arthur Evett (1982: 7–8). See also Ronald W. Clark (1973: 90–91).

[42] Consider in this context the attempts to overcome this situation: (1) G. F. Fitzgerald's suggestion in 1892 that all moving objects shorten along the axis of their movement, thereby enabling him to think that the contraction in the case of the earth would be quite negligible given its slow speed, and considerable in the case of objects moving at a speed approximating to that of light itself. (2) Lorentz similarly suggested, first in 1895 and then again in 1906, an explanation for the null result in terms of his contraction hypothesis, viz.:

The form of a solid body depends on the electromagnetic forces between its molecules as electrically charged bodies such that, when set in motion, these forces themselves would so change as to produce Lorentz-Fitzgerald contraction.

[43] Cf. A. Einstein (1960: 13). See also Ronald W. Clark (1973: 93–94).

[44] Cf. A. Einstein (1960: 49). See also Gerald Holton (1969: 197).

[45] See A. Einstein (1960: 53–54). See also Abraham Pais (1982: 141).

[46] See A. Einstein (1960: 13).

[47] To quote Einstein (1905: 891) in the German original: ". . . sondern daß vielmehr für alle Koordinatensysteme, für welche die mechanischen Gleichungen gelten, auch die gleichen elektrodynamischen und optischen Gesetze gelten . . .".

[48] A. Einstein (1960: 43).

[49] Arthur Evett (1982: 8) states it thus: "*No criterion exists for giving preference to one inertial reference frame over other inertial reference frame*". It is interesting to contrast this with the restricted Galilean principle of relativity, in the words of Evett (1982: 9): "No experiment based solely on Newton's laws of motion (i.e. the branch of physics known as mechanics) would yield a criterion for giving preference to one inertial reference frame over any other inertial reference frame".

[50] What Einstein here calls V is to be understood as the canonical symbol c for the velocity of light. While speaking of the "Principle of Relativity" and the principle of the velocity of light V(= c) in empty space as the two postulates of his STR, Einstein (1905: 891–892) wrote: "Wir wollen diese Vermutung (deren Inhalt im folgenden ,,Prinzip der Relativität" genannt werden wird) zur Voraussetzung erheben und außerdem die mit ihm nur scheinbar unverträgliche Voraussetzung einführen, daß sich das Licht im Leeren Raume stets mit einer bestimmten, vom Bewegungszustande des emittierenden Körpers unabhängigen Geschwindigkeit V fortpflanze. Diese beiden Voraussetzungen genügen, um zu einer einfachen und widerspruchsfreien Elektrodynamik bewegter Körper zu gelangen unter zugrundelegung der Maxwellschen Theorie für ruhende Körper. Die Einführung eines ,,Lichtäthers" wird sich insofern als überflüssig erweisen, als nach der zu entwickelnden Auffasung weder ein mit besonderen Eigenschaften ausgestatteter ,,absolut ruhender Raum" eingeführt, noch einem Punkte des leeren Raumes, in welchen elektromagnetische Prozesse stattfinden, ein Geschwindigkeitsvektor zugeordnet wird".

[51] A. Einstein (1960: 42).

[52] A. Einstein (1905: 891). According to STR the velocity of light c is the limiting velocity such that nothing can be *displaced* in space-time nor any signals or interactions of forces *transmitted* at a velocity $v > c$. Any serious attempt to answer the questions 'Whether faster than light signals/particles exist?' and 'In case they do, whether their velocity could be deccelerated to a velocity $v<c$?' would help us in understanding the STR better, if only because of the possibility of newer complexities and theoretical identifications that this would entail.

[53] Controversies have arisen in the history and philosophy of science as to the exact role, if any, of the Michelson-Morley experiment in the genesis of Einstein's STR. Einstein's own account of the development of physical theory from the time of Newton to his own does not assign any role to it. This is understandable since what fascinated Einstein most was, I believe, the physical theory's employment of the field concept (giving it a unique resolving power) without falling in the trap of mechanical models. Einstein must have, on the contrary, considered the ether drift hypothesis an awkward complication of the

background physical field theory of the electromagnetic phenomena and the wave theory of light.

I don't agree with Lakatos' view (1974b: 318) that "prior to the emergence of relativity theory the Michelson-Morley experiment was in no 'logical' sense a 'negative result' for classical physics". I think that there could have been perhaps no better experimental verdict against the customary resort to mechanical models everywhere. I find it equally difficult to accept the view of Adolf Grünbaum (1973: Chapter 12) that the negative result embodied in that experiment played a crucial logical role in the genesis of the STR.

[54] See G. L. Pandit (1983: 15, 162, 166, 182).

[55] We may say that the $T_{RP}(T_2) > T_{RP}(T_1)$ when T_2 covers more ground in the following sense: Besides explaining the motion of the planets, free fall and tides, Newtonian dynamics could also predict the motion of comets and bullets. Einstein's GTR similarly predicted the bending of light rays and redshift essentially inconceivable under the Newtonian dynamics. The resolving power of a theory progresses in two directions: greater accuracy and wider applicability of a theory.

[56] Cf. G. L. Pandit (1983: 153–155).

[57] In the process of making this distinction I stumbled across Popper's (1972a: 117) distinction between "two kinds of scientific prediction": between the "prediction of *events of a kind which is known*, such as eclipses or thunderstorms on the one hand and, on the other hand, the prediction of *new kinds of events* (which the physicist calls 'new effects') such as the prediction which led to the discovery of wireless waves, or of zero point energy, or to the artificial building up of new elements not previously found in nature". According to Popper such a distinction is not possible under instrumentalism. "It seems to me clear", argues Popper (1972a: 118), "that instrumentalism can account only for the first kind of prediction: if theories are instruments for prediction, then we must assume that their purpose must be determined in advance, as with other instruments. Predictions of the second kind can be fully understood only as discoveries". However, all that this seems to capture is the familiar doctrine of the pragmatical asymmetry of explanation and prediction otherwise considered as structurally identical. In any case, Popper's distinction is less sharp than the one I have drawn above. For further discussion on the nature of novel prediction see § 9.4. below.

[58] See A. Einstein (1948: 320–324).

[59] A. Einstein (1948: 321), as rendered from the German original in Don Howard (1986).

[60] See Don Howard (1985: 179).

[61] See G. L. Pandit (1983: 48–75).

[62] I intend to discuss this aspect of the Einstein separability elsewhere.

[63] See Don Howard (1985: 172–179).

[64] Don Howard (1985: 173).

[65] Owing mainly, I believe, to the traditional dogmatic and one-sided view of the role of initial conditions in physical explanation, it is not surprising at all if one does not come across in the literature explicit accounts of the problem of theoretical investigation of initial conditions in the context of the grand unified theories or even cosmology. The one brief discussion that I have just stumbled across is by Andrei Linde (1985a: 279–291), who considers this problem in the context of the recent cosmological model(s) known as the inflationary universe scenario. Himself an architect of this new model, his lucid account is available in Adrei Linde (1985b: 14–18).

[66] For a discussion on these see Tong Bor Tang and Li Zhi Fang (1984: 7).
[67] For a detailed discussion see Jack O. Burns (1986: 33–35).
[68] See Neil Turok (1986: 111–112).
[69] See Tong Bor Tang and Li Zhi Fang (1984: 2).
[70] For example, this is the view held by Eugene P. Wigner (1980: 14–15).
[71] *Ibid.*, p. 14.
[72] It seemed to me so largely as a result of the introduction of the concept of the resolving power of a scientific theory in my (1976). Thus for an explanation-theorist, for whom the appraisals of a theory T remain restricted to $T_{EP}(T)$, the question of a physical theory's methodological role in cosmology does not arise at all. For, obviously, whatever be the role of physical theory in cosmology, it cannot reside in its explanatory power.
[73] L. Woltjer (1978: 147–148).
[74] For details see A. Einstein (1960: 123–132).
[75] Cf. A. H. Klotz (1982: 12). The fact that the galaxies are electrically uncharged and that "the only force influencing their motion is gravity" – cf. D. F. Lawden (1982: 174) – makes it all the more clear why a physical theory such as the GTR can be expected to play a fundamental role in generating cosmological models capable of explaining and identifying the observable features of the universe.
[76] Resolving power in this sense is excellently illustrated in the Hubble law of expansion (extragalactic recession), viz.: the theoretical identification of redshift of the light from distant galaxies as a Doppler effect. Among other things, this identification utilizes the assumption that atoms behave in the same way everywhere in the universe. Cf. A. H. Klotz (1982: 103).
[77] See A. Einstein (1960: 156–157).
[78] Cf. A. H. Klotz (1982: 10). It could be shown how the GTR and the STR themselves give rise to this problem.
[79] The Hubble law says: The velocities at which galaxies are receding are directly proportional to their distances – i.e. their redshifts are proportional to their distances. As Hubble (1936: 120) puts it: "The velocity-distance relation is not merely a powerful aid to research, it is also a general characteristic of our sample of the universe – one of the very few that are known . . . if it could be fully interpreted, the relation would probably contribute an essential clew to the problem of the structure of the universe".
[80] Cf. *ibid.*, p. 122. Redshifts are generally interpreted as velocities (Doppler effects). Thus interpreted, they indicate "velocities of recession, implying an expanding universe". Cf. R. Berendzen *et al.* (1976: 206). The Doppler principle, which this interpretation uses, says that "for a shift $\Delta\lambda$ of any particular wavelength λ, the velocity v is found from the relation

$$c\,\frac{\Delta\lambda}{\lambda} = v$$

where c = velocity of light". (Cf., *ibid.*, p. 206). What is interesting theoretically and methodologically as well is Hubble's own view that the theoretical identification of redshifts as Doppler effects and hence as velocity-shifts must not be taken as the final word on the subject. "Meanwhile", says Hubble (1936: 123), "redshifts may be expressed on a scale of velocities as a matter of convenience. They behave as velocity-shifts behave and they are very simply represented on the same familiar scale, regardless of the ultimate interpretation. The term 'apparent velocity' may be used in carefully considered statements . . ." The term 'apparent velocity' is intended by him to emphasize "the empirical features of the correlation". Cf. R. Berendzen *et al.* (1976: 207).

The problem of the theoretical identification of redshifts thus remains even for Hubble who does not rule out other interpretations for the following reason: "Given the basic relation between energy and wavelength for radiation proposed by Einstein in 1905,
Energy *x* wavelength = constant,
an increase in the wavelength for distant nebulae (i.e. a redshift) would indicate a loss of energy in the radiation. This loss could occur either in the nebulae itself (the Doppler effect) or in the long path through space that the light must travel." Cf. R. Berendzen *et al.* (1976: 208–9). Hubble envisages observational tests to decide between the rival interpretations. Should the theoretical identification of redshifts as velocity-shifts (i.e. Doppler effect) prove wrong, it will be a step of great methodological significance, showing that velocity-distance relationship cannot represent an expanding universe and thereby leading to a radical reformulation of the cosmological problem.

[81] For example, physicists are at present inclined to suspect any theory that leads to a cosmological model which conflicts with what are taken to be the established 'facts' in cosmology: the overall isotropy of the universe and the Hubble law of expansion.

[82] Cosmology has come a long way from the earlier static models to non-static dynamical models of the universe. Relativistic cosmology began in 1917 with a basis in Einstein's GTR which he formulated in 1915, giving relativistic treatment of gravitation and applying non-Euclidean geometry to space-time continuum. It was around this time that two different static cosmological models were proposed, one by Einstein himself and the other by W. de Sitter (1872–1935). The former described a universe containing matter but static, while the latter described one allowing motion but devoid of matter.

[83] It was only in 1930 that Einstein withdrew the 'cosmological constant' in the wake of Hubble's discovery and increasing evidence in favour of the recession of the distant galaxies and the continual expansion of the universe. The cosmological constant, denoted by λ, was equivalent to a repulsive force which Einstein had introduced, in an *ad hoc* manner as it were, to overcome the difficulties of Newtonian gravitation theory when applied to the entire universe. This theory requires the universe to be unstable, while Einstein took it for granted that universe must be stable and unchangeable in time. Cf. Einstein (1960: 105–7, 133–4).

[84] We can say that the resolving power of NGT could resolve nothing beyond the solar system. The theory is notable for its difficulties when extended as a basis for cosmology. The main difficulties have to do with its basic assumptions which conceive gravity as an action-at-a-distance force and its influence or transmission as instantaneous. That in this sense the theory is based on an over-simplification is what became clear with Einstein's GTR appearing on the scene around 1915.

[85] Cf. Ronald W. Clark (1973: 210–11). Given Newton's gravitation theory and its resolving power, all that one can require or assume as a basis for cosmology is, in the words of Einstein (1960: 106), that the "stellar universe ought to be a finite island in the infinite ocean of space". Today we know better how unsatisfactory this island-universe theory is. One of its inherent difficulties is the implication of a maximal stellar density as the centre of the universe, whose invalidity is demonstrated by GTR and modern relativistic cosmology owing mainly to assumptions of greater resolving power. "It is still

less satisfactory", writes Einstein (1960: 106), "because it leads to the result that the light emitted by the stars and also individual stars of the stellar system are perpetually passing out into infinite space, never to return, and without ever again coming into interaction with other objects of nature. Such a finite material universe would be destined to become gradually but systematically impoverished".

[86] In Einstein's model the curvature of space is related, in the words of J. Narlikar (1977: 122) "to the density of matter filling up the space, and this emphasizes the central theme of the general theory of relativity that matter distribution and space-time geometry are related in an intimate fashion".

[87] For technical details of this discovery see R. W. Wilson (1979: 433–444).

[88] P. A. M. Dirac (1980: 1–11).

[89] *Ibid.*, p. 10.

[90] *Ibid.*, p. 10.

[91] For a discussion of the general methodological situation in sciences see G. L. Pandit (1983: 44–80).

[92] This is generally true of a majority of philosophers of science. See, e.g., C. G. Hempel (1965); and J. G. Kemeny and Paul Oppenheim (1970: 307–18).

[93] P. A. M. Dirac (1980: 6–9).

[94] *Ibid.*, p. 6.

[95] *Ibid.*, p. 7.

[96] *Ibid.*, p. 8.

[97] *Ibid.*, p. 9.

[98] This account, incidentally, does not clash with Einstein's (1960: 75) own account of what GTR predicts; they imply each other. Cf. J. M. Weisberg, J. H. Taylor and L. A. Fowler (1981: 68).

[99] In the words of Iain Nicolson (1981: 115): "What is a black hole? Essentially it is a region of space into which matter has fallen and from which nothing can escape; within a black hole, gravity is so powerful that not even light can move outwards".

[100] See J. M. Weisberg, J. H. Taylor and L. A. Fowler (1981: 66, 68).

[101] Such an application of GTR is envisaged by Einstein himself (1960: 132). See also J. M. Weisberg, J. H. Taylor and L. A. Fowler (1981: 68, 74).

[102] Cf. *ibid.*, p. 74.

[103] Cf. Arthur Evett (1982: 122–124); and John Saunders and John Norton (1982: 101–127).

[104] See Graham Nerlich (1982: 361–388).

[105] In the words of J. Narlikar (1977: 157): "The results so far show a deflection ranging from 1.43″ of arc to 2.7″ compared to the general relativistic value of 1.75″. If we adopt Newtonian gravitation and assume that photons, the light carriers, are attracted to the Sun like ordinary particles, we come up with half the relativistic value".

[106] In fact, the 'Queries' (1–31) in Newton's *Opticks* have received very little attention in the philosophical discussions on science. I myself stumbled across these as most interesting cases of methodological variance only recently.

[107] Its popularity can be gauged from the more recent writings of scientists and of philosophers of science and language. For example, see W. S. Cooper (1978: 51–53).

[108] See K. R. Popper (1968: 315).

[109] K. R. Popper (1972a: 241–244).

[110] *Ibid.*, p. 244. Insistence on *verifications* of novel predictions of a scientific theory is a common practice among scientists and philosophers alike. For example, see A. H. Klotz (1982: 1, 4–5, 12, 103, 139).
[111] See K. R. Popper (1968: 32–33).
[112] I. Lakatos (1971a: 91–136).
[113] *Ibid.*, p. 100.
[114] I. Lakatos (1973: 7).
[115] I. Lakatos (1978: 177).

CHAPTER 10

[1] As to the (biologically neutral) concept of *NFCPS* systems (Pandit, 1983: 62–79), I take the theory-problem interactive systems in the special sciences as their best examples. Lest they be mistaken for systems of an essentially incomplete nature because they are abstract, they are not only interactive systems in their own right, quite separate from those who discover and build them up, but interacting systems in constant interaction with relevant scientific communities. Thus they have a certain self-sufficiency (= ontological independence) which does not derive from those other types of systems (such as their relevant scientific communities and their language-frameworks) in which they may be embedded for their historical growth. Their chief difference from these systems, whether structural-hierarchical or control-hierarchical systems of a cognitive or natural order, lies in their capacity for *recurrent transcendence* from entanglement in their own rules by a process of methodological variance already discussed above. Understanding an earlier theory-problem interactive system in the light of a later one in their field is an important aspect of the same process.
[2] For example, K. R. Popper (1968, 1972a, 1972b, 1976, 1983), Bas C. van Fraassen (1980), and others for whom the rational acceptance of a theory from among a number of alternative theories involves the belief that it is the best possible step we can take towards that which science aims at. For a brief discussion see G. L. Pandit (1988a). A more detailed discussion is to be found in G. L. Pandit (1988b).
[3] For a discussion see G. L. Pandit (1988a, 1989).
[4] P. K. Feyerabend (1987).
[5] See G. L. Pandit (1983: 111).
[6] See G. L. Pandit (1989).
[7] The same idea has found its way into Feyerabend's (1987: 36, 281, 316) doctrine of the so-called self-sufficiency of the individual sciences as if they were just languages on the model of Wittgenstein's model language-games in his *Philosophical Investigations*.
[8] I discuss this point in G. L. Pandit (1988a, 1988b).
[9] In the order of new priorities set here, it is the question of the method and structure which dominates the question of aim of science.
[10] Among these, I find Ian Hacking's (1983) and C. A. Hooker's (1987) realistic approaches in certain important respects similar to the general interaction-theoretic approach I have adopted here and in my (1983) on important problems of rationality, method and scientific change. For limitations of space, their discussion must await another occasion. For a critical-historical review of the present scenario see John Passmore (1985).

[11] K. R. Popper (1968: 50).
[12] K. R. Popper (1968: 50).
[13] For the distinction that must be drawn between method and strategy, see G. L. Pandit (1983: 144–183).
[14] For a discussion see G. L. Pandit (1987).
[15] G. L. Pandit (1983: 41); see also p. 43.
[16] G. L. Pandit (1983: 144–183).
[17] G. L. Pandit (1983: 111).
[18] K. R. Popper (1968: 49).
[19] J. Watkins (1984: 124).
[20] For discussion, see G. L. Pandit (1986: 141–148).
[21] J. Watkins (1984: 123–125, 164, 279–280, 304–306, 335–336).
[22] J. Watkins (1984: 164).
[23] J. Watkins (1984: 279).
[24] J. Watkins (1984: 124).
[25] For a discussion, see G. L. Pandit (1986).
[26] J. Watkins (1984: 166–221).
[27] J. Watkins (1984: 159); see also pp. 164, 304.
[28] For a brief discussion, see G. L. Pandit (1988a: 125–138).
[29] See also G. L. Pandit (1988a).
[30] A detailed discussion is to be found in G. L. Pandit (1988b).
[31] K. R. Popper (1972a: 223–231, 402).
[32] To my knowledge, no philosopher of truth has come close enough to recognizing its complexities from which arises the possibility of a successor to the traditional conception of it. We have not looked deeper inside the structural-organizational complexities of our knowledge sufficiently enough to recognize how it moves back and forth between ontologically primordial $truth_1$ and $truth_2$ as a K-state of disclosedness. A fundamental distinction in this sense is being hinted at by Martin Heidegger (1927: 227) when he answers his own question: *Why must we presuppose truth*? – (Warum müssen wir voraussetzen, daß es Wahrheit gibt?). With Heidegger (1927; 226) we may, then, say that we must presuppose $truth_1$ if we are at all to understand $truth_2$ not as agreement/correspondence with objects/facts but as K-states of disclosedness (Erschlossenheit) or uncoveredness (Entdecktheit). What this suggests is not just a dualism of some sort or other but a dynamic interplay between them. They are the structural-dynamical states of knowledge always intertwined with each other. To quote Heidegger (1927: 226): " . . . Nicht die Aussage ist der primäre 'Ort' der Wahrheit, sondern *umgekehrt*, die Aussage als Aneignungsmodus der Entdecktheit und als Weise des In-der-Welt-seins gründet im Entdecken, bzw. der *Erschlossenheit* des Daseins. Die ursprünglichste 'Wahrheit' ist der 'Ort' der Aussage und die ontologische Bedingung der Möglichkeit dafür, daß Aussagen wahr order falsch (entdeckend oder verdeckend) sein können".
[33] See D. Miller (1974a, 1974b) and Pavel Tichý (1974).
[34] See K. R. Popper (1972a: 235).
[35] G. Schurz and P. Weingartner (1987: 47–77).
[36] For details see I. Niiniluoto (1984: 75–110, 159–192: 1987). See also David Pearce (1987a)/(1987b).
[37] My presentation here is mainly based on I. Niiniluoto (1984) and a talk by him in the

Freie Universität Berlin in the Fall of 1985. A more detailed discussion is to be found in G. L. Pandit (1988b).
[38] K. R. Popper (1972a: 233).
[39] I think that Niiniluoto (1984: 95) himself admits this.
[40] For example, by R. C. Jennings (1983, 1987) whose claims I discuss briefly in G. L. Pandit (1988a).
[41] See G. L. Pandit (1989).
[42] K. R. Popper (1972a: 226).
[43] For a discussion, see G. L. Pandit (1988a, 1988b).
[44] O. Neurath (1983: 92).
[45] See O. Neurath (1983: 61).
[46] O. Neurath (1983: 61).
[47] For my discussion see G. L. Pandit (1988a, 1988b).
[48] H. Putnam (1981a: 118).
[49] H. Putnam (1981a: 118; 1983: 167–169).
[50] N. Goodman (1978: 4, 17, 119).
[51] N. Goodman (1978: 4).
[52] G. L. Pandit (1988a, 1988b).
[53] See G. L. Pandit (1983: 75); see also § 8.2. above.
[54] Larry Laudan (1977: 66).
[55] Larry Laudan (1977: 68).
[56] Larry Laudan (1977; 1981: 144–155).
[57] Larry Laudan (1977: 14).
[58] Larry Laudan (1977: 68).
[59] Larry Laudan (1977: 22).
[60] Larry Laudan (1977: 16–17).
[61] Larry Laudan (1977: 60).
[62] Larry Laudan (1977: 17).
[63] Larry Laudan (1977: 66).
[64] Larry Laudan (1987).
[65] Larry Laudan (1977: 155–170).
[66] Larry Laudan (1977: 169).
[67] Larry Laudan (1977: 168).
[68] Larry Laudan (1977: 168).
[69] Larry Laudan (1987: 27–29).
[70] Larry Laudan (1987: 27–28).
[71] G. L. Pandit (1983: 111).
[72] See reference 13. Limitations of space do not permit me to discuss here this important point.
[73] See W. Stegmüller (1976, 1979), J. D. Sneed (1979) and P. Suppes (1969, 1970).
[74] See W. Stegmüller (1976: 36).
[75] For a detailed critical discussion of structuralism see David Pearce (1987b).
[76] See W. Stegmüller (1979: 39). For the ambiguity of the statement view see W. Stegmüller (1979).
[77] See reference 43 above.
[78] W. Stegmüller (1979: 69).

[79] " Linguistically this means", clarifies Stegmüller (1976: 128), " that *to one particular description of a physical system* in terms of the vocabulary of the reduced theory there correspond, as a rule, *various descriptions of the same physical system* in the language of the reducing theory". And philosophically, as Stegmüller (1979: 36) puts it, "What is, from the standpoint of T, one and the same state of affairs, may split up into many different states of affairs, if viewed from T′". See W. Stegmüller (1979: 33–40).
[80] See W. Stegmüller (1976: 270).
[81] See W. Stegmüller (1976: 133; 1979: 12).
[82] See Stegmüller's (1979: 1–2, 12, 25–28, 86–88) own account to this effect.
[83] P. K. Feyerabend (1977: 351–369).
[84] W. Stegmüller (1979: 87).
[85] W. Stegmüller (1979: 87–88). See also W. Stegmüller (1976: 132–133).
[86] W. Stegmüller (1979: 79).
[87] W. Stegmüller (1979: 12).
[88] W. Stegmüller (1979: 4).
[89] Cf. W. Stegmüller (1979: 4); see also pp. 1–7.
[90] W. Stegmüller (1976: 132–133; 1979: 26–27).
[91] W. Stegmüller (1979: 26–28).
[92] See W. Stegmüller (1979: 29–32).
[93] W. Stegmüller (1979: 26).
[94] W. Stegmüller (1976, 1979).
[95] W. Stegmüller (1979: 28).
[96] W. Stegmüller (1979: 35).
[97] See W. Stegmüller (1979: 35).
[98] See W. Stegmüller (1979: 35).
[99] W. Stegmüller (1979: 37).
[100] See W. Stegmüller (1979: 35, 39, 71, 77, 82).
[101] W. Stegmüller (1979: 37).
[102] See W. Stegmüller (1979) and W. Balzer et al. (1984).
[103] See W. Stegmüller (1979). A recent discussion of theory-reduction in this sense is to be found in David Pearce and V. Rantala (1985: 126–138).
[104] For limitations of space, I shall here refrain from any discussion of reduction. Interesting critical discussions of the relevant intertheoretic relations in this context are to be found in Erhard Scheibe (1976) and Lorenz Kruger (1980a, 1980b).
[105] See S. Toulmin (1972: 478–508) and C. Dilworth (1986).
[106] See G. L. Pandit (1989).
[107] W. Stegmüller (1979: 87).
[108] W. Stegmüller (1979: 88).
[109] W. Stegmüller (1979: 8–24).

BIBLIOGRAPHY

Agassi, J. and Cohen, R. S. (eds.): 1982, *Scientific Philosophy Today*, D. Reidel, Dordrecht, Holland.

Allen, G. E.: 1978, *Life Sciences in the Twentieth Century*, Cambridge University Press, Cambridge.

Ammerman, R. R. (ed.): 1965, *Classics of Analytic Philosophy*, Tata McGraw-Hill, New York.

Baigrie, B. and Hattiangadi, J. N. : 1981, 'Laudan's Problems', *Metaphilosophy* **12** 85–95.

Balzer, W. *et al.* (eds.): 1984, *Reduction in Science*, D. Reidel, Dordrecht, Holland.

Bell, J. S.: 1964, 'On the Einstein-Podolsky-Rosen Paradox', *Physics* **1** 195–200.

Bell, J. S.: 1966, 'On the Problem of Hidden Variables in Quantum Mechanics', *Reviews of Modern Physics* **38** 447–452.

Ben, S. I. and Mortimore, G. W.(eds.): 1976, *Rationality and the Social Sciences*, Routledge and Kegan Paul, London.

Ben, S. I. *et al.*: 1976, 'Technical Models of Rational Choice', in Ben and Mortimore (1976), pp. 157–195.

Bergmann, P. G.: 1970, 'Ernst Mach and Contemporary Physics', in Cohen and Seeger (1970), pp. 69–78.

Bernstein, J.: 1973, *Einstein*, Fontana/Collins, London.

Berendzen, R., Hart, R. and Seeley, D.: 1976, *Man Discovers Galaxies*, Science History Publications, New York.

Bohm, D.: 1957, *Causality and Chance in Modern Physics*, Routledge and Kegan Paul, London.

Borger, R. and Cioffi, F. (eds.): 1970, *Explanation in the Behavioural Sciences*, Cambridge University Press, Cambridge.

Brody, B. A. (ed.): 1970, *Readings in the Philosophy of Science*, Prentice-Hall, INC., Englewood Cliffs, New Jersey.

Buck, R. C. and Cohen, R. S. (eds.): 1971, PSA 1970 *In Memory of Rudolf Carnap*, D. Reidel, Dordrecht, Holland.

Bulmer, M. G.: 1980, *The Mathematical Theory of Quantitative Genetics*, Clarendon Press, Oxford.

Burns, J. O.: 1986, 'Very Large Structures in the Universe', *Scientific American* **255** (1), 30–39.

Carlson, L.: 1983, *Dialogical Games: An Approach to Discourse Analysis*, D. Reidel, Dordrecht, Holland.

Carnap, R.: 1937, *Logical Syntax of Language*, Routledge and Kegan Paul, London.

Carnap, R.: 1950, *The Logical Foundations of Probability*, University of Chicago Press, Chicago.

Carnap, R.: 1956, *Meaning and Necessity*, University of Chicago Press, Chicago.

Carnap, R.: 1963, 'Intellectual Autobiography', in Schilpp, P. A. (1963), pp. 1–43.

Carnap, R.:1965, 'Testability and Meaning', in Ammerman, R. R. (1965), pp. 130–195. Reprinted from *Philosophy of Science* **3** (1936) and **4** (1937).

Carnap, R.: 1974, *An Introduction to the Philosophy of Science*, Basic Books Inc., New York (The Ist ed. of 1966 appeared as *Philosophical Foundations of Physics*).

Chandrasekhar, S.: 1974, 'Development of General Relativity', *Nature* **252** 15–17.

Chomsky, N.: 1967, 'The Formal Nature of Language', in Lenneberg (1967), pp. 397–442.

Chomsky, N.: 1972, *Language and Mind*, Harcourt Brace Jovanovich, Inc. (enlarged edition).

Chomsky, N.: 1976, *Reflections on Language*, Fontana/Collins, London.

Chomsky, N.: 1977, *Essays on Form and Interpretation*, North-Holland Publ. Co., Amsterdam.

Chomsky, N. and Walker, E.: 1978, 'Introduction: The Linguistic and the Psycholinguistic Background', in Walker (1978), pp. 15–26.

Clark, R. W.: 1973, *Einstein: The Life and Times*, Hodder and Stroughton, London.

Clauser, J. F. and Shimony, A.: 1978, 'Bell's Theorem: Experimental Tests and Implications', *Reports on Progress in Physics* **41** 1881–1927.

Cohen, L. J.: 1980, 'Some Comments on Third World Epistemology', *British Journal for the Philosophy of Science* **31** 175–180.

Cohen, R. S. and Seeger, R. J. (eds.): 1970, *Ernst Mach: Physicist-Philosopher*, D. Reidel, Dordrecht, Holland.

Cohen, R. S. *et al.* (eds.): 1976, *Essays in Memory of Imre Lakatos*, D. Reidel, Dordrecht, Holland.

Cohen R. S. and Wartofsky, M. W. (eds.): 1983, *Language, Logic and Method*, D. Reidel, Dordrecht, Holland.

Cooper, W. S.: 1978, *Foundations of Logico-Linguistics*, D. Reidel, Dordrecht, Holland.

Davidson, D. *et al.* (eds.): 1975, *Words and Objections*, D. Reidel, Dordrecht, Holland (revised ed.).

d'Espagnat, B. (ed.): 1971, *Foundations of Quantum Mechanics*, Academic Press, New York/London.

Dilworth, C.: 1986, *Scientific Progress*, D. Reidel, Dordrecht, Holland (2nd ed.).

Dirac, P. A. M.: 1980, 'Why We Believe in the Einstein Theory', in Gruber and Millman (1980), pp. 1–11.

Duhem, P.: 1954, *The Aim and Structure of Physical Theory*, Princeton University Press, Princeton. (translated by P. P. Wiener from the French 2nd ed. of 1914.)

Einstein, A.: 1905, 'Zur Elektrodynamik bewegter Körper', *Annalen der Physik*, Band **17** 891–921.

Einstein, A.: 1934, *Mein Weltbild*, Querida Verlag, Amsterdam. (paperback ed.).

Einstein, A.: 1948, 'Quanten-Mechanik und Wirklichkeit', *Dialectica* **2** 320–324.

Einstein, A.: 1949, 'Autobiographical Notes', in Schilpp, P. A. (ed.): *Albert Einstein: Philosopher-Scientist*, Open Court, Illinois, Lasalle (3rd ed. 1970), pp. 1–95; and 'Reply to Criticisms', *ibid*, pp. 665–688.

Einstein, A: 1960, *Relativity: The Special and the General Theory*, Methuen and Co. Ltd., London (University paperback ed.). Translated by Robert W. Lawson and first published by Methuen and Co. Ltd. in 1920.

Einstein, A., Podolsky, B., and Rosen, R.: 1935, 'Can Quantum Mechanical Description of Physical Reality Be Considered Complete?', *The Physical Review* **47** 777–780.

Eisen, C. and Eisen, M.: 1979, *Finite Mathematics*, Glencoe Publishing Co. Inc. Encino, California/Collier Macmillan Publishers, London.

Evett, A: 1982, *Understanding the Space-Time Concepts of Special Relativity*, Publishing Creative Services Inc.

Fanofsky, W. K. H. and Phillips, M.: 1955, *Classical Electricity and Magnetism*, Addison-Wesley, Reading, Massachussets.

Farrel, B. A.: 1970, 'A Psychological Look at Some Problems of Perception', in *Royal Institute of Philosophy Lectures*, Vol 3, 1968/1969, Macmillan and Co. Ltd., New York, pp. 51–72.

Fetzer, J. H.: 1981, *Scientific Knowledge*, D. Reidel, Dordrecht, Holland.

Fetzer, J. H.: 1983a, 'Critical Study: Wolgast's Paradoxes of Knowledge', in *Philosophia* **12** 404–421.

Fetzer, J. H.: 1983b, 'Transcendent Laws and Empirical Procedures', in Rescher, N. (ed.): 1983, *The Limits of Lawfulness*, University Press of America, Lanham, Maryland, pp. 25–32.

Fetzer, J. H.: 1983c, 'Probability and Objectivity in Deterministic and Indeterministic Situations', *Synthese* **57** 367–386.

Feyerabend, P. K.: 1965, 'Problems of Empiricism', in Colodny (1965), pp. 145–260.

Feyerabend, P. K.: 1977, 'Changing patterns of reconstruction', *The British Journal for the Philosophy of Science* **28** 351–369.

Feyerabend, P. K.: 1978a, *Against Method*, Verso edition, London/ New York.

Feyerabend, P. K.: 1978b, *Science in A Free Society*, London.

Feyerabend, P. K.: 1980, 'Zahar on Mach, Einstein and Modern Science', *British Journal for the Philosophy of Science* **31** 273–282.

Feyerabend, P. K.: 1981a, *Realism, Rationalism and Scientific Method* (Philosophical Papers Vol. 1), Cambridge University Press, Cambridge.

Feyerabend, P. K.: 1981b, *Problems of Empiricism* (Philosophical Papers Vol. 2), Cambridge University Press, Cambridge.

Feyerabend, P. K.: 1981c, 'How to Defend Society Against Science', in Hacking (1981), pp. 156–167.

Feyerabend, P. K.: 1987, *Farewell to Reason*, Verso ed., London/New York.

Godelier, M.: 1972, *Rationality and Irrationality in Economics*, Monthly Review Press, New York/London, (translated by Brain Pearce).

Good, I. J. and McMichael, A. F.: 1984, 'A Pragmatic Modification of Explicativity For the Acceptance of Hypotheses', *Philosophy of Science* **51** 120–127.

Goodman, N.: 1966, *The Structure of Appearance*, Bobbs-Merril, Indianapolis (2nd ed.).

Goodman, N.: 1978, *Ways of Worldmaking*, Hackett Press, Indianapolis.

Gottlieb, F. J.: 1966, *Developmental Genetics*, Chapman and Hall, London.

Gruber, B. and Millman, R. S. (eds.): 1980, *Symmetries in Science*, Plenum Press, New York/London, (Proceedings of the Einstein Centennial Celebration Science Symposium on Symmetries in Science, held at Southern Illinois University, Carbondale, Illinois: February 23–March 2, 1979).

Grünbaum, A.: 1973, *Philosophical Problems of Space and Time*, D. Reidel, Dordrecht, Holland (2nd enlarged edition).

Gustafson, D. F. *et al.* (eds.): 1979, *Body, Mind, and Method: Essays in Honour of Virgil C. Aldrich*, D. Reidel, Dordrecht, Holland.

Guttenplan, S. (ed.): 1975, *Mind and Language*, Clarendon Press, Oxford.
Hacking, I. (ed.): 1981, *Scientific Revolutions*, Oxford University Press, Oxford.
Hacking, I.: 1983, *Representing and Intervening*, Cambridge University Press, Cambridge.
Harman, G.: 1975, 'An Introduction to 'Translation and Meaning' Chapter Two of Word and Object', in Davidson *et al.* (1975), pp. 14–26.
Harre, R. (ed.): 1975, *Problems of Scientific Revolution*, Clarendon Press, Oxford.
Heath, A. F. (ed.): 1981, *Scientific Explanation*, Clarendon Press, Oxford.
Heidegger, M.: 1927, *Sein und Zeit*, Max Niemeyer Verlag, Tübingen, Sechzehnte Auflage 1986.
Hempel, C. G.: 1965, *Aspects of Scientific Explanation*, The Free Press, New York.
Hempel, C. G.: 1966, *Philosophy of Natural Science*, Prentice-Hall Inc.: Englewood Cliffs, New York.
Hempel, C. G.: 1980, 'Comments on Goodman's Ways of Worldmaking', *Synthese* **45** 193–199.
Hintikka, J.: 1965, 'Cogito, Ergo Sum: Inference or Perfomance?', in Sesonske, A. and Fleming, N. (1965), pp. 50–76. Reprinted from *The Philosophical Review* LXXI (3–32), 1962.
Hintikka, J.: 1979, 'Language-Games', in Sarrinen (1979), pp. 1–26.
Hintikka, J.: 1982, 'A Dialogical Model of Teaching', *Synthese* **51** 39–59.
Holton, G.: 1969, 'Einstein, Michelson, and the "Crucial" Experiment', *Isis* **60** 133–197.
Hooker, C. A.: 1987, *A Realistic Theory of Science*, State University of New York Press, Albany/New York.
Howard, D.: 1985, 'Einstein on Locality and Separability', *Studies in History and Philosophy of Science* **16** 171–201.
Howard, D.: 1986, 'Locality, Separability, and the Physical Implications of the Bell Experiments: A New Interpretation', pp. 1–39 (unpublished version).
Hubble, E.: 1936, *The Realm of the Nebulae*, Yale University Press, New Haven.
Hume, D.: 1739, *A Treatise of Human Nature*, Book I, D. G. C. Macnabb (ed.) 1962, Fontana Library.
Jaki, S. L.: 1978, 'Johann Georg von Soldner and the Gravitational Bending of Light, With an English Translation of His Essay on It Published in 1801', *Foundations of Physics* **8** 927–950.
Jammer, M.: 1974, *The Philosophy of Quantum Mechanics*, Wiley, New York.
Jeffreys, H.: 1973, *Scientific Inference*, Cambridge University Press, Cambridge (3rd ed.).
Jennings, R. C.: 1983, 'Popper, Tarski and Relativism' *Analysis* **43** 118–123.
Jennings, R. C.: 1987, 'Tarski – A Dilemma', *Inquiry* **30** 155–172.
Jones, A.J.: 1980, *Game Theory: Mathematical Models of Conflict*, Ellis Horwood Ltd.
Kant, I.: 1781, *Kritik der reinen Vernunft*, Riga (2. verbesserte Auflage 1787).
Kemeny, J. G. and Oppenheim P.: 1970, 'On Reduction', in Brody (1970), pp. 307–318.
Kirchhoff, G.: 1897, *Vorlesungen über Mechanik*, Leipzig (4. Auflage).
Klotz, A. H.: 1982, *Macrophysics and Geometry*, Cambridge University Press, Cambridge.
Knorr, K. D., Strasser, H. and Zilian, H. G. (eds.): 1975, *Determinants and Controls of Scientific Development*, D. Reidel, Dordrecht, Holland.
Kuhn, T. S.: 1970a, *The Structure of Scientific Revolutions*, University of Chicago Press, Chicago (2nd ed.).

Kuhn, T. S.: 1970b, 'Logic of Discovery or Psychology of Research', in Lakatos and Musgrave (1970), pp. 1–23.

Kuhn, T. S.: 1971, 'Notes on Lakatos', in Buck and Cohen (1971), pp. 137–146.

Kuhn, T. S.: 1974, 'Second Thoughts on Paradigms', in Suppe, F. (ed.): *The Structure of Scientific Theories*, University of Illinois Press, Urbana, pp. 459–482.

Kuhn, T. S.: 1983, 'Commensurability, Comparability, Communicability', *PSA 1982*, Vol. 2, pp. 669–688.

Krüger, L.: 1980a, 'Intertheoretic Relations as a Tool For the Rational Reconstructions of Scientific Development', *Studies in History and Philosophy of Science* **11** 89–101.

Krüger, L.: 1980b, 'Reduction as a Problem', in Hintikka, J., et al. (eds.): 1980, *Pisa Conference Proceedings* II, D. Reidel, Dordrecht, Holland, pp. 147–174.

Lakatos, I.: 1970, 'Falsification and the Methodology of Scientific Research Programmes', in Lakatos and Musgrave (1970), pp. 91–195.

Lakatos, I.: 1971a, 'History of Science and Its Rational Reconstructions', in Buck and Cohen (1971), pp. 91–136.

Lakatos, I.: 1971b, 'Replies to Critics', in Buck and Cohen (1971), pp. 174–182.

Lakatos, I.: 1973, 'Science and Pseudoscience', *BBC Talk*, Open University, pp. 1–9.

Lakatos, I.: 1974a, 'Popper on Demarcation and Induction', in Schilpp (1974), pp. 241–273.

Lakatos, I.: 1974b, 'The Role of Crucial Experiments in Science', *Studies in History and Philosophy of Science* 309–325.

Lakatos, I.: 1978, *Mathematics, Science and Epistemology* (Philosophical Papers Vol. 2, edited by J. Worral and Gregory Currie), Cambridge University Press, Cambridge.

Lakatos, I. and Musgrave, A. (eds.): 1970, *Criticism and the Growth of Knowledge*, Cambridge University Press, Cambridge.

Laudan, L.: 1977, *Progress and Its Problems*, Routledge and Kegan Paul, London.

Laudan, L.: 1981, 'A Problem-Solving Approach to Scientific Progress', in Hacking (1981), pp. 144–155.

Laudan, L.: 1987, 'Progress Or Rationality? The Prospects For Normative Naturalism', *American Philosophical Quarterly* **24** 19–31.

Lawden, D. F.: 1982, *An Introduction to Tensor Calculus, Relativity and Cosmology*, John Wiley and Sons Ltd. (3rd ed.).

Lebovitz, N. R. *et al.* (eds.): 1978, *Theoretical Principles in Astrophysics and Relativity*, University of Chicago Press, Chicago.

Lenard, P.: 1921, 'Über die Ablenkung eines Lichtstrahls von seiner geradlinigen Bewegung durch die Attraktion eines Weltkörpers, an welchem er nahe vorbeigeht; von J. Soldner, 1801. Mit einer Vorbemerkung von P. Lenard', *Annalen der Physik* **65** 593–604.

Lenneberg, E. H.: 1967, *Biological Foundations of Language*, John Wiley and Sons Inc, New York.

Levi, I.: 1983, 'Truth, Fallibility and the Growth of Knowledge', in Cohen et al. (1983), pp. 153–174.

Linde, A.: 1985a, 'Particle Physics and Cosmology', *Progress of Theoretical Physics* Supplement No. 85, 279–291.

Linde, A.: 1985b, 'The Universe: Inflation Out of Chaos', *New Scientist* March 7, 1985 14–18.

Locke, J.: 1690, *An Essay Concerning Human Understanding* J. M. Dent and Sons Ltd. (Vol. I, edited by J. W. Yolton 1961).

Locke, M. (ed.): 1966, *Major Problems in Developmental Biology*, New York: Academic Press.

Mach, E.: 1976, *Knowledge and Error: Sketches on Psychology of Enquiry*, Vienna Circle Collection 3: translated from the original German edition 1905 by T. J. McCormack and Paul Foulkes, D. Reidel, Dordrecht, Holland.

Margenau, H.: 1982, 'Physics and the Doctrine of Reduction', in Agassi and Cohen (1982), pp. 187–199.

Marx, K.: 1859, *Zur kritik der Politischen Ökonomie*. (Karl Marx Friedrich Engels Werke 13), Dietz Verlag, Berlin 1964.

McElroy, W. D. and Swanson, C. P.: *Modern Cell Biology*, Prentice-Hall Inc: Englewood Cliffs, New York.

McLaughlin, R. (ed.): 1982, *What? Where? When? Why?*, D. Reidel, Dordrecht, Holland.

Michel, L.: 1981, 'The Description of the Symmetry of Physical States, and Spontaneous Symmetry Breaking', in Boccara, N. (ed.) (1981), *Symmetries and Broken Symmetries in Condensed Matter Physics*, I. D. S. E. T. Paris, pp. 21–28.

Miller, D.: 1974a, 'Popper's Qualitative Theory of Verisimilitude' *The British Journal for the Philosophy of Science* **25** 166–177.

Miller, D.: 1974b, 'On the Comparison of False Theories by Their Bases', *The British Journal for the Philosophy of Science* **25** 178–188.

Monod, J.: 1977, *Chance and Necessity*, Collins/Fount Paperbacks, London.

Mountcastle, V. B.: 1975, 'The View From Within: Pathways to the Study of Perception', *John Hopkins Medical Journal* **136** 109–131.

Mundy, B.: 1986, 'The Physical Content of Minkowski Geometry', *The British Journal for Philosophy of Science* **37** 25–54.

Narlikar, J.: 1977, *The Structure of the Universe*, Oxford Universe Press, Oxford.

Nelkowski, H. *et al.* (eds.): 1979, *Einstein Symposium Berlin*, Springer Verlag, Berlin.

Nerlich, G.: 1982, 'Special Relativity is not Based on Causality', *The British Journal for the Philosophy of Science* **33** 361–388.

Neurath, O.: 1983, *Philosophical Papers 1913–1946*, Vienna Circle Collection 16: edited and translated by Robert S. Cohen and Marie Neurath, D. Reidel, Dordrecht, Holland.

Newton, I.: 1704, *Opticks*, Dover edition 1952, an unabridged and an unaltered republication of the 1931 edition of G. Bell and Sons Ltd., New York.

Nicolson, I.: 1981, *Gravity, Black Holes and the Universe*, David and Charles Newton Abbot, London.

Niiniluoto, I.: 1984, *Is Science Progressive*?, D. Reidel, Dordrecht, Holland.

Niiniluoto, I.: 1987, *Truthlikeness*, D. Reidel, Dordrecht, Holland.

Oddie, G. 1986, *Likeness to Truth*, D. Reidel, Dordrecht, Holland.

Pandit, G. L.: 1971, 'Two Concepts of Psychologism', *Philosophical Studies* **22** 85–91.

Pandit, G. L.: 1972, 'From Traditional Monism to Traditional Pluralism: A Characterization', *Anviksiki* **5** 50–59.

Pandit, G. L.: 1976, 'Epistemology and an Interactive Model of the Growth of Knowledge', *Indian Philosophical Quarterly* **3** 409–436.

Pandit, G. L.: 1981a, 'Towards a Limit Concept of the Unit of Empirical Content', *Visva - Bharati Journal of Philosophy* **12** 14–42.

Pandit, G. L.: 1981b, 'The Cartesian Questions of Methodology and Against Psychologism: A Rejoinder', *Indian Philosophical Quarterly* **8** 299–304.

Pandit G. L.: 1983, *The Structure and Growth of Scientific Knowledge: A Study in the Methodology of Epistemic Appraisal*, D. Reidel Dordrecht, Holland.

Pandit, G. L.: 1986, 'Rationality of an Optimum Aim For Science', *The Journal of Indian Council for Philosophical Research* **3** 141–148.

Pandit, G.L.: 1987, 'Epistemological Ontology and the Special Sciences: An Interaction-theoretic Argument Against Relativism', *The Journal of Indian Council of Philosophical Research* **4** 35–45.

Pandit, G. L.: 1988a, 'Science and Truthlikeness', *The Journal of Indian Council of Philosophical Research* **5** 125–138.

Pandit, G. L.: 1988b, 'Problems of Scientific Realism', (unpublished).

Pandit, G. L.: 1989, 'Scientific Change: The Possibility of a Unified Approach', in *Grenzfragen Zwischen Philosophie und Naturwissenschaft*, Herausgeber: Paul Weingartner und Gerhard Schurz, Hölder-Pichler-Tempsky, pp. 168–179.

Pais, A.: 1982, *'Subtle is the Lord . . .': The Science and the Life of Albert Einstein*, Oxford University Press, Oxford.

Passmore, J.: 1985, *Recent Philosophers* (a supplement to *A Hundred Years of Philosophy*) Duckworth, London.

Pearce, D.: 1982, 'Stegmüller on Kuhn and Incommensurability', *The British Journal for the Philosophy of Science* **33** 389–396.

Pearce, D. and Rantala, V.: 1985, 'Approximative Explanation is Deductive-Nomological', *Philosophy of Science* **52** 126–140.

Pearce, D.: 1987a, 'Critical Realism in Progress: Reflections on Ilkka Niiniluoto's philosophy of Science', *Erkenntnis* **22** 147–171.

Pearce, D.: 1987b, *Roads to Commensurability*, D. Reidel, Dordrecht, Holland.

Pears, D.: 1972, *Russell's Logical Atomism*, Fontana/Collins, London.

Planck, M.: 1949, *Vorträge und Erinnrungen*, Darmstadt.

Popper, K. R.: 1934, *Logik der Forschung*, Wien.

Popper, K. R.: 1961, *The Poverty of Historicism*, Routledge and Kegan Paul, London.

Popper, K. R.: 1968, *Logic of Scientific Discovery*, Hutchinson, London, revised paperback ed.

Popper, K. R.: 1972a, *Conjectures and Refutations*, London: Routledge and Kegan Paul, 4th edition.

Popper, K. R.: 1972b, *Objective Knowledge: An Evolutionary Approach*, Clarendon Press, Oxford.

Popper, K. R.: 1975, 'Rationality of Scientific Revolutions', in Harré (1975), pp. 72–101.

Popper, K. R.: 1976, *Unended Quest: An Intellectual Autobiography*, Fontana, London, (revised ed.), first published in Schilpp, (1974).

Popper, K. R. and Eccles, J. C.: 1977, *The Self and Its Brain*, Springer International.

Popper, K. R.: 1983, *Realism and the Aim of Science*, Rowman and Littlefield, New Jersey.

Putnam, H.: 1981a, 'Philosophers and Human Understanding', in Heath (1981), pp. 99–120.

Putmam, H.: 1981b, 'The Corroboration of Theories', in Hacking (1981), pp. 60–79.

Putmam, H.: 1983, *Realism and Reason* (*Philosophical Papers* Vol. 3), Cambridge University Press, Cambridge.

Quine, W. V. O.: 1960, *Word and Object*, MIT Press, Cambridge, Mass.

Quine, W. V. O.: 1963a, *From a Logical Point of View*, Harper and Row (2nd ed. paperback).
Quine, W. V. O.: 1963b, 'Two Dogmas of Empiricism', in Quine (1963a), pp. 20–46.
Quine, W. V. O.: 1969, 'Epistemology Naturalized', in Quine (1969), pp. 69–90.
Quine, W. V. O.: 1969, *Ontological Relativity and Other Essays*, Columbia University Press, New York.
Quine, W. V. O.: 1970, 'On the Reasons for Indeterminancy of Translation', *The Journal of Philosophy* **67** 178–183.
Quine, W. V. O.: 1971, 'Translation and Meaning', in Rosenberg, *et al.* (1971), pp. 290–324, reprinted from Quine (1960).
Quine, W. V. O.: 1975a, 'Mind and Verbal Dispositions,' in Guttenplan (1975), pp. 83–95.
Quine, W. V. O.: 1975b, 'Replies', in Davidson, *et al.* (1975), pp. 292–351.
Quine, W. V. O.: 1979, 'Facts of the Matter', in Shahan and Swoyer (1979), pp. 155–169.
Quinton, A. M.: 1955, 'The Problems of Perception', *Mind* **LXIV** 28–51.
Radnitzky, G.: 1976, 'Popperian Philosophy of Science as an Antidote Against Relativism', in Cohen *et al.* (1976), pp. 505–546.
Reichenbach, H.: 1938, *Experience and Prediction*, University of Chicago Press, Chicago.
Reichenbach, M. and Cohen, R. S. (eds.): 1978, *Hans Reichenbach: Selected Writings 1909–1953* Vol. 2: Vienna Circle Collection 4, D. Reidel, Dordrecht, Holland.
Rosenberg, J. F. *et al* (eds.): 1971, *Readings in Philosophy of Language*, Prentice-Hall Inc.: Englewood Cliffs, New York.
Rosen, R. (ed.): 1972, *Foundations of Mathematical Biology* Vol. 1, Academic Press, New York/London.
Rosen, R.: 1972, 'Quantum Genetics', in Rosen (ed.) (1972), Chapter 3, pp. 215–252.
Russel, B.: 1914, *Our Knowledge of the External World*, The Open Court Publishing Co.
Ruwet, N.: 1973, *An Introduction to Generative Grammar*, North-Holland, Amsterdam, translated by N. S. H. Smith.
Rychlak, J. F.: 1977, *The Psychology of a Rigorous Humanism*, John Wiley and Sons.
Salmon, W. C.: *The Foundations of Scientific Inference*, University of Pittsburgh Press, Pittsburgh, first ed. 1966.
Sarrinen, E. (ed.): 1979, *Game-Theoretical Semantics*, D. Reidel, Dordrecht, Holland.
Saunders, J. and Norton, J.: 1982, 'Einstein, Light Signals and the ε- Decision', in McLaughlin (1982), pp. 101–127.
Scheibe, E.: 1976, 'Conditions of Progress and the Comparability of Theories', in Cohen *et al.* (eds.) (1976), pp. 547–568.
Scheibe, E.: 1986, 'What Kind of Hidden Variables are Excluded by Bell's Inequality?', in Weingartner, P. and Dorn, G. (eds.) (1986), *Foundations of Physics*, Verlag Hölder-Pichler-Tempsky, Vienna, pp. 251–271.
Schilpp, P. A. (ed.): 1963, *The Philosophy of Rudolf Carnap*, Open Court, La Salle, Illinois.
Schilpp, P. A. (ed.): 1974, *The Philosophy of Sir Karl Popper*, Open Court, La Salle, Illinois.
Schurz, G. and Weingartner, P.: 1987, 'Verisimilitude defined by relevant consequence-elements', in A. F. Kuipers (ed.): 1987, *What is Closer-To-The-Truth*? Editions Rodopi B. V., Amsterdam, pp. 47–77.

Sesonske, A. and Fleming, N. (eds.): 1965, *Meta–Meditations: Studies in Descartes*, Wadsworth Publishing Co., Inc,. Belmont, California.

Shahan, W. R. and Swoyer, C. (eds.): 1979, *Essays on the Philosophy of W. V. Quine*, The Harvester Press.

Shankland, R. S.: 1963, 'Conversations With Albert Einstein', *American Journal of Physics* **31** 47–57.

Shimony, A.: 1976, 'Comments on Two Epistemological Theses of Thomas Kuhn', in Cohen *et al.* (eds.) (1976), pp. 569–588.

Shoemaker, S.: 1963, *Self-Knowledge and Self-Identity*, Cornell University Press, Cornell.

Sircello, G.: 1979, 'Beauty and Sex', in Gustafson *et al.* (1979), pp. 225–239.

Sneed, J. D.: 1979, *The Logical Structure of Mathematical Physics*, D. Reidel, Dordrecht, Holland, 2nd ed.

Sober, E.: 1975, *Simplicity*, Clarendon Press, Oxford.

Stachel, J.: 1986a, 'Einstein and the Quantum: Fifty Years of Struggle', in Colodny, R. (ed.) (1986), *From Quarks to Quasars*, University of Pittsburgh Press, Pittsburgh, pp. 349–385.

Stachel, J.: 1986b, 'What a Physicist Can Learn From the Discovery of General Relativity', Proceedings of the Fourth Marcel Grossmann Meeting on General Relativity (forthcoming).

Stegmüller, W.: 1976, *The Structure and Dynamics of Theories*, Springer-Verlag, New York /Heidelberg/Berlin (German ed. 1973).

Stegmüller, W.: 1979, *The Structuralist View of Theories*, Springer-Verlag, Berlin.

Stöckler, M.: 1986, 'EPR, Relativity and Realism', in Weingartner, P. and Dorn, G. (eds.) (1986), *Foundations of Physics*, Verlag Hölder-Pichler-Tempsky, Vienna pp. 301–315.

Strawson, P. F.: 1964, *Individuals*, Methuen, London.

Suppes, P.: 1969, *Studies in the Methodology and Foundations of Sciences*, D. Reidel, Dordrecht, Holland.

Suppes, P.: 1970, *Set-theoretical Structures in Science*, mimeo reprint, Stanford.

Tang, T. B. and Fang, L.Z.: 1984, 'The Cosmic Asymmetry in Matter-Antimatter', *Vistas in Astronomy* ***27*** 1–23.

Tichý, P.: 1974, 'On Popper's Definition of Verisimilitude' *The British Journal for the Philosophy of Science* **25** 155–160.

Tondl, L.: 1973, *Scientific Procedures*, D. Reidel, Dordrecht, Holland.

Toulmin, S.: 1972, *Human Understanding* I, Princeton Press, New Jersey.

Toulmin, S.: 1976 'History, Praxis and the 'Third World'', in Cohen *et al.* (eds.) (1976), pp. 655–675.

Turok, N.: 1986, 'Particles and the Universe', *Nature* **322** 111–112.

van Fraassen, B. C.: 1980, *The Scientific Image*, Clarendon Press, Oxford.

von Neumann, J. and Morgenstern, O.: 1944, *Theory of Games and Economic Behaviour*, Princeton University Press, Princeton.

von Soldner, J. G.: 1801, 'Über die Ablenkung eines Lichtstrahls von seiner geradlinigen Bewegung durch die Attraktion eines weltkörpers, an welchem er nahe vorbeigeht', Astronomisches Jahrbuch für das Jahr 1804, Berlin 1801 (161–172), hersg. von J. E. Bode.

Waddington, C. H.: 1966, 'Fields and Gradients', in Locke (1966) pp. 105–230.

Waddington, C. H.: 1977, *Tools for Thought*, Jonathan Cape Ltd, London.

Walker, E. (ed.): 1978, *Explorations in the Biology of Language*, Harvester Press, Hassocks, Sussex.

Watkins, J. W. N.: 1970, 'Imperfect Rationality', in Borger *et al.* (1970), pp. 167–217.

Watkins, J. W. N.: 1975, 'The Popperian Approach' *Criteria of Scientific Progress: A Critical Rationalist View*, pp. 1/1–31 (L. S. E. Workshop No. 1 – Position Paper).

Watkins, J. W. N.: 1978, 'The Popperian Approach to Scientific Knowledge', in Radnitzky, G. *et al.* (eds.) (1978), *Progress and Rationality in Science*, D. Reidel, Dordrecht, Holland, pp. 23–38.

Watkins, J. W. N.: 1984, *Science and Scepticism*, Hutchinson, London.

Watkins, J. W. N.: 1987, 'A Note on Incongruent Counterparts and Verisimilitude', *Erkenntnis* **26** 295–300.

Watson, J. D. and Crick, F. H. C.: 1953, 'A Structure for Deoxyribose Nucleic Acid', *Nature* **171** 737–738.

Watson, J. D.: 1968, *The Double Helix: A Personal Account of the Discovery of the Structure of DNA*, Weidenfeld and Nicolson, London.

Weinberg, S.: 1979, 'Is Nature Simple?', In Huff, D. and Prewelt, O. (eds.) (1979), *The Nature of the Physical Universe*, John Wiley & Sons, New York, pp. 47–62 (1976 Nobel Conference organized by Gustavus Adolphus College, St. Peter, Minnesota).

Weinberg, S.: 1980, 'Unified Theories of Elementary-Particle Interaction', *Readings From Scientific American: Particles and Fields*, W.H. Freeman and Co.

Weisberg, J. M., Taylor, J. H. and Fowler, L. A.: 1981, 'Gravitational Waves From an Orbiting Pulsar', *Scientific American* **245** (4) 66–74.

Wettersten, J.: 1982, 'The Place of Mario Bunge', in Agassi and Cohen (1982), pp. 465–486.

Wigner, E. P.: 1980, 'The Role and Value of the Symmetry Principles and Einstein's Contribution to their Recognition', in Gruber and Millman (eds.) (1980), pp. 13–21.

Williams, B.: 1973, *Problems of the Self*, Cambridge University Press, Cambridge.

Wilson, R.W.: 1979, 'The Cosmic Microwave Background Radiation', *Reviews of Modern Physics* **51** (3) 433–445.

Wittgenstein, L.: 1958, *The Blue and Brown Books* (Preliminary Studies for the "Philosophical Investigations"), Basil Blackwell, Oxford.

Wittgenstein, L.; 1968, *Philosophical Investigations*, Basil Blackwell, Oxford, reprinted from 3rd ed. of 1967.

Wittgenstein, L.: 1974a, *Tractatus Logico-Philosophicus*, Routledge and Kegan Paul, London, paperback ed. (First English edition 1922).

Wittgenstein, L.: 1974b, *Philosophical Grammar*, Basil Blackwell, Oxford.

Woltjer, L.: 1978, 'Relativistic Astrophysics and Observation', in Lebovitz, *et al.* (1978), pp. 141–148.

Worrall, J.: 1975, 'The Ways in which the Methodology of Scientific Research Programmes improves on Popper's Methodology', in *Criteria of Scientific Progress: A Critical Rationalist View*, pp. 2/1–35 (L. S. E. Workshop No. 1 – Position Paper).

Zahar, E.: 1982, 'Feyerabend on Observation and Empirical Content', *The British Journal for the Philosophy of Science* **33** 397–433.

Ziman, J.: 1978, *Reliable Knowledge*, Cambridge University Press, Cambridge.

INDEX OF SYMBOLS

(References are here given mainly to places where a symbol is either introduced or explained).

INDEX OF NAMES

INDEX OF SUBJECTS

(Index of subjects here is intended to supplement and not to duplicate the Contents).

Boston Studies in the Philosophy of Science

Editor: Robert S. Cohen, *Boston University*

1. M.W. Wartofsky (ed.): *Proceedings of the Boston Colloquium for the Philosophy of Science, 1961/1962.* [Synthese Library 6] 1963 ISBN 90-277-0021-4
2. R.S. Cohen and M.W. Wartofsky (eds.): *Proceedings of the Boston Colloquium for the Philosophy of Science, 1962/1964.* In Honor of P. Frank. [Synthese Library 10] 1965 ISBN 90-277-9004-0
3. R.S. Cohen and M.W. Wartofsky (eds.): *Proceedings of the Boston Colloquium for the Philosophy of Science, 1964/1966.* In Memory of Norwood Russell Hanson. [Synthese Library 14] 1967 ISBN 90-277-0013-3
4. R.S. Cohen and M.W. Wartofsky (eds.): *Proceedings of the Boston Colloquium for the Philosophy of Science, 1966/1968.* [Synthese Library 18] 1969 ISBN 90-277-0014-1
5. R.S. Cohen and M.W. Wartofsky (eds.): *Proceedings of the Boston Colloquium for the Philosophy of Science, 1966/1968.* [Synthese Library 19] 1969 ISBN 90-277-0015-X
6. R.S. Cohen and R.J. Seeger (eds.): *Ernst Mach, Physicist and Philosopher.* [Synthese Library 27] 1970 ISBN 90-277-0016-8
7. M. Čapek: *Bergson and Modern Physics.* A Reinterpretation and Re-evaluation. [Synthese Library 37] 1971 ISBN 90-277-0186-5
8. R.C. Buck and R.S. Cohen (eds.): *PSA 1970.* Proceedings of the 2nd Biennial Meeting of the Philosophy and Science Association (Boston, Fall 1970). In Memory of Rudolf Carnap. [Synthese Library 39] 1971 ISBN 90-277-0187-3; Pb 90-277-0309-4
9. A.A. Zinov'ev: *Foundations of the Logical Theory of Scientific Knowledge (Complex Logic).* Translated from Russian. Revised and enlarged English Edition, with an Appendix by G.A. Smirnov, E.A. Sidorenko, A.M. Fedina and L.A. Bobrova. [Synthese Library 46] 1973 ISBN 90-277-0193-8; Pb 90-277-0324-8
10. L. Tondl: *Scientific Procedures.* A Contribution Concerning the Methodological Problems of Scientific Concepts and Scientific Explanation.Translated from Czech by D. Short. [Synthese Library 47] 1973 ISBN 90-277-0147-4; Pb 90-277-0323-X
11. R.J. Seeger and R.S. Cohen (eds.): *Philosophical Foundations of Science.* Proceedings of Section L, 1969, American Association for the Advancement of Science. [Synthese Library 58] 1974 ISBN 90-277-0390-6; Pb 90-277-0376-0
12. A. Grünbaum: *Philosophical Problems of Space and Times.* 2nd enlarged ed. [Synthese Library 55] 1973 ISBN 90-277-0357-4; Pb 90-277-0358-2

13. R.S. Cohen and M.W. Wartofsky (eds.): *Logical and Epistemological Studies in Contemporary Physics*. Proceedings of the Boston Colloquium for the Philosophy of Science, 1969/72, Part I. [Synthese Library 59] 1974
ISBN 90-277-0391-4; Pb 90-277-0377-9
14. R.S. Cohen and M.W. Wartofsky (eds.): *Methodological and Historical Essays in the Natural and Social Sciences*. Proceedings of the Boston Colloquium for the Philosophy of Science, 1969/72, Part II. [Synthese Library 60] 1974
ISBN 90-277-0392-2; Pb 90-277-0378-7
15. R.S. Cohen, J.J. Stachel and M.W. Wartofsky (eds.): *For Dirk Struik*. Scientific, Historical and Political Essays in Honor of Dirk J. Struik. [Synthese Library 61] 1974 ISBN 90-277-0393-0; Pb 90-277-0379-5
16. N. Geschwind: *Selected Papers on Language and the Brains*. [Synthese Library 68] 1974 ISBN 90-277-0262-4; Pb 90-277-0263-2
17. B.G. Kuznetsov: *Reason and Being*. Translated from Russian. Edited by C.R. Fawcett and R.S. Cohen. 1987 ISBN 90-277-2181-5
18. P. Mittelstaedt: *Philosophical Problems of Modern Physics*. Translated from the revised 4th German edition by W. Riemer and edited by R.S. Cohen. [Synthese Library 95] 1976 ISBN 90-277-0285-3; Pb 90-277-0506-2
19. H. Mehlberg: *Time, Causality, and the Quantum Theory*. Studies in the Philosophy of Science. Vol. I: *Essay on the Causal Theory of Time*. Vol. II: *Time in a Quantized Universe*. Translated from French. Edited by R.S. Cohen. 1980 Vol. I: ISBN 90-277-0721-9; Pb 90-277-1074-0
Vol. II: ISBN 90-277-1075-9; Pb 90-277-1076-7
20. K.F. Schaffner and R.S. Cohen (eds.): *PSA 1972*. Proceedings of the 3rd Biennial Meeting of the Philosophy of Science Association (Lansing, Michigan, Fall 1972). [Synthese Library 64] 1974
ISBN 90-277-0408-2; Pb 90-277-0409-0
21. R.S. Cohen and J.J. Stachel (eds.): *Selected Papers of Léon Rosenfeld*. [Synthese Library 100] 1979 ISBN 90-277-0651-4; Pb 90-277-0652-2
22. M. Čapek (ed.): *The Concepts of Space and Time*. Their Structure and Their Development. [Synthese Library 74] 1976
ISBN 90-277-0355-8; Pb 90-277-0375-2
23. M. Grene: *The Understanding of Nature*. Essays in the Philosophy of Biology. [Synthese Library 66] 1974 ISBN 90-277-0462-7; Pb 90-277-0463-5
24. D. Ihde: *Technics and Praxis*. A Philosophy of Technology. [Synthese Library 130] 1979 ISBN 90-277-0953-X; Pb 90-277-0954-8
25. J. Hintikka and U. Remes: *The Method of Analysis*. Its Geometrical Origin and Its General Significance. [Synthese Library 75] 1974
ISBN 90-277-0532-1; Pb 90-277-0543-7
26. J.E. Murdoch and E.D. Sylla (eds.): *The Cultural Context of Medieval Learning*. Proceedings of the First International Colloquium on Philosophy,

Science, and Theology in the Middle Ages, 1973. [Synthese Library 76] 1975 ISBN 90-277-0560-7; Pb 90-277-0587-9

27. M. Grene and E. Mendelsohn (eds.): *Topics in the Philosophy of Biology.* [Synthese Library 84] 1976 ISBN 90-277-0595-X; Pb 90-277-0596-8
28. J. Agassi: *Science in Flux.* [Synthese Library 80] 1975 ISBN 90-277-0584-4; Pb 90-277-0612-3
29. J.J. Wiatr (ed.): *Polish Essays in the Methodology of the Social Sciences.* [Synthese Library 131] 1979 ISBN 90-277-0723-5; Pb 90-277-0956-4
30. P. Janich: *Protophysics of Time.* Constructive Foundation and History of Time Measurement. Translated from the 2nd German edition. 1985 ISBN 90-277-0724-3
31. R.S. Cohen and M.W. Wartofsky (eds.): *Language, Logic, and Method.* 1983 ISBN 90-277-0725-1
32. R.S. Cohen, C.A. Hooker, A.C. Michalos and J.W. van Evra (eds.): *PSA 1974.* Proceedings of the 4th Biennial Meeting of the Philosophy of Science Association. [Synthese Library 101] 1976 ISBN 90-277-0647-6; Pb 90-277-0648-4
33. G. Holton and W.A. Blanpied (eds.): *Science and Its Public.* The Changing Relationship. [Synthese Library 96] 1976 ISBN 90-277-0657-3; Pb 90-277-0658-1
34. M.D. Grmek, R.S. Cohen and G. Cimino (eds.): *On Scientific Discovery.* The 1977 Erice Lectures. 1981 ISBN 90-277-1122-4; Pb 90-277-1123-2
35. S. Amsterdamski: *Between Expeience and Metaphysics.* Philosophical Problems of the Evolution of Science. Translated from Polish. [Synthese Library 77] 1975 ISBN 90-277-0568-2; Pb 90-277-0580-1
36. M. Marković and G. Petrović (eds.): *Praxis.* Yugoslav Essays in the Philosophy and Methodology of the Social Sciences. [Synthese Library 134] 1979 ISBN 90-277-0727-8; Pb 90-277-0968-8
37. H. von Helmholtz: *Epistemological Writings.* The Paul Hertz / Moritz Schlick Centenary Edition of 1921. Translated from German by M.F. Lowe. Edited with an Introduction and Bibliography by R.S. Cohen and Y. Elkana. [Synthese Library 79] 1977 ISBN 90-277-0290-X; Pb 90-277-0582-8
38. R.M. Martin: *Pragmatics, Truth and Language.* 1979 ISBN 90-277-0992-0; Pb 90-277-0993-9
39. R.S. Cohen, P.K. Feyerabend and M.W. Wartofsky (eds.): *Essays in Memory of Imre Lakatos.* [Synthese Library 99] 1976 ISBN 90-277-0654-9; Pb 90-277-0655-7
40. B.M Kedrov and V. Sadovsky (eds.): Current Soviet Studies in the Philosophy of Science. (In prep.) ISBN 90-277-0729-4
41. M. Raphael: Theorie des geistigen Schaffens aus marxistischer Grundlage. (In prep.) ISBN 90-277-0730-8

Boston Studies in the Philosophy of Science

42. H.R. Maturana and F.J. Varela: *Autopoiesis and Cognition*. The Realization of the Living. With a Preface to 'Autopoiesis' by S. Beer. 1980
ISBN 90-277-1015-5; Pb 90-277-1016-3
43. A. Kasher (ed.): *Language in Focus: Foundations, Methods and Systems*. Essays in Memory of Yehoshua Bar-Hillel. [Synthese Library 89] 1976
ISBN 90-277-0644-1; Pb 90-277-0645-X
44. T.D. Thao: *Investigations into the Origin of Language and Consciousness*. 1984 ISBN 90-277-0827-4
45. A. Ishmimoto (ed.): Japanese Studies in the History and Philosophy of Science. (In prep.) ISBN 90-277-0733-3
46. P.L. Kapitza: *Experiment, Theory, Practice*. Articles and Addresses. Edited by R.S. Cohen. 1980 ISBN 90-277-1061-9; Pb 90-277-1062-7
47. M.L. Dalla Chiara (ed.): *Italian Studies in the Philosophy of Science*. 1981
ISBN 90-277-0735-9; Pb 90-277-1073-2
48. M.W. Wartofsky: *Models*. Representation and the Scientific Understanding. [Synthese Library 129] 1979 ISBN 90-277-0736-7; Pb 90-277-0947-5
49. T.D. Thao: *Phenomenology and Dialectical Materialism*. Edited by R.S. Cohen. 1986 ISBN 90-277-0737-5
50. Y. Fried and J. Agassi: *Paranoia*. A Study in Diagnosis. [Synthese Library 102] 1976 ISBN 90-277-0704-9; Pb 90-277-0705-7
51. K.H. Wolff: *Surrender and Cath*. Experience and Inquiry Today. [Synthese Library 105] 1976 ISBN 90-277-0758-8; Pb 90-277-0765-0
52. K. Kosík: *Dialectics of the Concrete*. A Study on Problems of Man and World. 1976 ISBN 90-277-0761-8; Pb 90-277-0764-2
53. N. Goodman: *The Structue of Appearance*. [Synthese Library] 1977
ISBN 90-277-0773-1; Pb 90-277-0774-X
54. H.A. Simon: *Models of Discovery* and Other Topics in the Methods of Science. [Synthese Library 114] 1977 ISBN 90-277-0812-6; Pb 90-277-0858-4
55. M. Lazerowitz: *The Language of Philosophy*. Freud and Wittgenstein. [Synthese Library 117] 1977 ISBN 90-277-0826-6; Pb 90-277-0862-2
56. T. Nickles (ed.): *Scientific Discovery, Logic, and Rationality*. 1980
ISBN 90-277-1069-4; Pb 90-277-1070-8
57. J. Margolis: *Persons and Mind*. The Prospects of Nonreductive Materialism. [Synthese Library 121] 1978 ISBN 90-277-0854-1; Pb 90-277-0863-0
58. G. Radnitzky and G. Andersson (eds.): *Progress and Rationality in Science*. [Synthese Library 125] 1978 ISBN 90-277-0921-1; Pb 90-277-0922-X
59. G. Radnitzky and G. Andersson (eds.): *The Structure and Development of Science*. [Synthese Library 136] 1979
ISBN 90-277-0994-7; Pb 90-277-0995-5
60. T. Nickles (ed.): *Scientific Discovery*. Case Studies. 1980
ISBN 90-277-1092-9; Pb 90-277-1093-7

61. M.A. Finocchiaro: *Galileo and the Art of Reasoning*. Rhetorical Foundation of Logic and Scientific Method. 1980 ISBN 90-277-1094-5; Pb 90-277-1095-3
62. W.A. Wallace: *Prelude to Galileo*. Essays on Medieval and 16th-Century Sources of Galileo's Thought. 1981 ISBN 90-277-1215-8; Pb 90-277-1216-6
63. F. Rapp: *Analytical Philosophy of Technology*. Translated from German. 1981 ISBN 90-277-1221-2; Pb 90-277-1222-0
64. R.S. Cohen and M.W. Wartofsky (eds.): *Hegel and the Sciences*. 1984 ISBN 90-277-0726-X
65. J. Agassi: *Science and Society*. Studies in the Sociology of Science. 1981 ISBN 90-277-1244-1; Pb 90-277-1245-X
66. L. Tondl: *Problems of Semantics*. A Contribution to the Analysis of the Language of Science. Translated from Czech. 1981 ISBN 90-277-0148-2; Pb 90-277-0316-7
67. J. Agassi and R.S. Cohen (eds.): *Scientific Philosophy Today*. Essays in Honor of Mario Bunge. 1982 ISBN 90-277-1262-X; Pb 90-277-1263-8
68. W. Krajewski (ed.): *Polish Essays in the Philosophy of the Natural Sciences*. Translated from Polish and edited by R.S. Cohen and C.R. Fawcett. 1982 ISBN 90-277-1286-7; Pb 90-277-1287-5
69. J.H. Fetzer: *Scientific Knowledge*. Causation, Explanation and Corroboration. 1981 ISBN 90-277-1335-9; Pb 90-277-1336-7
70. S. Grossberg: *Studies of Mind and Brain*. Neural Principles of Learning, Perception, Development, Cognition, and Motor Control. 1982 ISBN 90-277-1359-6; Pb 90-277-1360-X
71. R.S. Cohen and M.W. Wartofsky (eds.): *Epistemology, Methodology, and the Social Sciences*. 1983. ISBN 90-277-1454-1
72. K. Berka: *Measurement*. Its Concepts, Theories and Problems. Translated from Czech. 1983 ISBN 90-277-1416-9
73. G.L. Pandit: *The Structure and Growth of Scientific Knowledge*. A Study in the Methodology of Epistemic Appraisal. 1983 ISBN 90-277-1434-7
74. A.A. Zinov'ev: *Logical Physics*. Translated from Russian. Edited by R.S. Cohen. 1983 ISBN 90-277-0734-0
    See also Volume 9.
75. G-G. Granger: *Formal Thought and the Sciences of Man*. Translated from French. With and Introduction by A. Rosenberg. 1983 ISBN 90-277-1524-6
76. R.S. Cohen and L. Laudan (eds.): *Physics, Philosophy and Psychoanalysis*. Essays in Honor of Adolf Grünbaum. 1983 ISBN 90-277-1533-5
77. G. Böhme, W. van den Daele, R. Hohlfeld, W. Krohn and W. Schäfer: *Finalization in Science*. The Social Orientation of Scientific Progress. Translated from German. Edited by W. Schäfer. 1983 ISBN 90-277-1549-1
78. D. Shapere: *Reason and the Search for Knowledge*. Investigations in the Philosophy of Science. 1984 ISBN 90-277-1551-3; Pb 90-277-1641-2

79. G. Andersson (ed.): *Rationality in Science and Politics*. Translated from German. 1984 ISBN 90-277-1575-0; Pb 90-277-1953-5
80. P.T. Durbin and F. Rapp (eds.): *Philosophy and Technology*. [*Also* Philosophy and Technology Series, Vol. 1] 1983 ISBN 90-277-1576-9
81. M. Marković: *Dialectical Theory of Meaning*. Translated from Serbo-Croat. 1984 ISBN 90-277-1596-3
82. R.S. Cohen and M.W. Wartofsky (eds.): *Physical Sciences and History of Physics*. 1984. ISBN 90-277-1615-3
83. É. Meyerson: *The Relativistic Deduction*. Epistemological Implications of the Theory of Relativity. Translated from French. With a Review by Albert Einstein and an Introduction by Milić Čapek. 1985 ISBN 90-277-1699-4
84. R.S. Cohen and M.W. Wartofsky (eds.): *Methodology, Metaphysics and the History of Science*. In Memory of Benjamin Nelson. 1984 ISBN 90-277-1711-7
85. G. Tamás: *The Logic of Categories*. Translated from Hungarian. Edited by R.S. Cohen. 1986 ISBN 90-277-1742-7
86. S.L. De C. Fernandes: *Foundations of Objective Knowledge*. The Relations of Popper's Theory of Knowledge to That of Kant. 1985 ISBN 90-277-1809-1
87. R.S. Cohen and T. Schnelle (eds.): *Cognition and Fact*. Materials on Ludwik Fleck. 1986 ISBN 90-277-1902-0
88. G. Freudenthal: *Atom and Individual in the Age of Newton*. On the Genesis of the Mechanistic World View. Translated from German. 1986 ISBN 90-277-1905-5
89. A. Donagan, A.N. Perovich Jr and M.V. Wedin (eds.): *Human Nature and Natural Knowledge*. Essays presented to Majorie Grene on the Occasion of Her 75th Birthday. 1986 ISBN 90-277-1974-8
90. C. Mitcham and A. Hunning (eds.): *Philosophy and Technology II*. Information Technology and Computers in Theory and Practice. [*Also* Philosophy and Technology Series, Vol. 2] 1986 ISBN 90-277-1975-6
91. M. Grene and D. Nails (eds.): *Spinoza and the Sciences*. 1986 ISBN 90-277-1976-4
92. S.P. Turner: *The Search for a Methodology of Social Science*. Durkheim, Weber, and the 19th-Century Problem of Cause, Probability, and Action. 1986. ISBN 90-277-2067-3
93. I.C. Jarvie: *Thinking about Society*. Theory and Practice. 1986 ISBN 90-277-2068-1
94. E. Ullmann-Margalit (ed.): *The Kaleidoscope of Science*. The Israel Colloquium: Studies in History, Philosophy, and Sociology of Science, Vol. I. 1986 ISBN 90-277-2158-0; Pb 90-277-2159-9
95. E. Ullmann-Margalit (ed.): *The Prism of Science*. The Israel Colloquium: Studies in History, Philosophy, and Sociology of Science, Vol. II. 1986 ISBN 90-277-2160-2; Pb 90-277-2161-0

96. G. Márkus: *Language and Production.* A Critique of the Paradigms. Translated from French. 1986 ISBN 90-277-2169-6
97. F. Amrine, F.J. Zucker and H. Wheeler (eds.): *Goethe and the Sciences: A Reappraisal.* 1987 ISBN 90-277-2265-X; Pb 90-277-2400-8
98. J.C. Pitt and M. Pera (eds.): *Rational Changes in Science.* Essays on Scientific Reasoning. Translated from Italian. 1987 ISBN 90-277-2417-2
99. O. Costa de Beauregard: *Time, the Physical Magnitude.* 1987 ISBN 90-277-2444-X
100. A. Shimony and D. Nails (eds.): *Naturalistic Epistemology.* A Symposium of Two Decades. 1987 ISBN 90-277-2337-0
101. N. Rotenstreich: *Time and Meaning in History.* 1987 ISBN 90-277-2467-9
102. D.B. Zilberman: *The Birth of Meaning in Hindu Thought.* Edited by R.S. Cohen. 1988 ISBN 90-277-2497-0
103. T.F. Glick (ed.): *The Comparative Reception of Relativity.* 1987 ISBN 90-277-2498-9
104. Z. Harris, M. Gottfried, T. Ryckman, P. Mattick Jr, A. Daladier, T.N. Harris and S. Harris: *The Form of Information in Science.* Analysis of an Immunology Sublanguage. With a Preface by Hilary Putnam. 1989 ISBN 90-277-2516-0
105. F. Burwick (ed.): *Approaches to Organic Form.* Permutations in Science and Culture. 1987 ISBN 90-277-2541-1
106. M. Almási: *The Philosophy of Appearances.* Translated from Hungarian. 1989 ISBN 90-277-2150-5
107. S. Hook, W.L. O'Neill and R. O'Toole (eds.): *Philosophy, History and Social Action.* Essays in Honor of Lewis Feuer. With an Autobiographical Essay by L. Feuer. 1988 ISBN 90-277-2644-2
108. I. Hronszky, M. Fehér and B. Dajka: *Scientific Knowledge Socialized.* Selected Proceedings of the 5th Joint International Conference on the History and Philosophy of Science organized by the IUHPS (Veszprém, Hungary, 1984). 1988 ISBN 90-277-2284-6
109. P. Tillers and E.D. Green (eds.): *Probability and Inference in the Law of Evidence.* The Uses and Limits of Bayesianism. 1988 ISBN 90-277-2689-2
110. E. Ullmann-Margalit (ed.): *Science in Reflection.* The Israel Colloquium: Studies in History, Philosophy, and Sociology of Science, Vol. III. 1988 ISBN 90-277-2712-0; Pb 90-277-2713-9

    See also Volumes 94 and 95.
111. K. Gavroglu, Y. Goudaroulis and P. Nicolacopoulos (eds.): *Imre Lakatos and Theories of Scientific Change.* 1989 ISBN 90-277-2766-X
112. B. Glassner and J.D. Moreno (eds.): *The Qualitative- Quantitative Distinction in the Social Sciences.* 1989 ISBN 90-277-2829-1
113. K. Arens: *Structures of Knowing.* Psychologies of the 19th Century. 1989 ISBN 0-7923-0009-2

Boston Studies in the Philosophy of Science

114. A. Janik: *Style, Politics and the Future of Philosophy*. 1989 ISBN 0-7923-0056-4
115. F. Amrine (ed.): *Literature and Science as Modes of Expression*. With an Introduction by S. Weininger. 1989 ISBN 0-7923-0133-1
116. J.R. Brown and J. Mittelstrass (eds.): *An Intimate Relation*. Studies in the History and Philosophy of Science. Presented to Robert E. Butts on His 60th Birthday. 1989 ISBN 0-7923-0169-2
117. F. D'Agostino and I.C. Jarvie (eds.): *Freedom and Rationality*. Essays in Honor of John Watkins. 1989 ISBN 0-7923-0264-8
118. D. Zolo: *Reflexive Epistemology*. The Philosophical Legacy of Otto Neurath. 1989 ISBN 0-7923-0320-2
119. M. Kearn, B.S. Philips and R.S. Cohen (eds.): *Georg Simmel and Contemporary Sociology*. 1989 ISBN 0-7923-0407-1
120. T.H. Levere and W.R. Shea (eds.): *Nature, Experiment and the Science*. Essays on Galileo and the Nature of Science. In Honour of Stillman Drake. 1989 ISBN 0-7923-0420-9
121. P. Nicolacopoulos (ed.): *Greek Studies in the Philosophy and History of Science*. 1990 ISBN 0-7923-0717-8
122. R. Cooke and D. Costantini (eds.): *Statistics in Science*. The Foundations of Statistical Methods in Biology, Physics and Economics. 1990 ISBN 0-7923-0797-6
123. P. Duhem: *The Origins of Statics*. Translated from French by G.F. Leneaux, V.N. Vagliente and G.H. Wagner. With an Introduction by S.L. Jaki. 1991 ISBN 0-7923-0898-0
124. H. Kamerlingh Onnes: *Through Measurement to Knowledge*. The Selected Papers, 1853-1926. Edited and with an Introduction by K. Gavroglu and Y. Goudaroulis. 1991 ISBN 0-7923-0825-5
125. M. Čapek: *The New Aspects of Time: Its Continuity and Novelties*. Selected Papers in the Philosophy of Science. 1991 ISBN 0-7923-0911-1
126. S. Unguru (ed.): *Physics, Cosmology and Astronomy, 1300- 1700*. Tension and Accomodation. 1991 ISBN 0-7923-1022-5
127. Z. Bechler: *Newton's Physics on the Conceptual Structure of the Scientific Revolution*. 1991 ISBN 0-7923-1054-3
128. É. Meyerson: *Explanation in the Sciences*. Translated from French by M-A. Siple and D.A. Siple. 1991 ISBN 0-7923-1129-9
129. A.I. Tauber (ed.): *Organism and the Origins of Self*. 1991 ISBN 0-7923-1185-X
130. F.J. Varela and J-P. Dupuy (eds.): *Understanding Origins*. Contemporary Views on the Origin of Life, Mind and Society. 1991 ISBN 0-7923-1251-1
131. G.L. Pandit: *Methodological Variance*. Essays in Epistemological Ontology and the Methodology of Science. 1991 ISBN 0-7923-1263-5

Boston Studies in the Philosophy of Science

132. G. Munévar (ed.): *Beyond Reason.* Essays on the Philosophy of Paul Feyerabend. 1991 ISBN 0-7923-1272-4
133. T.E. Uebel (ed.): *Rediscovering the Forgotten Vienne Circle.* Austrian Studies on Otto Neurath and the Vienna Circle. Partly translated from German. 1991 ISBN 0-7923-1276-7
134. W.R. Woodward and R.S. Cohen (eds.): *World Views and Scientific Discipline Formation.* Science Studies in the [former] German Democratic Republic. Partly translated from German by W.R. Woodward. 1991 ISBN 0-7923-1286-4
135. P. Zambelli: *The Speculum Astronomiae and Its Enigma.* Astrology, Theology and Science at Albertus Magnus' Time. (forthcoming) ISBN 0-7923-1380-1

Also of interest:

R.S. Cohen and M.W. Wartofsky (eds.): *A Portrait of Twenty-Five Years Boston Colloquia for the Philosophy of Science, 1960-1985.* 1985 ISBN Pb 90-277-1971-3

Previous volumes are still available.

KLUWER ACADEMIC PUBLISHERS – DORDRECHT / BOSTON / LONDON